TRANSPORT REVOLUTIONS

MOVING PEOPLE AND FREIGHT WITHOUT OIL

Richard Gilbert & Anthony Perl

NEW SOCIETY PUBLISHERS

Cataloging in Publication Data:
A catalog record for this publication is available
from the National Library of Canada.

Cover design by Diane McIntosh.
Cover images: Highway— © iStock/Maciej Noskowski;
Electric Train— © iStock/htng

Printed in Canada by Friesens.
First printing March 2010.

New Society Publishers acknowledges the support of the
Government of Canada through the Book Publishing Industry
Development Program (BPIDP) for our publishing activities.

Paperback ISBN: 978-0-86571-660-5

To order directly from the publishers,
please call toll-free (North America) 1-800-567-6772,
or order online at www.newsociety.com

Any other inquiries can be directed by mail to:

New Society Publishers
P.O. Box 189, Gabriola Island, BC V0R 1X0 Canada
(250) 247-9737

New Society Publishers' mission is to publish books that contribute in fundamental ways to building an ecologically sustainable and just society, and to do so with the least possible impact on the environment, in a manner that models this vision. We are committed to doing this not just through education, but through action. This book is one step toward ending global deforestation and climate change. It is printed on Forest Stewardship Council-certified acid-free paper that is **100% post-consumer recycled** (100% old growth forest-free), processed chlorine free, and printed with vegetable-based, low-VOC inks, with covers produced using FSC-certified stock. New Society also works to reduce its carbon footprint, and purchases carbon offsets based on an annual audit to ensure a carbon neutral footprint. For further information, or to browse our full list of books and purchase securely, visit our website at: www.newsociety.com

NEW SOCIETY PUBLISHERS
www.newsociety.com

Mixed Sources
Cert no. SW-COC-001271
© 1996 FSC

Contents

Preface to the First Edition

Transport Revolutions was begun during the summer of 2006, from two apartments on the south shore of Vancouver's Burrard Inlet, which provides oceangoing vessels with access to Canada's busiest port complex. The wealth of transport activity visible from our buildings inspired reflections on what has enabled human mobility to attain its current performance and what the future may hold.

The rooftop patio of the Gastown building that houses Richard Gilbert's rental apartment—his temporary home away from Toronto—provides a panoramic view of Vancouver's major container-ship berths, where many thousands of containers from Asia are unloaded each week. Across the water, on the north shore of Burrard Inlet, raw materials are loaded onto bulk carriers for shipment to Asia. Separating the container quays from Gastown are the marshalling yards of the Canadian Pacific Railway, among the world's oldest and largest freight carriers. Trains several kilometers in length are assembled here to carry containers and other cargo to destinations across Canada and the United States.

At the west edge of the rail yard lies the West Coast Express terminal, where diesel-fueled, double-decked commuter trains arrive from Vancouver's eastern suburbs on weekday mornings and depart in the afternoons. Between the rail tracks and the quays is a highway mostly used by trucks moving things to and from the container ships. At the side of this highway is a large parking area used mainly by tourist coaches (buses), and another parking area used by a car rental company.

A smaller dock just west of the container-ship piers is used by tugboats that move the huge freighters in and out of the harbor. Next to the tugboat dock is a heliport where frequent flights carry Vancouver's business and political elite a hundred kilometers southwest across the Georgia Straight to Victoria, a much smaller city that is the province of British Columbia's capital. West of the heliport is the Vancouver terminus of the Seabus, a ferry that carries an average of 15,000 passengers a day between Vancouver and North Vancouver, three kilometers across the harbor.

Where Gastown meets Vancouver's central business district is Canada Place, visible from both authors' apartments. This is a large and growing convention center, and office and hotel complex. It is also a cruise ship terminal. Berthing at the terminal, often three at a time, are the floating resorts that carry a million people each year on weeklong trips to view Alaska's glaciers. More noticeable from Anthony Perl's Coal Harbour apartment is the steady stream of float planes arriving from and departing to Vancouver Island, Whistler and communities in the Gulf Islands. Both buildings are beneath the flight path of the 20 or so large jet aircraft that leave for Europe each day from Vancouver International Airport, 25 kilometers to the south west. Farther away but visible are aircraft of many sizes arriving from and en route to numerous places in the Americas and Asia.

What these diverse transport modes have in common is their use of one form or another of processed crude oil: the bunker oil used in ships, gasoline, diesel fuel, jet kerosene and others. Oil products fuel more than 95 percent of the world's transport. Without a steady flow of this energy source, all the motorized mobility visible in, around and above Burrard Inlet would come to a halt.

On the city side of the Gastown and Coal Harbour buildings that provide such extraordinary views of Burrard Inlet lies a local transport panorama dominated by cars and buses. It is a familiar urban traffic scene except in one respect. Most of the buses in sight are trolley buses, which have electric motors powered through overhead wires rather than the internal combustion engines that propel the world's much more numerous fleet of diesel-fueled buses. The trolley buses move almost silently through Vancouver's streets, responsible for essentially no pollution in the city and little elsewhere because most of Vancouver's electricity is generated from falling water.

Many of these trolley buses are old, in service for 25 years or more, and for the most part rely on technology developed in the century before last. Nevertheless, they are popular, and TransLink, the regional transport authority, is upgrading the fleet with 228 state-of-the-art trolley buses purchased from a Winnipeg manufacturer and powered by German propulsion technology.

The trolley buses are popular because they are quiet and odorless and thus relatively inconspicuous, fitting well into the urban fabric. They could well become popular for another reason: they do not rely on oil. As world oil production peaks, supply fails to keep up with potential demand, and prices rise dramatically as a consequence, use of oil products as transport fuels could become prohibitively expensive.

Similarly popular but less noticeable in the city center—because there it is below ground—is Vancouver's Skytrain, a 33-station electrified light-rail system running on guideways that are mostly above ground, sometimes at ground level, and sometimes below. The Skytrain system provides a substantial part of the Vancouver region with rail capacity approaching that of the more familiar heavy-rail systems known in the UK as the "Tube" or "Underground," in North America as "subways," and elsewhere often as "metros." Skytrains are fully automated: no operator is required. Thus, they could serve as a technical bridge to future automated systems in which passengers can direct small vehicles on guideways to particular destinations.

The popular alternatives provided by Vancouver's trolley bus fleet and Skytrain system beg questions as to the roles electric traction could play in maintaining the mobility of a world challenged by declining oil production. We make a case in this book that electric vehicles are the most important viable alternative to vehicles moved by internal combustion engines, and that they could quite quickly begin to replace oil-fueled mobility on land. Their power could come mostly from overhead wires or rails, which can deliver renewable energy in a remarkably efficient manner. With some loss of efficiency, the power could also come from batteries. With much less efficiency, the power could come from fuel cells (which we do not find promising).

What is now a nearly invisible feature of the world's transport could become the dominant form, much as early mammals, scurrying inconspicuously at the feet of dinosaurs, adapted better to imperatives of cosmogeology and climate some 65 million years ago.

This book begins the exploration of a future in which mostly renewably produced electricity will increasingly replace oil as a transport fuel. In the course of this change, there will likely be several transport revolutions, transforming the movement of people and freight and the organization and use of the various transport modes.

Today's transport systems are little different from those of 30 years ago except in the amount of transport activity they support, which has increased by about a factor of three. Tomorrow's systems, those of 30 or more years hence, promise to be radically different, and support substantially different patterns of trade and social activity. In the following pages, we set out to explain why and how these changes will occur, and what they might imply for the human condition.

Such an investigation would have been much more arduous without the support we received from key people and organizations. Canada's Auto 21 Network of Centres of Excellence funded part of our collaboration through grants in its Societal Issues theme area for the project "Policy Options for Alternative Automotive Futures." Simon Fraser University supported this work through a President's Research Grant and Dean's Research Grants. We benefitted from research assistance provided by Ruby Arico, Graham Senft, Liu Xiaofei and Michaela Rollingson. Many friends and colleagues provided valuable feedback on early drafts of this manuscript. Al Cormier, Susan Dexter and Neal Irwin read drafts of all the chapters and had much influence on the book's final form and content. John Adams, David Gurin, Brendon Hemily, Tony Hiss, Catherine O'Brien, Bob Oliver, Judith Patterson and Linda Sheppard each commented on a substantial part of the emerging book and influenced it greatly. Emily Gilbert made numerous invaluable contributions to the identification of appropriate sources. Errors, omissions and analytical weaknesses in this book remain our full responsibility. There would have been many more of them without this input, for which we are truly grateful.

The forbearance of our spouses, Rosalind Gilbert and Andrea Banks, deserves special recognition, reflected in the book's dedication. Our transcontinental collaboration consumed much of the time we would have otherwise spent with them during 2006 and 2007.

— Richard Gilbert & Anthony Perl
Toronto and Vancouver, July 2007

Preface to the Second Edition

We finished writing the second edition of *Transport Revolutions* in August 2009, but were able to add this preface in early January 2010. It allows a very brief update of the information and analysis assembled in mid 2009.

The first edition of *Transport Revolutions* was published early in 2008, which went on to be among the most tumultuous years in the histories of transport and energy. We know of nobody who anticipated that the price of a barrel of oil—just over $90 in January 2008—would rise above $145 in July and fall almost to $30 in December.[1] We are unaware of anyone who anticipated that sales in the US of cars and light trucks would fall by a remarkable 36 percent during 2008,[2] or—perhaps more astonishing—that the Baltic Dry Index, which reflects the price of shipping dry raw materials by sea, would plummet by more than 90 percent.[3]

In many ways, 2009 was even more tumultuous. The price of a barrel of oil nearly doubled, ending the year at just under $80. World automobile production and sales fell for the second year in a row.[4] China passed Japan to become the world's top automobile producer, having passed the US as the second largest producer in 2008. China also led the world in domestic auto sales.[5] Meanwhile, India passed Japan as the leading producer of the smallest category of car—known as super-compact or sub-sub-compact—whose retail price is usually much less than the equivalent of $8,000.[6] Aviation suffered in 2009 what the International Civil Aviation Organization described as its "worst ever performance."[7] This seemed particularly true of air freight, down 15 percent over 2008, and highly profitable premium travel—first and business class—down 18 per cent.[8]

* Superscript numbers refer to reference and other notes that are arranged by chapter and begin on page 317.

The thesis of our first edition, continued here, is that we are entering an era of revolutions in how people and freight move, caused by the inevitable decline in world production of oil and anticipation of the decline. We define transport revolutions in Chapter 1. We don't yet have evidence that transport revolutions, as defined, actually occurred in 2008 and 2009 with the possible exception noted below. However, the fluxes of 2008 and 2009 provide signs that awareness of oil depletion has begun to trigger major changes in organization and technology that will yield transport revolutions.

At the beginning of 2010, world production of petroleum liquids appears to be at an uneven plateau after rising almost continuously for decades. Production may have declined in 2007 and did in 2009, but it rose in 2008 and may do so again in 2010.[9] Some believe 2008 was the year of peak production.[10] We anticipate that a continuing decline in production will begin soon after 2012.[11] Others believe the peak will be later in the current decade.[12] Yet others, notably the International Energy Agency (IEA), assert the peak will be much later.[13]

IEA is the main adviser to oil-importing countries. Although its projections for future oil production continue to speak to production increases, the extent of the increases—as reported in *World Energy Outlook* (*WEO*), IEA's flagship publication—declines almost every year. IEA's first projection of production in 2030 was in *WEO 2002*: 120.0 million barrels a day (mb/d). In *WEO 2009*, it was 103.0 mb/d. Shortly before the November publication of *WEO 2009*, IEA suffered embarrassing revelations by present and past employees that "it has been deliberately underplaying a looming shortage for fear of triggering panic buying."[14]

In this edition of *Transport Revolutions*, we favor the argument that the economic turmoil of 2008 and 2009 was caused mostly by the rising price of oil. This argument appeared to gain support in the second half of 2009.[15] We note another growing realization: that an oil price above $80 per barrel may be sufficient to cause economic distress; it does not have to go above $145.[16] As the price hovers around $80 again, and expansion of supply seems challenging, the imperative for governments to mandate oil conservation to avoid a further price rise becomes ever more urgent.

Governments are indeed moving to get transport off oil. The preferred strategy of most governments of countries with auto industries—as revealed during 2009—is to electrify motorized land transport, now

95-percent powered by oil products. The almost exclusive focus has been on developing battery-electric and hybrid automobiles. China, Germany and the US are vying to lead the production of these vehicles, a position presently held by Japan. Korea and France also have ambitious plans.[17]

Electrification of land transport is among the important transport revolutions we anticipate, but not necessarily with battery-powered vehicles as its centerpiece. We favor grid-connection while in motion. For this means of powering mobility, only China appears to have clearly embarked on a revolutionary trajectory as it implements what we note in Chapter 6 is the "largest railway expansion in history," most of it powered from the electrical grid. A late-2009 landmark was inauguration of the world's fastest long-distance passenger train service, across the 968 kilometers between Guangzhou and Wuhan, with 56 trains per day traveling at average running speeds of more than 300 kilometers per hour.[18] Buttressing expansion of electric traction is China's massive investment in the national electrical grid, "an area where China has actually moved ahead of the US."[19]

The only 21st-century transport revolution to have fully emerged so far, in our terms, has likely been in China. It is the recent decade's massive expansion in use of electric bicycles, described as the "pack mules of the nation's economic boom."[20] Some 2,000 Chinese manufacturers produce 21 million of these vehicles a year (compared with China's annual production of a few thousand electric cars). A late-2009 controversy concerned whether the largest of them should be licensed as motorcycles. It proved such a hot potato the national government left it up to provincial and municipal governments to enforce the rules, should they want to. This rapid rise of electric bicycles augurs well for many more transport revolutions that will shift transport substantially away from oil during the next few decades.

— Richard Gilbert & Anthony Perl
Toronto and Vancouver, January 2010

Introduction: Transport Revolutions Ahead

What Is in This Book

This book examines the kinds of change that motorized transport around the world could undergo during the next few decades. Today, 95 percent of this transport is fueled by products of petroleum oil, chiefly gasoline, diesel fuel and jet kerosene. We believe world oil production is on the verge of peaking and the amount available for use will decline progressively after about 2012. Meanwhile, demand or potential demand for oil could continue to increase, chiefly to fuel growing motorized movement of people and freight. A shortfall between what people want to use and what is available could cause petroleum prices to rise, perhaps steeply.

High oil prices, or even the anticipation of them, could cause at least four kinds of transport revolution:

1. *Now,* almost all transport is propelled by internal combustion engines. *In the future,* transport could be propelled increasingly by electric motors, using electricity that is increasingly generated from renewable resources.
2. *Now,* almost all land transport is by vehicles that carry their fuel on board: gasoline or diesel fuel. *In the future,* much land transport could be in electric vehicles that are grid-connected; that is, they are powered while in motion, from wire or rails or in other ways, because of the efficiency inherent in this means of providing electricity to vehicles.
3. *Now,* almost all marine transport is propelled by diesel engines. Their use will continue but with considerable assistance from wind via sails and kites.

4. *Now,* air travel and air freight movement have been the fastest growing transport activities. Soon, they could begin a continuing decline because there will be no adequate substitute for increasingly expensive aviation fuels refined from petroleum. Air travel and air freight movement will continue, but at lower intensities and mostly in large, more fuel-efficient aircraft flying a limited number of well-patronized routes, also with some use of partially solar-powered airships (dirigibles).

Four other factors could support and shape these revolutions:

1. Concern about pollution in cities, today caused mainly by the burning of petroleum fuels in vehicles, as we note in Chapter 4. Electric motors produce no such pollution at the vehicle.
2. Concern about how human activity, particularly transport, may be contributing to climate change, also touched on in Chapter 4. Vehicles using electric motors can be readily fueled from renewable resources that make no such contribution.
3. Concern with achieving sustainability, so that succeeding generations can have a reasonable measure of well-being. Sustainability requires reliance on renewable resources that can be as available in the future as they are today. Oil is not a renewable resource, but electricity can be.
4. Avoidance of international conflict over energy resources, which will become more intense as oil production declines unless strong steps are taken to reduce oil consumption, particularly for transport.

The four revolutions we explore in this book may not be inevitable, at least not within the next few decades. Enough oil *could* be found to maintain incremental growth in today's forms of transport activity. Petroleum-based transport fuels *could* be replaced by a to-be-developed liquid fuel that can be renewably produced in sufficient quantities. However, in our view, neither is likely to happen. After about 2012, as we will explain in Chapter 3, the world will enter an era of oil depletion characterized by progressively declining oil production. As we will explain in the same chapter, biofuels, liquids from coal and other products will nowhere near make up for the decline.

A more likely impediment to these revolutions will be lack of timely preparation. High oil prices will cause change, but the change will be destructive if it is not anticipated, as occurred in 2008. In one of many dismal scenarios, car-dependent suburban residents who can no longer afford to refuel their cars, and have no alternative means to travel to work or buy essential goods, will have to abandon their homes or live at a subsistence level on what they can produce from their backyards. If a region dependent on food imports by truck can no longer afford the transport costs, and there is no alternative means of moving food, residents will have to rely on what can be produced in the region, which may be too little for the numbers of people who live there. There have been hints of such a scenario in part of the US over the last few years, as we shall note in Chapters 3 and 6 in discussions about how high oil prices may have contributed to housing woes.

Economic and social collapse is a real prospect if our oil-dependent societies do not prepare and implement workable plans to anticipate oil depletion. Then, there will be another kind of transport revolution, resulting in very much less motorized transport activity than humans now enjoy. This would be a transport revolution to avoid. Again, there have been hints of such a decline in transport activity and loss of amenity as a result of the rising prices of transport fuels during the last decade.

The four transport revolutions noted above could allow humanity to continue with at least the comfort and convenience of present arrangements—and quite possibly more. How people and goods move will be different, but they will still move, with all the benefits of such movement. Moreover, there will be fewer of the costs we accept today as being the price of progress such as transport-related poor air quality.

In Chapters 5 and 6, we discuss how transport-related preparations for oil depletion could unfold, with a focus on the US and China. These are respectively the most challenging among richer and poorer countries. We propose a process for initiating the required transport revolutions and offer some suggestions as to how matters might transpire during the first stage of these revolutions.

Four chapters prepare the ground for Chapters 5 and 6. Chapter 1 sets out what we mean by a transport revolution and looks back at five earlier examples to gain perspectives on what transport revolutions bring about.

Chapter 2 reviews current transport worldwide, including the movement of both people and freight. Much more information is publicly available on the movement of people, and we spend more time discussing this matter. However, we note that in many respects the movement of freight is as important and deserves more consideration than we and most others have been able to give it. For both people and freight we discuss local movement and movement among cities, countries and continents, considering differences between richer and poorer places. We look at recent trends and current projections, and discuss some of the causes of the transport activity. Almost all of our discussion concerns motorized transport, but we do touch on cycling and walking.

Chapter 3 focuses on transport and energy. We begin by explaining why we believe oil production will soon reach a peak and then decline progressively, noting the likely role of rising oil prices in the 2008 recession. Next we consider alternatives to oil as a transport fuel, focusing on electricity, which we believe to be the most viable alternative. Different kinds of electric vehicle and delivery system are assessed, and we conclude that grid-connected systems offer the most promise in an era of energy constraints. Finally, we consider how enough electricity might be generated to support widespread replacement of internal combustion engines by electric motors.

In Chapter 4, we discuss transport's adverse impacts, beginning with consideration of the global impacts. Currently, the most newsworthy potential impact is climate change, but we also consider other global impacts including stratospheric ozone depletion, dispersion of persistent organic pollutants, and ocean acidification. Then we move on to local and regional impacts, including air pollution and noise. Finally, we discuss what might loosely be identified as the adverse social and economic impacts of transport, the most salient of which are the outcomes of transport crashes and collisions. We note in several places in Chapter 4 that many of the impacts of transport would be reduced were electric motors to replace internal combustion engines as the prime means of traction.

Chapter 5 is the core of the book. There we look ahead to 2025 and show how for the US and China high levels of transport activity can be maintained while substantially reducing oil use. The overall framework for Chapter 5 is the amount of oil we believe will be available in 2025, based on the analysis in Chapter 3. This will be about 17 percent below what was produced in 2007. We expect that the US, as the example of a richer country, will achieve a greater reduction—in the order of 40 percent. China, as the example of a poorer country, will increase its oil consumption from the 2007 level, but by much less than current trends suggest. Moreover, by 2025, after reaching a peak in about 2020, China's consumption will be falling with further declines to come.

For both the US and China, and for the movement of both people and freight, we set out mode by mode how transport activity could change between 2007 and 2025. We stress that the detailed proposals implied by our scenarios for 2025 are of much less significance than the process used to figure out, for example, how much of China's movement of people might be by grid-connected high-speed train, or how much movement of freight in the US might continue to be moved by diesel-fueled trucks. We stress too the advantage of beginning transitions to new transport arrangements as soon as possible compared to waiting for additional episodes of very high prices and their accompanying turmoil to trigger change.

In Chapter 6, we expand the discussion we begin in Chapter 3 of the remarkable happenings with respect to oil and transport before, during, and after 2008. We sketch out a "vicious cycle" of high oil prices, economic recession, collapse in oil prices, economic recovery, and high oil prices again, a cycle that if repeated could become a spiral into economic crisis. We suggest that the US and other countries may be in such a vicious cycle and propose that a way out of it is massive investment in the kind of agenda we have proposed in Chapter 5. Moves have been made in the right direction in the US, but they are not evidently sufficient. The vicious cycle touched China but major timely investments toward electrification of transport appear to have contributed to early economic recovery and perhaps even turned the vicious cycle into a virtuous cycle. China's Achilles' heel could be the parallel continued growth in transport activity that depends on oil. We conclude by stressing the role of leadership in securing necessary transport revolutions, in the US, in China and elsewhere.

Who Could Benefit from This Book, and How

The information and analysis presented in this book could be of interest to many people who would like to know more about today's transport activity, its energy use and impacts, and how these things could change. Some readers will have a general interest in this profoundly important aspect of modern societies. Other readers will have a professional perspective, perhaps that of transport or land-use planner, policy adviser or traffic engineer—or as a student in the process of gaining a professional perspective. Here we discuss what these two groups of readers—general and professional—could gain from spending time with *Transport Revolutions.*

General readers could learn from this book how to be better prepared for and how to influence what happens during oil depletion: that indefinite period after the world peak in oil production when the oil that fuels our present transport will be less and less available. If we are prepared, the transport revolutions ahead could be relatively painless, and even provide for an era of peace and prosperity. At the other extreme, lack of preparation and difficulty in keeping modern mobility functioning during oil depletion could trigger massive social unrest, economic decline and international conflict.

People in richer countries often spend more on mobility than they realize, with direct expenditures on transport usually being second only to housing costs in the typical household budget. Transport's contribution to costs rises further when the transport component of other costs (e.g., food and clothing) is factored in. Transport's contribution rises even further when the influence on asset values such as real property, equities and pension plans is also taken into account. The overall result is that much personal welfare is at stake in transport system performance beyond being able to move with ease from one place to another. In richer countries, the remarkable effectiveness of our complex transport systems is mostly taken for granted. But these systems could fail in the event that oil becomes costly or scarce, or both.

Transport that does not adapt to oil depletion could make most things much more expensive, including food and medicine, and lower the value of homes and savings. There were hints of the turmoil that can be unleashed by disruptions in the supply of transport fuels during the oil crises of the 1970s, during the protests against high fuel prices that occurred in the UK and elsewhere in Europe during September 2000, and the 2007 riots when gasoline rationing was introduced in Iran. People of many countries, notably the US, are still coming to terms with the economic turmoil unleashed by the 2008 oil shock.

Knowing the challenges in developing transport options that can function without oil or with less oil, and initiating efforts before oil depletion becomes acute, could motivate demands for workable programs of transport redesign from governments. Understanding what may lie ahead could also encourage some people to make different decisions about their own living arrangements in anticipation of the coming transport revolutions. Where one lives and works could be of profound importance during oil depletion. Life in a car-dependent neighborhood could be much harder than one where most places to be reached are a walk or a bicycle ride away, or a short journey by public transport.

Quality of life can be vastly improved or greatly diminished by the way transport works, at local, national or global scales. Safe and welcoming communities, healthy cities and peaceful countries are all facilitated by successful transport systems. Success in achieving these ends will depend increasingly upon the ability to stop using oil as a fuel. Learning that it is possible to move large numbers of people and high volumes of goods without oil could be a surprise to many people. Understanding why this is so and how to make it happen will help general readers hold leaders accountable for delivering on the promises of transport revolutions.

Tomorrow's transport professionals will have important roles to play in implementing transport revolutions, but the roles will differ from those of today. For example, there will be less need for the technical skills that go into building roads, airports and motor vehicles in a world in which cars and aircraft are used less and grid-connected vehicles—many operating on rails or other guideways—are used more. Much current work is guided by the models of economists and others that predict demand for mobility. These models will become unreliable as oil depletion becomes a major influence, but it is not yet clear where planners will be able to turn for help in anticipating changes in transport demand.

There will be a need for new skills in both the technical and strategic domains of transport system development. This book offers insights into both. In Chapters 4, 5 and 6, there is a focus on what we have come to call *energy redesign,* that is, refashioning transport systems to accommodate new energy realities. This will become a core planning skill. Electrical engineering will become a much more central feature of the technical expertise needed to develop land transport systems.

Transport professionals who can see the changes ahead will be better prepared to assume new responsibilities in the coming transport revolutions. Some transport careers will flourish because of abilities to deliver mobility without oil. Other careers will stagnate or even terminate after oil-dependent mobility begins to decline. Retraining to plan, build and manage systems that deliver mobility without oil may be a challenge for some of today's transport professionals, but their future careers could depend upon it.

Government officials will be held accountable for society's ability to make a smooth transition to transport systems that require less oil. This book could stimulate them to look for more and possibly even better advice in their formulation of the policies that guide us. Much of what is presently received as wisdom in government circles deserves to be challenged. We have heard, for example, that there is no need to worry about oil depletion because of all the oil in Alberta's tar sands. We have heard too that if there are transport-related challenges, they can be satisfactorily addressed with vehicles that use hydrogen fuel cells or biofuels, or both. In Chapter 3, we provide reasons for being skeptical about the efficacy of such "cures."

The most important feature of coming oil depletion we want to convey to government officials is its imminence. In respect to oil production, the world appears to be in a similar situation to that of the US in 1970. In that year, the US produced more than ever before (or since) and the high levels of production were comforting. The decline began in 1971 and by 1973 US production was down by three percent. This was enough, with the actions of the Organization of the Petroleum Exporting Countries (OPEC), to precipitate the first "oil crisis" in late 1973, leading to large price increases for transport fuels, fuel rationing (as noted above) and major declines in stock values. Today, the world is at or near a peak in oil production, again with some of the comfort brought by high production levels. That production is inexorably declining may not be apparent until about 2015 or later. If moves toward transport revolutions are not made before then, too many years will have been wasted.

There appear to us to be some government officials who understand something about oil depletion but believe the challenges to be so far in the future that early, effective action is not warranted. Such a disposition favors the promotion of uncertain, long-term solutions such as hydrogen fuel cells. "If they don't work," the attitude seems to be, "there will be time to try something else; and in any case I will be long retired and therefore absolved from responsibility." An understanding of the potential imminence of oil depletion could be a stimulus to timely and effective policy making.

There will be huge amounts of money to be made, and lost, in the coming transport revolutions. In the business world, the difference between riches and ruin is often a matter of timing. An owner who sells a shopping center that will not be well served by transport once oil depletion sets in could avoid a large loss. Similarly, purchase of a presently depressed property that will be well served by electric transport could result in considerable gains.

Comparable opportunities and risks from oil depletion await business leaders in manufacturing, trade and finance. This book could help some get ahead of others in profiting from the considerable adjustments to come. The return on business leaders' time and money spent on this book could be among their better investments.

Lastly, we commend our book to students in the academic fields that fill the ranks of the transport professions. They will find it differs considerably from most of the books and articles assigned during their training to become transport engineers, managers and planners. Few, if any, of the tools, models and techniques presented to today's students will be helpful in making the most of the coming transport revolutions. Most current university transport programs prepare students to continue supporting transport arrangements that will not survive the coming oil depletion crisis.

Ideally, of course, we would like this book, or similar analyses, to become assigned reading for students in transport courses, including courses in engineering, business, economics, geography, political science and other programs that have some focus on transport. Short of that, we urge students to use this book to help them challenge what they are taught. Such challenges should always be offered judiciously, in the recognition that much of what we have to say is about the future, which will surely produce surprises.

We hope this book has another kind of value to students and to researchers generally. It brings together in one place an unusually large and varied amount of information about the state of transport today, and transport's energy use and impacts. We have taken pains to be accurate and balanced in our presentation and, even more, to provide readily accessible sources where the information can be checked and further analyzed.

For all these reasons, we believe this book will be of value even if our basic argument turns out to be wrong. We could be wrong in several ways. The peak in world oil production could be far ahead rather than "just around the corner." In that case, transport revolutions could well be driven more by a perceived need to avoid climate change, and much of what we say could apply. The peak in oil production could be in 2012 as we expect, but not trigger steep increases in the prices of transport fuels. Steep price increases could occur, but with little impact on transport activity. Even if we are wrong in one or more of these ways, we hope there is enough in the book to provide a solid resource about transport's present and to stimulate thinking about transport's future.

We believe our basic argument—oil depletion, high energy prices, serious impacts—is the best conclusion that can now be made about the future. The prospect of depletion of the main energy source for systems we rely on completely should be truly alarming. It is as if the power for our life-support system were about to be cut off by a blackout—or at least limited by a brownout. We are alarmed, but we also are confident that solutions exist to deal with our predicament. These solutions involve redesign of the transport sector, which need not wait for breakthroughs in technology and could begin today with good planning and effective leadership. We hope this book will help you participate in these transport revolutions and inspire you to make them happen soon.

Sources and Terminology

We have tried to be scrupulous in providing pointers to sources for all the information provided in this book. Where no source is given, it is an oversight, or the information is common knowledge or it is the result of an analysis by the authors during the book's preparation. *Superscript numbers provide the links between the text of each chapter and source and other notes at the end of the book.* The sources of material in boxes, figures and tables are also given in these notes, referenced by the superscript numbers at the end of their captions.

Sources are usually substantial documents such as government reports, books and articles in academic journals. Such sources are compiled in a reference list at the end of each chapter. They are referred to in the source notes using the Harvard system, e.g. Gilbert and Perl (2009). Some sources are more ephemeral; for example, newspaper articles.

These sources are listed fully in the source notes, usually with the uniform resource locator (URL) that points to the web page where the source is available. URLs are also given for sources in the reference lists, where they are available. *All sources we give were available during July or August 2009. All the URLs we give worked during that period.*

The notes at the end of the book are chiefly about sources, but a note can also contain additional information relevant to the part of the text to which it is linked. We urge readers to consult the notes, where there are often answers to questions begged by the text. Some of the notes may be considered to be interesting reading in their own right. Material is there rather than in the text because it is more technical or because it would interrupt the flow of the text.

This edition of *Transport Revolutions* is being both written and published in Canada where metric units increasingly prevail. We follow Canadian practice in using kilometers and liters rather than miles and gallons. A kilometer is roughly six-tenths of a mile. A liter is roughly a quarter of a US gallon. Vehicle speeds are in kilometers per hour (km/h). Mass is usually in kilograms (roughly 2.2 pounds) or (metric) tonnes. Each tonne is 1,000 kilograms, or about 2,200 pounds, or about 1.1 (short) tons.

Energy terms, too, are metric: for example, joules rather than British Thermal Units (BTUs) and exajoules rather than quads. A thousand joules or kilojoule (kJ) are roughly the same as a BTU. An exajoule (EJ) is a billion billion (10^{18}) joules and is roughly the same as a quad (which is a million billion BTU).

Perhaps the most confusing presentation for readers who use miles and gallons is that for vehicle fuel consumption. We use liters per 100 kilometers (L/100km) rather than miles per gallon. In this metric representation of fuel use, lower numbers mean lower fuel use. Five L/100km is roughly equivalent to 47 miles per US gallon (m/g); 9 L/100km is about 26 m/g; 16 L/100km is about 15 m/g.

By way of compensation for using units and terminology that many people in the US (and some elsewhere) are unfamiliar with, we have put all monetary values in US dollars unless otherwise indicated.

The only non-metric measure we use often is "barrel," because in Chapter 3 in particular we say much about oil production and consumption and so many of the available data are in barrels. A barrel is about 159 liters, or 42 US gallons.

For exact equivalents for the above units and many others, we recommend among several good sources the relevant part of the website of the US National Institute of Standards and Technology (NIST) at physics.nist.gov/cuu/Reference/unitconversions.html.

Finally, we should apologize for the large number of abbreviations used in the book, for measures, organizations and countries. They have been used chiefly to reduce the book's length, although they are sometimes an aid to reading. We apologize particularly for the ungainly "US," used for the United States of America. The more usual abbreviations are U.S. and U.S.A., but these look odd against the standard abbreviation for the United Kingdom—UK—and odd in combination with certain kinds of punctuation. We use US often, partly because the book has a focus on this most transport-intensive country and partly because we have often stayed away from using America or American(s), at least until Chapter 5, in order not to confuse or irritate people in other countries of the Americas. Unfamiliar abbreviations are defined quite often in the text and the notes, and there is a compilation of all abbreviations used, and what they stand for, at the end of the book.

1

Learning from Past Transport Revolutions

Overview

There have been two kinds of change in human mobility since hominids began exploring the African savannah: incremental change and revolutionary change. For much of history, people made incremental improvements to their inherited technology and practices for moving about. Tinkering with wheels, sails and engines produced real transport advances, but these gradual changes do not provide understanding of what makes a transport revolution occur and where it can lead. This chapter focuses on revolutionary changes: the more dramatic instances of rapid shifts from prevailing to new mobility patterns. These sudden changes were disruptive. They broke patterns of how people relied on technology for enabling mobility and they quickly changed expectations of what the norm in trade and travel was. These revolutions thus show how transport alternatives can reshape society.

We discuss past revolutionary changes in this chapter to help readers of this book think about the revolutionary changes in transport that we expect will occur during the early part of the 21st century. With the possible exceptions of riding a new high-speed train in France or receiving confirmation that their first overnight express package arrived, few readers will have personally experienced a transport revolution at around the time it was launched. Our five vignettes of past transport revolutions should help evoke those dynamics of change that, sooner or later, will become a lived experience for those reading this book.

What do we mean by a revolutionary change in transport? We need a definition that provides a clear, measurable distinction between an incremental change and a revolutionary change. Here is our proposal: A transport evolution is a substantial change in a society's transport activity—moving people or freight, or both—that occurs in less than 25 years. By "substantial change" we mean one or both of two things. Either something that was happening before increases or decreases dramatically, say by 50 percent; or a new means of transport becomes prevalent to the extent that it becomes a part of the lives of ten percent or more of the society's population. The two key features of our definition are these: First, there is a change in how people or freight move; the mere availability of a new technology does not constitute a revolution. Second, the change occurs relatively quickly; by our definition, horseback riding would not qualify as revolutionary because its extensive adoption likely took hundreds or even thousands of years.

A transport revolution is a substantial change in a society's transport activity that occurs in less than 25 years.

A new technology such as the unicycle or the Segway is not revolutionary until it results in a significant shift in the way people travel. This could take the form of a large number of new trips using the new mode or a shift to the new mode from an existing mode such as bicycling or walking. Even "big" technological advances such as the Boeing 747 or the Airbus 380 aircraft do not count as revolutionary unless they result in large, rapid changes in transport activity.

Our concept of a transport revolution is thus behavioral, and differs from the usual way of characterizing transport revolutions in terms of availability of transport modes or technologies. An example of a more conventional characterization is in Table 1.1.

In this chapter, we present five examples of transport revolutions that expose the common and uncommon elements of major mobility change. Through these examples we identify some of the factors and forces that precipitate revolutionary rather than evolutionary mobility change. Our examples are meant to be illustrative rather than exhaustive of the range of factors associated with a significant reconfiguration of transport technology and socio-economic organization.

We begin with a transport revolution motivated by the belief that Britain's industrial revolution was generating more goods movement than existing roads and canals could accommodate. Britain's emerging railway entrepreneurs believed that the steam locomotive offered a technology that could deliver a faster and cheaper mobility option and thus generate considerable profit while meeting growing demand. A belief that existing transport is inadequate and that major improvements are required can thus be a key factor spurring the investment and risk-taking required to launch revolutionary new mobility.

A different kind of transport revolution occurred during the Second World War, when the US suddenly restricted the production and use of automobiles and the expansion of its road network in order to accelerate military mobilization. This revolution highlights the role that governmental authority can play in reorganizing mobility when national security is perceived to be at stake. In this case, the reorganization was achieved through the imposition of gasoline rationing and industrial planning, used as tools to radically redesign the way people moved locally and between cities. By 1942, the private automobile had lost its place at the forefront of America's mobility growth. Intercity trains and local public transport were filled as they had never been before and mostly have never been since. This transport revolution ended as suddenly as it began, with a quick downsizing of military production and a rush back to car production that set the stage for a great suburban expansion.

TABLE 1.1 Transport revolutions in human history[1]

Era	Approximate Date	Ways of moving people and goods
Paleolithic	From ca. 700,000 BP	First migrations of hominids from Africa
	From ca. 100,000 BP	First migrations of modern humans from Africa
	From ca. 60,000 BP	First migrations by sea to Australasia
Agrarian	From ca. 4000 BCE	Animal-powered transport
	From ca. 3500 BCE	Wheeled transport
	From ca. 1500 BCE	Long-distance ships in Polynesia
	1st millennium BCE	State-built roads and canals
Modern	1st millennium CE	Improvements in shipbuilding, navigation
	From early 19th century	Railways and steamships
	From late 19th century	Internal combustion engines
	From early 20th century	Air travel
	From mid 20th century	Space travel

Note: BP = before the present; BCE = before the common era (i.e., before Year 1 in the Christian dating system); CE = common era.

Between 1950 and 1975, the third transport revolution we describe involved a profound transformation in the way people traveled over long distances. The rapid replacement of ocean liners by aircraft as the main means of traveling across the Atlantic represented a revolution in the intercontinental movement of people. A key element of this change was the adaptation of transport technology invented for military use—jet aircraft—to yield dramatic performance improvements in an existing mode. This example also shows how a revolution can trigger the subsequent reinvention of an apparently obsolete transport mode: in this case, the reincarnation of the ocean liner as a cruise ship.

Our fourth example concerns another approach to adaptation where the innovation in technology occurs under public sector initiative with a civilian focus. The reinvention of the passenger train began with the introduction of high-speed rail in Japan in 1964 and in Europe by 1982. Limitations of existing train technology and enterprise structure prompted innovators to "go back to the drawing board" and develop a new railway system that had little in common with its predecessors. The result was a major change in the way that people traveled between cities 300–800 kilometers apart. This transport revolution reinforces the concept that mobility options can be reinvented after a period in which they experience decline in the face of competition.

From 1980 onward, the movement of cargo by aircraft underwent a transport revolution that rounds out our consideration of these upheavals. Before this revolution, air cargo was being moved almost entirely in the holds of passenger aircraft, with limited integration into ground transport networks. Entrepreneurs at Federal Express applied "hub and spoke" routing to flights carrying only cargo, integrated these with door-to-door delivery, and launched a revolutionary expansion in freight movement. From being an exceptional and expensive proposition, next-day delivery times became commonplace. This transformation in air freight service levels made it possible to develop global logistics networks that could support production and distribution on an unprecedented scale. It shows how organizational changes can be as important for a transport revolution as changes in technology.

We chose to explore these five transport revolutions because they illustrate a range of dynamics that could be expected to occur in coming transport revolutions. In their impact, they have not necessarily been the most important revolutions. Moreover, only some of their attributes will appear in the coming round of mobility changes. With an understanding of how things have happened in the past gained through review of these revolutions, there could be less surprise at the scenarios for revolutionary change presented later in the book and possibly even less surprise at the changes that will eventually occur.

Had we been seeking out transport revolutions of the greatest magnitude, rather than those that illustrate a broad spectrum of change dynamics, we might well have included the transformation of global logistics and manufacturing enabled by widespread adoption of standardized shipping containers. According to one economist, "The shipping container may be a close second to the Internet in the way it has changed the international economy, and in that way, our lives."[2] Box 1.1 provides excerpts from a book that elaborates this point. We provide current data about the movement of shipping containers in Chapter 2.

Britain's Railway Revolution of 1830 to 1850

By the 1820s, Britain's industrial activity and global trade required the movement of large volumes of raw and finished materials between her cities. Most of this movement was on the extensive network of canals built up over the previous 50 years. Despite the significant profits generated for canal owners, canal capacity was not expanding as fast as the demand to move goods, and something more was needed. Some of the most intensive freight movement occurred between Liverpool, which was England's major Atlantic seaport of the 19th century, and Manchester, a rapidly growing industrial center located 50 kilometers inland.

The considerable growth of freight transport made conditions ideal for developing new transport capacity. Between 1820 and 1825, the number of ocean-going vessels docking annually at Liverpool rose from 4,746 to 10,837, with a commensurate increase in the weight of goods shipped.[4] Henry Booth, one of the entrepreneurs behind the ensuing railway revolution, characterized the Liverpool to Manchester mobility needs as ripe for a breakthrough. In 1824, 409,670 bags of American cotton arrived in Liverpool, much of it destined for Manchester's textile factories. The wealth generated by industrial development and colonial trade had spurred population growth in both cities, with Liverpool counting 135,000 inhabitants and Manchester 150,000 in 1824.[5] Each day in 1825,

BOX 1.1 Shipping containers[3]

On April 26, 1956, a crane lifted fifty-eight aluminum truck bodies aboard an aging tanker ship moored in Newark, New Jersey. Five days later, the *Ideal-X* sailed into Houston, where fifty-eight trucks waited to take on the metal boxes and haul them to their destinations. Such was the beginning of a revolution.

A soulless aluminum or steel box held together with welds and rivets, with a wooden floor and two enormous doors at one end: the standard container has all the romance of a tin can. The value of this utilitarian object lies not in what it is, but how it is used. The container is at the core of a highly automated system for moving goods from anywhere, to anywhere, with a minimum of cost and complication on the way.

Merchant mariners, who had shipped out to see the world [now have only] a few hours ashore at a remote parking lot for containers, their vessel ready to weigh anchor the instant the high-speed cranes finish putting huge metal boxes off and on the ship.

...at a major container terminal, the brawny longshoremen carrying bags of coffee on their shoulders [are] nowhere to be seen.

A 35-ton container of coffeemakers can leave a factory in Malaysia, be loaded aboard a ship, and cover the 9,000 miles to Los Angeles in 16 days. A day later, the container is on a unit train to Chicago, where it is transferred immediately to a truck headed for Cincinnati. The 11,000-mile trip from the factory gate to the Ohio warehouse can take as little as 22 days, a rate of 500 miles a day, at a cost lower than that of a single first-class air ticket. More than likely, no one has touched the contents, or even opened the container, along the way.

The container, combined with the computer, made it practical for companies like Toyota and Honda to develop just-in-time manufacturing... Such precision...has led to massive reductions in manufacturers' inventories and corresponding huge cost savings.

Fewer than one third of the containers imported through southern California in 1998 contained consumer goods. Most of the rest were links in global supply chains, carrying what economists call "intermediate goods," factory inputs that have been partially processed in one place and will be processed further someplace else.

22 horse coaches made the three-hour journey between the two cities, with up to seven more added when demand warranted.[6] They provided a total daily capacity of 688 passengers in each direction.

In 1820, three canal routes carried cargo between Liverpool and Manchester in circuitous routes of up to 80 kilometers. Journey time for the flat-bottomed canal boats, each carrying up to 29 tonnes (about 32 short tons), was 12 to 18 hours. Even with three alternative routes, canals were a profitable investment during their heyday. One estimate noted an annual return of over 50 percent on the Duke of Bridgewater's investment in his privately held canal.[7] Another study documented that shares in the publicly traded Mersey & Irwell Navigation Company grew in value from the initial issue price of £70 in 1736 to £1,250 in 1825, and paid an annual dividend of £35 per share.[8]

These profits, and the canals' limited capacity, led to discontent among Liverpool merchants and Manchester industrialists. Some were keen to invest in an alternative that promised strong competition for the canals in terms of cost and speed. They noted that Britain's first steam-hauled railway, the Stockton and Darlington, had produced an immediate reduction in the cost of transporting coal, and thus its price. One year after the railway opened in 1825, the price of coal at Stockton had fallen by more than 50 percent.[9]

The first public prospectus for the "Liverpool and Manchester Rail-Road" [*sic*] was issued in October 1824. Seeking to both raise investment capital and build public support for this project, the prospectus suggested that a new means of mobility was needed to serve the public interest by improving on the *status quo*:

> ...the [rail] transit of merchandise will be effected in four or five hours, and the charge to the merchant will be reduced at least one-third. Nor must we estimate the value of this saving merely by its nominal amount, whether in money or in time: it will afford a stimulus to the productive industry of the country...
>
> It is not that the water companies have not been able to carry goods on more reasonable terms, it is that they have not thought proper to do so. Against the most arbitrary exactions the public have hitherto had no protection, and against indefinite continuance or recurrence of the evil, they have but one security: IT IS COMPETITION THAT IS WANTED.[10]

This argument for meeting mobility needs by a wholly new means attracted support from shippers and manufacturers. The railway revolution also faced vigorous opposition from skeptics who perceived it as a threat. Canal and turnpike operators suggested that the railway could wipe out the value of their investments and thus undermine the British economy. Simon Garfield quotes Cornwall member of parliament James Loch, who served as auditor of the Bridgewater Canal, as demanding a high threshold of proof before Parliament chartered any more railways because "...it can never be advantageous to a country that much of its capital should be unnecessarily annihilated and a vast number of persons dependent on the existence of that capital reduced to poverty, except [when] such a sacrifice is demanded on the clearest public necessity, founded on incontrovertible general principles."[11]

Such proof was hard to establish. There was little commercial experience to draw from. Moreover, accurate surveys for future rail lines were vigorously resisted by landowners who saw the insertion of rail infrastructure across their property as a costly disruption. Initial surveys of the Liverpool and Manchester railway had to be conducted at night, sometimes under false pretext, to overcome landowners' refusals to allow railway survey parties access to their property.

The Liverpool and Manchester's proponents failed to secure enabling legislation from Parliament in 1825. Opponents characterized the uncertainties associated with rail transport as presenting excessive risk, which, at the outset of transport revolutions, can appear especially formidable. Financial risk could ruin investors in canals and turnpikes. Safety risk could injure or kill passengers, employees and bystanders in horrific accidents. Environmental risk could undermine agriculture and wildlife once the English countryside was penetrated by machines generating enormous quantities of noise and smoke. At the outset of some transport revolutions, the risks of breaking away from existing technology and established organizational arrangements can be magnified by the uncertainty of what might unfold. But the growing demand for mobility did not ease up, and may have even been stimulated by a pre-emptive 25 percent reduction in canal charges.[12]

By 1826, when Parliament did approve the Liverpool and Manchester's charter, some key skeptics had been won over. The Marquess of Stafford, who then controlled the Bridgewater Canal, was enticed to become a major shareholder in the railway. He was offered 1,000 shares in the Liverpool and Manchester at £100 per share. This was one quarter of the railway's initial capitalization and offered the opportunity to appoint three members to the company's board of directors.[13]

Major landowners along the rail line became convinced that their natural resources and agricultural produce would be worth more if trains could carry them to new markets. Not all the advances in support for this project came through a clearer understanding of what the railway could do. The 1825 legislative proposal had been definite about the intended use of steam locomotives. The 1826 act and debate surrounding it left the matter of propulsion intentionally unspecified. If steam locomotives did not prove themselves, the Liverpool and Manchester's promoters would rely upon other means. The uncertainty over how much a new technology would become a part of this new venture appeared to facilitate skeptics' acceptance of this transport revolution.

Inter-city horse coaches could not compete with the railway on either speed or price.

The technical innovation behind this revolution was introduced to a large audience at the now famous Rainhill steam locomotive trials of 1829, held well into the Liverpool and Manchester railway's construction. George Stephenson's "Rocket" achieved immortality for itself and its designer by hauling heavy loads and reaching a top speed of 48 kilometers per hour (reported as 30 miles per hour). This performance made quite an impression on those who saw it and on the many more who read reports about the event in the UK and US newspapers. The Rocket's tour de force set a new benchmark for inland transport because,

> Rainhill demonstrated that man was now freed from the constraints of animal power for land transport, that he was going to be able to travel two to three times faster than animal power had ever permitted, with consequent reductions in journey times... This was widely understood at the time (though not invariably accepted immediately). All the improvements made to transport over the preceding century with much effort were insignificant by comparison...[14]

The Liverpool and Manchester Railway's commercial results proved to be similarly dramatic following inauguration on September 15, 1830. Performance during the first six months is shown in Table 1.2 in comparison with what the railway's architects had anticipated five years previously. The expectation that more would be earned from freight than from passengers turned out to be wholly wrong.

Faced with new competition, canal companies lowered their rates and thus attracted shipments that were both less time sensitive and more price sensitive than the goods moving by rail. Ransom comments that profits on the Bridgewater canal stopped growing after the Liverpool and Manchester railway was inaugurated, and

> though goods traffic on the L & M railway increased over the ensuing years, so, in general, did traffic on the Bridgewater Canal... Even the canal's passenger traffic continued and was not decimated as the road coaches were—it seems that although the waterway journey between Liverpool and Manchester took longer than the rail journey, it was also cheaper. The eventual ascendancy of railways over canals was a much more gradual affair than their ascendancy over roads...[16]

Canal operators thus adopted a new commercial strategy following the railway revolution that has kept their business going to this day. Those in the road transport business also made adjustments to accommodate the much greater efficiency and speed of railway transport in moving people. These changes required embracing a wholesale substitution of rail for coach in moving passengers between cities.

Inter-city horse coaches could not compete with the railway on either speed or price. Before the railway, coach journeys from Liverpool to Manchester took three hours at a price of 12 shillings for an inside seat

TABLE 1.2 Expected and actual performance of the first trains to carry passengers and freight[15]

	Expected (1st year)		Actual (1st 6 months)	
	Number	Net revenue	Number	Net revenue
Passengers	90,000	£10,000	188,726	£43,600
Freight (tonnes)	305,000	£70,000	36,400	£22,000

and seven shillings for an outside perch. The train cut the travel time to two hours. A seat cost seven shillings for first class and five shillings for second class, which was still more comfortable than the interior of a horse coach. The full capacity of the coach services between these two cities had peaked at 108,000 travelers per year before the railway, but the Liverpool and Manchester carried 460,000 passengers in its first year.[17] Faced with such a competitor, horse coach operators did one or both of two things: they joined the rail revolution as investors and partners, and they shifted their operations to provide much more local mobility.

Coach operators quickly shifted their intercity operations to much more local "omnibus" service, carrying people on short journeys that railways did not serve. They also began moving freight to and from the nearest railway station. Once the railways reached London, the city's two largest coach operators, William Chaplin and B. W. Horne, merged their businesses in 1846 and became major carriers for the railways.[18] Horne also bought into the London & South Western Railway and became its chairman. Horse-powered coaches quickly disappeared from Britain's intercity roads, but they remained a fixture of local travel for decades following the railway revolution.

The discrepancy between pre-construction expectations and the actual performance highlights two outcomes often arising from transport revolutions. First, new transport revolutions often produce results that differ considerably from the predictions made about their future. Second, established organizations that were delivering mobility before the revolution rarely quit the business. Instead, they attempt to refashion their business model to take account of new competition and reposition their service.

New transport revolutions often produce results that differ considerably from the predictions made about their future.

Britain's railway revolution marked a major milestone in expanding the efficiency, speed and capacity of overland transport. The growing demand for goods and passenger movement during the Industrial Revolution encouraged the integration of available technologies such as iron rails and steam engines to produce faster, more reliable and cheaper inland mobility over long distances. Once the potential of this breakthrough was evident, existing transport technology and techniques were adapted to provide other forms of mobility, filling niches that the steam railway could not serve well (at least in its infancy). As a result, British society experienced a quantum leap in mobility, while the canal and horse coach continued to move people and goods over and above what the growing railway network could deliver.

Britain's railway revolution ushered in a new era of mobility that was incompletely anticipated by both its architects and those who were skeptical of such massive change to the transport system. The new mobility was eagerly embraced by those who could subsequently afford to travel or to buy products that were shipped using the new transport capacity.

The Great Wartime Pause in Motorization in the US

The great pause in the growth of mass motorization that occurred during the Second World War receives scant attention from most accounts of transport achievements in the US. The dramatic abatement in the production and use of motor vehicles is typically viewed as an anomaly compelled by the necessity to shift manufacturing from civilian to military production, and the rationing of key resources including rubber and gasoline. Driving less was thus seen as just another of the many sacrifices demanded from Americans during the war, like meatless Tuesdays and drinking coffee without cream or sugar. Yet this understanding of temporary sacrifice captures only part of the story behind the most ambitious effort in the US to date to restrain personal mobility. This effort deserves closer attention in order to understand the full story behind the transport revolution that put the growth in mass motorization on hold so decisively and so effectively.

As the war expanded, the stakes involved in curtailing automotive production were very high.

All belligerents in the Second World War limited civilian automobile production and restricted the use of cars, but the scale of motorization in the US going into the war put its effort to restrict the automobile in a class by itself.[19] In 1941, Americans owned almost 30 million cars, about three quarters of the world total.[20] Those cars were driven a total of more than 440 billion kilometers. Annual vehicle production in 1941 was 3.8 million cars, an output that had been exceeded only in 1928, 1929 and 1937. Indeed, as wartime destruction spread across Europe, civilian auto production in America picked up because more Americans were working in military production and thus earning the means to purchase cars.

Automakers were profiting greatly from this indirect, but quite powerful, boost to their sales produced by the start of hostilities in Europe.[21]

As the war expanded, the stakes involved in curtailing automotive production were very high. Gropman wrote that in 1941, automobile manufacturing in the US "was equal to the total industry of most of the countries in the world" and went on to explain the implications for industrial production:

> ...[it] was spread over 44 states and 1,375 cities... More than 500,000 workers produced autos and trucks when the United States entered the war—one out of every 260 Americans. And 7 million others—one out of every 19 Americans—were indirectly employed in the industry. Automobiles consumed 51 percent of the country's annual production of malleable iron, 75 percent of plate glass, 68 percent of upholstery leather, 80 percent of rubber, 34 percent of lead, 13 percent of copper, and about 10 percent of aluminium.[22]

With auto production taking up so much manpower and resources, the US could not meet its military production needs at the speed demanded when it entered the war in December 1941 without reducing the number of cars being built. The key question was how far to go in converting car production capacity to military manufacturing. Government and industry failed to answer the question before Pearl Harbor was bombed, setting the stage for a transport revolution.

On May 29, 1940, after Holland and Belgium capitulated to invading German forces, and less than one month before France's surrender, the US became serious about industrial mobilization. President Roosevelt signed an administrative order establishing the National Defense Advisory Commission (NDAC). The Commission was charged with developing a voluntary plan to reorient American industry toward military production. NDAC's transport division was chaired by Ralph Budd, president of the Association of American Railroads. Budd was strongly averse to government intervention. Under Budd's leadership, the transport planning efforts undertaken by NDAC were "usually preliminary or perfunctory" because "protecting the transportation sector from the regulatory hand of government remained Budd's first priority."[23]

Railroads were not the only industry to resist closer links to defense mobilization, and by 1941, NDAC had achieved so little that the Roosevelt administration brought the effort to mobilize industry within government. The Office of Production Management (OPM) was created to orchestrate military manufacturing and the Office of Price Administration and Civilian Supply (OPACS) was launched to regulate non-military production at a level that could support optimal military output.

Both OPM and OPACS negotiated with the automotive industry, yielding rival plans for converting civilian vehicle production to military output. OPM shared the auto industry's perspective on manufacturing priorities, likely the result of its two chief negotiators' backgrounds in the automotive sector. OPM's negotiations were led by William Knudsen, who had been a vice president of General Motors Corporation, and John Biggers, a former executive at the Libbey-Owens-Ford Glass Company, a major automotive supplier. They accepted industry's contention that not all auto production facilities could be converted to military manufacturing, and that there was nothing to be gained by stopping production at the plants that had no defense production capabilities. The OPM plan proposed a remarkably precise 20.15 percent reduction in the output of personal vehicles (i.e., cars) between August 1, 1941 and July 31, 1942.[24]

From the 3.8 million personal vehicles that rolled off US assembly lines in 1941, production shrank to just 143 vehicles in 1943.

OPACS had a different plan. Director Leon Henderson and assistant administrator Joseph Weiner were academics trained in economics and law. They approached the need to reduce auto production from a broader perspective that sought to balance inputs and outputs across sectors. Henderson and Weiner were particularly concerned that the auto industry's enormous appetite for raw materials would cripple other economic functions. For example, steel going into passenger cars would hamstring food processing plants that required steel for canning. Weiner was confident that sharply restricting the output of passenger cars would pose fewer difficulties than trying to maintain production. He claimed "That the civilian population can get along without them [passenger cars] in time of emergency and that such existing goods can be made to last longer was demonstrated during the depression and undoubtedly will be demonstrated in the present war."[25]

OPACS pressed for a 50 percent reduction in auto production, more than twice that proposed by OPM. In late August 1941, a plan was reached to reduce car output by 6.5 percent in the coming quarter, with cuts increasing each quarter thereafter to reach a 43.3 percent overall reduction by the summer of 1942. This plan was just starting to be implemented when Japan attacked the Hawaiian Islands on December 7, 1941.

The War Production Board (WPB) replaced OPM on January 16, 1942, and its chair Donald Nelson wasted no time in calling a halt to US car production. His first official act was to order a cessation of all passenger car production and light trucks by February 10, 1942.[26] The industry's claim that its manufacturing capacity could not be converted to military production was quickly disproved as the major manufacturers pulled car assembly lines apart, retrofitted as much as 75 percent of this machinery to produce war materiel, from anti-aircraft guns to heavy bombers, and literally threw the remaining material and equipment into scrap heaps. One month after WPB's stop-production order had been issued, automotive capacity was being converted with extraordinary rapidity: "...Discarded machinery is pushed out with such great haste that workmen do not even have time to apply a cover or a coating of grease to protect metal against the elements."[27]

From the 3.8 million personal vehicles that rolled off US assembly lines in 1941, production shrank to just 143 vehicles in 1943. This left more than 25 million civilian vehicles in operation for the war's duration,[28] but two key resources available to run them were in short supply. America's access to natural rubber was cut off in late 1941 by the Japanese occupation of Southeast Asia. Pre-war consumption had been about 600 million tonnes annually, and by 1941 a stockpile of a year's supply had been amassed.[29] By 1944, the stockpile had shrunk to slightly more than 100 million tonnes despite massive investment in synthetic rubber manufacturing. In the intervening years, new tires were strictly rationed. Noting that pre-war drivers had replaced an average of one tire per year, a journalist described tire rationing as a "slow paralysis" of America's car fleet and suggested that

> If businessmen and housewives gauge their driving against standards of absolute necessity, they will find very little of it is imperative. The choice is theirs to conserve tires or to become pedestrians within six months to two years.[30]

Gasoline rationing began in 17 states along the East Coast from the middle of May 1942 and was extended nationwide by December of that year. It was necessitated not by any lack of petroleum supply, which the US had in relative abundance at that time, but by constraints on transporting fuels from refineries along the Gulf and West coasts that had shipped the bulk of their output by ocean-going tankers. These ships had come under attack by submarines, and were being diverted to supply military operations in Europe and the Pacific. Government officials emphasized the grave risks facing those who moved oil by tanker to justify the utmost efforts to conserve gasoline (see Box 1.2).

Major war production plants joined in the effort by charging workers a fee of ten cents for each empty seat in their car.

BOX 1.2 Wartime appeal to reduce petrol consumption[31]

Washington, 23 April — Following is the joint appeal to automobile owners issued by Harold L. Ickes, Petroleum Coordinator; Donald Nelson, War Production Board chairman; the Price Administrator, Leon Henderson; J.B. Eastman, Defense Transportation Director, and the War Shipping Administrator, Emory S. Land:

It is not possible to transport enough petroleum to the seventeen Eastern States to meet both essential war needs and normal civilian demands. Very substantial reductions in gasoline consumption must be achieved immediately. Motoring-as-usual is out. Already hundreds of men have been lost at sea trying to bring in the oil needed for war. No patriotic American can or will ask men to risk their lives to preserve motoring-as-usual.

Since the sailors' lives are at stake every time a tanker plies between the Gulf Coast and the East, it is unthinkable that they be asked to take the risk of going down on a burning ship in order that some one may have gasoline to go to a bridge party or the ball game. In fact, it is unthinkable that they be asked to take this risk for any purpose except to fill those requirements which are absolutely essential.

If a motorist fills up the tank to go to a picnic, some defense worker may not be able to get to his job. If a man drives to work alone every day, instead of working out a car-sharing plan with his neighbors, he may take gasoline from a truck that is hauling for a war plant.

Motorists are asked by their government to reduce their gasoline consumption to the absolute minimum — not any specified percentage, but as much as they possibly can.

Government officials did not hesitate to regulate the ways in which civilians used cars during the Second World War. For most of 1943 a ban on "recreational" driving was in effect. Those caught by vigilant police officers were summoned before local ration boards and could be stripped of their gasoline coupons as punishment. The hearings were open to the public and covered by the press to drive the message home.[32]

As well as the stick of canceling the rations of wasteful drivers, there was the carrot of extra rations for participants in "car sharing clubs" that pooled travel by neighbors or co-workers in a single vehicle. News reports emphasized the positive and suggested that the car in wartime was becoming a "sociable drawing room on wheels." Reporting on this new phenomenon, Coan wrote,

> ...car sharing—also called the group riding plan or club and many other names—establishes a new outlook on motoring. That basic American institution, the automobile, is undergoing changes. Drivers are learning to know one another. Through proximity, they are taking the chips off their shoulders.[33]

Major war production plants joined in the transport demand management effort by charging workers a fee of ten cents for each empty seat in their car on entering their parking lots. Government advertising reinforced the message that car travel should be shared whenever possible. The poster shown in Figure 1.1 illustrates the graphic connections that were made between conservation and patriotism and between wasteful driving and aiding the enemy.

FIGURE 1.1 Poster promoting car-sharing produced by the US government during the Second World War[34]

Americans proved quite accepting of these changes to their mobility, perhaps because a majority did not yet consider cars to be essential to daily life.[35] A Gallup poll in January 1942 asked a nationwide sample, "If it were not possible for you to use your car, would this make any great difference to you?"

A majority of 54 percent answered there would be no great difference and 73 percent said they could get to work without a car.[36]

Another reason for the openness of the US to this transport revolution was its participatory implementation. In Canada and the UK, gasoline rationing was administered by civil servants. In the US, rationing boards were run largely by volunteers. Derber notes that "the board of neighbors idea facilitated the recruitment of many prominent and highly capable citizens who would have otherwise been unobtainable. This resulted in both securing community acceptance for rationing and in providing a very capable rationing board."[37]

The number of car vehicle-kilometers declined by 41 percent from 1941 to 1943, rising slightly in 1944 and a little more in 1945.[38] Car sharing met some of the mobility demand, but many travelers turned to public modes of transport including buses and streetcars, and local, regional and intercity trains. This turned out to be a "golden age" for public transport in the US, one that would be followed by a steep and long-lasting decline.

The success of local public transport and intercity railway carriers in providing for a surge in traffic while facing wartime restrictions on equipment, fuel and personnel was an extraordinary accomplishment. Passenger trains' share of intercity travel rose fourfold from 8 percent of total passenger-kilometers in 1941 to 32 percent in 1944, and intercity bus travel more than doubled, from 4 to 9 percent (see Figure 1.2).[39] Rail's share of freight movement increased from 61 percent of total tonne-kilometers (tkm) in 1940 to 72 percent of a much larger total in 1943.[40] Public transport ridership rose dramatically from 1940, when close to 13 billion trips were recorded, to 1946, when there were some 23 billion trips.[41]

Wartime traveling conditions were far from comfortable. Most trains and buses were crowded, schedules for long-distance trains were particularly unreliable, and railways and transit companies were perennially short staffed. Some sense of the conditions under which people traveled by train and bus during this transport revolution is offered by Cardozier, who wrote,

> ...the demand for transportation exceeded the supply and travelers became accustomed to long waits and discomfort. ...passenger trains...were always crowded; it was not unusual to board a train and find dozens of soldiers and sailors sleeping on the floor.

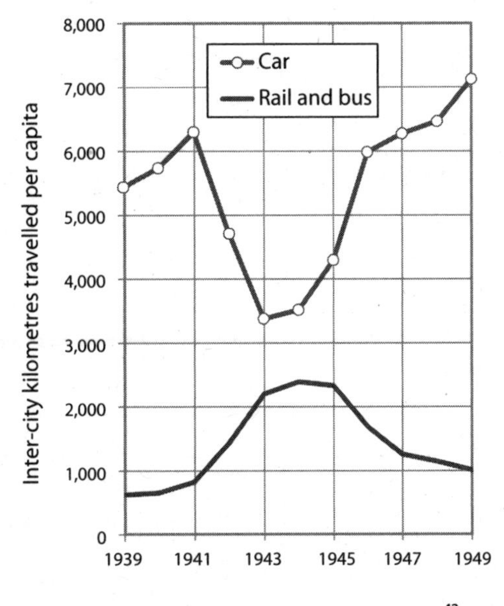

FIGURE 1.2 US intercity travel, 1939–1949[42]

> Any limitations on the number passengers allowed per car were ignored, and trains accepted passengers as long as there were places for them to sit, stand or lie down.[43]

Public transport providers, most of which remained privately owned, were aware that service quality was generally poor and that extra efforts would be needed to retain their new riders once wartime restrictions were lifted. In its 1945 *Transit Fact Book,* the American Transit Association adopted a defensive tone:

> A decline in traffic after the war was inevitable. Every management recognized that. The problem is to hold the decline to a minimum. That means that new equipment must be procured at the earliest possible moment and that maintenance of equipment must be restored to pre-war standards or better as fast as the manpower and materials to do it can be obtained.[44]

As the end of war drew within sight, government officials guiding the civilian economy shifted their priority from maximizing efficiency to stimulating production. Their primary goal was to avoid a steep recession, as had occurred following the First World War's rapid demobilization. Their solution was to boost civilian industries that could pick up as much as possible of the manufacturing capacity created over the previous three years. The winding down of military production was seen to require a parallel winding up of consumer purchasing in order to avoid a steep downturn.

The car took top billing as a product to fill the looming void in manufacturing that would arise once the armed forces were no longer in combat. All the steel, rubber, chrome, glass, leather, gasoline and other materials that could not be spared during wartime would rapidly glut the market unless these commodities could be absorbed into consumer production. And the defense workers skilled in building aircraft, ships and tanks would be readily employable in car manufacturing. This challenge demanded fast action, since "within hours after Japan surrendered, billions of dollars worth of government contracts were cancelled, including $1.5 billion in the Detroit area alone, and hundreds of thousands of people were thrown out of work."[45] Suddenly, rail's efficiency in providing high levels of mobility with low inputs became irrelevant, if not an outright liability. Trams and trains were entirely off the radar when government searched for means to stimulate postwar prosperity. WPB authorized car plants to resume vehicle production in the second half of 1945, and 69,500 cars rolled off the assembly lines that year. In 1946, production was back in full swing and over two million cars were built.[46]

The wartime experiment in putting the brakes on motorization revealed how quickly and dramatically the often characterized "love affair" with the automobile could be interrupted.

Intercity passenger train travel became the first casualty of this postwar return to mass motorization. Passenger-kilometers by train dropped by 30 percent between 1945 and 1946 and by another 30 percent from 1946 to 1947.[47] Ridership on local public transport reached a peak of some 23 billion trips in 1946, but then fell steeply such that by 1960 there were only about 9 billion trips and by 1971 only 6.5 million, the lowest post-war total.[48] Most intercity railways and many transit companies invested in new equipment after the war, but the pace of producing this more specialized transport technology lagged well behind the mass production of the car. During the several years that it took for new passenger trains, streetcars and subway cars to arrive in service, momentum had swung against public transport in the US. It would not be until the 1970s that the post-war decline in ridership was reversed. Then, massive government assistance was committed to saving the remnants of the country's passenger trains and public transport operations.

The wartime experiment in putting the brakes on motorization revealed how quickly and dramatically the often characterized "love affair" with the automobile could be interrupted. One key condition for success was the call to sacrifice in the name of a serious threat. Another seemed to be community participation in adjusting the burden imposed by the rationing of fuel and tires.

The halt to car production ensured that new mobility needs would have to be met by transport alternatives. For example, after 1941, people hired into wartime jobs could no longer buy a new car, but would go to work by public transport or by riding in somebody else's car. A year-long ban on recreational driving spurred the discovery of ways to enjoy free time without a vehicle. Even after the ban was lifted in 1943, the limited supply of tires and fuel reinforced the perception that a car no longer provided unlimited mobility benefits. The car's wartime mobility role was reshaped by a fixed ration of gasoline and by the unpredictability of having to deal with a flat tire. Either constraint could leave drivers stuck in a sudden state of immobility that was far more constrained than the public transport services available to those without cars.

What the US wartime experience shows above all is that in an emergency government can act to change transport activity dramatically and effectively. Supported by a clear rationale and sufficient but flexible authority, the US government was quickly able to reduce Americans' reliance on automobility. *Government can be a prime instigator of a transport revolution.*

The Big Switch in Transatlantic Travel in the 1950s

Over the past four centuries, the Atlantic Ocean has been the busiest avenue for very long-distance movement of goods and people. Movement between Europe and the Americas was initiated by the rise of empires, and then accelerated by commerce and migration. For most of the time that people have sought to move across the Atlantic, there was little choice about how to make the trip. Ships offered the only option until the 1930s, and remained the predominant means of making this journey during the 1940s and early 1950s. Transatlantic ship crossings had improved from being dangerous and uncomfortable[49] to offering luxury for some and widespread comfort for many travelers, while providing reliable and efficient movement of cargo.

During the 1950s, the balance of options for intercontinental mobility shifted decisively. As shown in Figure 1.3, between 1949 and 1973, air travel across the Atlantic exhibited the kind of unbroken growth that signals a transport revolution. The number of people crossing the Atlantic by air surpassed those making the journey by sea in 1957, and in 1958 the first commercial jet planes began flying across the Atlantic, simultaneously expanding the speed and reducing the cost of intercontinental air transport.

> Over the past four centuries, the Atlantic Ocean has been the busiest avenue for very long-distance movement of goods and people.

During subsequent decades, transatlantic air travel set the pace for transforming flight between continents from an expensive luxury service to a popular mode of transport. This transport revolution grew out of airlines' and aircraft manufacturers' success in commercializing the major advances in military aircraft design and propulsion achieved during the Second World War. Domestic air travel in North America and Europe during the late 1930s would be at least somewhat recognizable to those who have flown

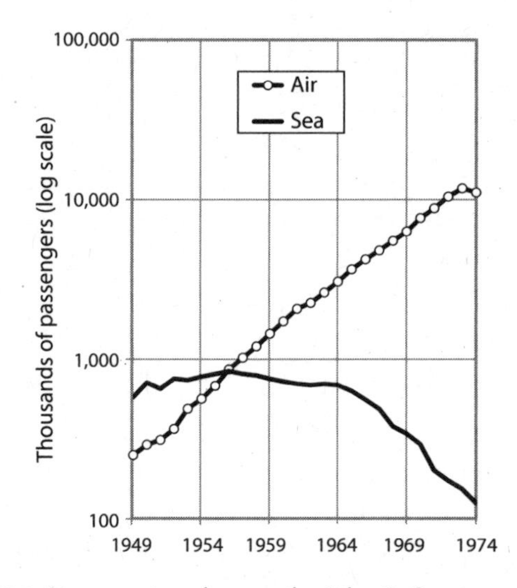

FIGURE 1.3 Air vs. sea travel across the Atlantic Ocean, 1949–1974[50]

recently aboard propeller aircraft, but flying between continents was another matter.

During the 1930s, flights linking Europe with Africa, North and South America and Asia were pioneered by carriers using "flying boats." The carriers included Imperial Airways in Britain, Pan American Airways (Pan Am) in the US and Australia's QANTAS. These massive propeller aircraft landed and took off on water because existing airport runways were too short to accommodate their take-off requirements. Flying boats were relatively slow and had to make frequent refueling stops, but still reached their destinations three to five times faster than ships. Pan Am's flying boat from San Francisco to Hong Kong was scheduled over five days and Imperial's Southampton to Sydney service took nine days. Flying boats rivaled the first class service on ocean liners with sleeping berths, dining rooms serving freshly cooked meals, bars and lounges. And the 20 to 70 passengers on these pioneering intercontinental flights paid fares that were up to six times the cost of passage on an ocean liner, roughly equaling what would be charged for a flight on the supersonic Concorde during the 1980s.[51]

Advances in aviation technology during the Second World War rendered flying boats obsolete and set the stage for an intercontinental transport revolution. Virtually all the elements found in today's aircraft were introduced to meet military needs during the period 1940–1945. Jet engines had propelled the German Messerschmitt ME 262 and the British Gloster Meteor in dogfights at Mach 0.8, the cruising speed of today's commercial jets. More importantly for the early post-war period, bombers such as the B-29 were designed to fly long distances above the range of ground-based anti-aircraft guns. These long-range bombers advanced airframe designs enough for later commercial flights to be able to fly across oceans and stay above bad weather. The bigger runways that were built for bombers could accommodate aircraft with the fuel payload needed to fly between continents. Government-led war production had yielded many key ingredients of modern air travel, but a post-war integration was needed to adapt these breakthroughs to a commercial framework that would make intercontinental aviation a part of the transport mainstream.

In 1950, transatlantic flying already represented the world's busiest air travel market.[52] Airlines were then carrying over one third of travelers crossing the Atlantic.[53] Large pressurized propeller aircraft such as the Lockheed Constellation Douglas DC-6 and Boeing 377 Stratocruiser had replaced flying boats.[54] Air France and Belgium's Sabena had made the least change from the flying boat business model by being the first to offer all-sleeper flights across the North Atlantic in 1947.[55] Given the flight time of 12 to 14 hours, with refueling stops in Gander, Newfoundland and Shannon, Ireland, providing each traveler with a bed was attractive. But traveling in a bed moving at almost 600 kilometers per hour was a luxury that could be afforded only by the elite. Business leaders, government officials, celebrities and dignitaries who flew across the Atlantic paid a premium fare in exchange for saving three or four days at sea. Many more could not afford the cost of flying. New and refurbished ocean liners served this broader market, thereby supporting a modest growth in business in the early 1950s, illustrated in Figure 1.3.

Ships such as Cunard's *Queen Mary* and *Queen Elizabeth* offered refined luxury and spacious accommodation in first class suites that were priced far above the transatlantic airfare. These great liners, and many lesser ships, also offered spartan accommodation in multi-berth tourist cabins for those who were traveling on a budget. Indeed, passenger ships had been the first low-cost carriers in international travel for a previous generation, facilitating the mass departure of Europeans to the new world at "immigrant fares" as low as £1.50 in 1904.[56]

North American airlines were especially eager to expand their market, and targeted the large US middle class in two ways. Carriers cut their transatlantic fares in the spring of 1952 and introduced "fly now pay later" credit arrangements to attract American leisure travelers to take their holidays in Europe.[57] The typical two-week vacation period in the US made taking a holiday in Europe by ship impractical. Bringing the price of flying within reach of the American middle class opened up a major market for intercontinental tourism. Transatlantic airlines embraced this opportunity, and their adoption of "tourist class" was so popular that it remains a synonym for economy airfare. Tourist travel contributed to an accelerated growth in air travel in 1953 and 1954. Meanwhile, Pan Am was pressing aircraft manufacturers to build a commercial jet that could cross the Atlantic without refueling.

Jet aircraft promised a major advance in expanding air travel through a combination of better performance, higher productivity and lower operating costs in an era where fuel was a near-negligible component of airline expenses.[58] Jet engines were much simpler and thus cheaper to maintain than propeller engines. One study cited airline estimates that a Boeing 707 would yield a 30 percent reduction in operating costs per seat mile compared to a Lockheed Constellation.[59] Aircraft productivity would be greatly improved because the seven- to eight-hour flight times between Europe and North America would enable a plane to make a return flight every 24 hours.

The British de Havilland Comet was the world's first commercial jet. In 1952, British Overseas Airways Corporation (BOAC) inaugurated Comet service between London and Johannesburg. Even with refueling stops in Rome, Beirut, Khartoum, Entebbe and Livingstone, the Comet made the trip in 23.5 hours compared to the 40 hours propeller planes had required.[60] However, the Comet lacked the range for a non-stop flight between New York and London or Paris and was beset by a series of initially mysterious crashes resulting from a design flaw that took several years to re-engineer. The Comet's tarnished safety record dealt a setback to the de Havilland's commercial jet development and opened the door for US manufacturers Boeing and Douglas to overtake the British in producing civilian jets. But it would take a push from the US airlines for the manufacturers to seize this opportunity.

Pan Am's competitors appeared satisfied to expand their fleets with large propeller planes, based on manufacturers' claims that commercial jets were not yet ready for the North Atlantic. But Pan Am's president Juan Trippe was intent on pushing Boeing and Douglas to meet his company's needs. To meet Trippe's demand, manufacturers would have to deliver major innovations in airframe and jet engine designs.

Boeing had become the US leader in jet airframe design because of an unusual energy advantage. Designing the first generation of jet aircraft had required wind tunnels producing tremendous airflows to test aerodynamics at close to the speed of sound. Rodgers pointed out that "wind tunnels drew vast amounts of current, and the cost of electricity to run them was a major consideration. Because of the newly built Grand Coulee Dam, power in the Pacific Northwest was cheap."[61] It is thus not a coincidence that America's first civilian jetliner was designed by a company with access to vast amounts of low-cost electricity to run a 13.4-megawatt wind tunnel.

The plane that emerged from this test bed was the Boeing 367-80, known as the "Dash 80," a dual purpose prototype that was intended to yield both a commercial passenger jet and a tanker used for mid-air refueling of military aircraft. This approach enabled Boeing to spread development and production costs across both military and civilian markets, but it did not offer a design that could carry 160 passengers across the Atlantic non-stop. The factor limiting the Dash 80's length (which determined passenger capacity) and fuel payload (which determined its operating range) was the power of its Pratt & Whitney J-T3C jet engines, the civilian variant of a design powering the first generation of B-52 bombers. However, Trippe knew that Pratt & Whitney was developing a more powerful engine for the military.

In a negotiating coup that became a legend in post-war airline history, Trippe persuaded Pratt & Whitney to move their prototype into production even before any aircraft had been purchased. Trippe threatened to go to the UK and buy from Rolls Royce if there were no American engine powerful enough to move Pan Am's planes from New York to London or Paris without refueling. He also committed to buying 120 of these powerful engines even before he had secured any aircraft that could make use of them.[62] After Pratt & Whitney agreed, Trippe played Boeing and Douglas off against each other to design planes that could use these powerful engines for the non-stop Atlantic crossing.

By the spring of 1959, the 707-320 was flying nonstop between New York and London and Paris at the speeds that are offered today.

On October 13, 1955, Trippe addressed the annual general meeting of the International Air Transport Association, suggesting that mass mobility by air would "prove to be more significant to world destiny than the atom bomb."[63] As a result of Trippe's negotiating acumen, the jet aircraft had come to serve aviation's biggest market faster than would have otherwise occurred. Aircraft manufacturers' incremental adaptation of military designs could have added years to the jet's emergence as the predominant means of trans-Atlantic travel.

The first commercial Boeing 707 flight was operated by Pan Am between New York and Paris on October 26, 1958, three weeks after a BOAC Comet 4 had flown the first paying passengers across the Atlantic by jet. Initially making a refueling stop in Gander, Pan Am reached Europe in

8 hours and 41 minutes. By the spring of 1959, the 707-320 was flying non-stop between New York and London and Paris at the speeds that are offered today, with Douglas's DC-8 entering the North Atlantic service in September 1959.[64] Over 90 percent of the seats were occupied, and Pan Am's transatlantic air passenger volumes doubled within five years of launching the 707.[65]

The airlines' bonanza came at a high cost for the aircraft manufacturers. Designing a jet for the transatlantic market had cost Boeing a great deal more than allocated for designing a joint platform civilian and military aircraft. According to one account, "...it wasn't until 1964, nine years after Boeing sold its first 707, and twelve years after the company started heavy spending specifically for the prototype, that Boeing recouped its entire investment in the jetliner... The expenditure on all aircraft delivered to the break-even point was over two billion dollars."[66] Aircraft manufacturers took over a decade to cash in on the jet age, but the passenger shipping industry would take even longer to develop a workable strategy that could adapt to aviation's progress.

Ocean traffic held steady for a few years following the introduction of jets, but steamship lines had to reduce fares to keep passengers sailing. Cunard introduced significant economy fare discounts in 1962, in direct response to competition from jet aircraft.[67] The shrinking revenues occurred at the same time as shipping lines faced growing operating costs. As the resulting losses from operating transatlantic liner service mounted, schedules were cut back steadily during the 1960s. Some ships were redeployed to warm water cruise itineraries during the winter months, while others were simply laid up and sold off, either for scrap or to new operators who could dramatically reduce operating costs.

Although shipping companies had been operating cruises as a sideline to passenger transport since 1844, when P&O Lines pioneered a "special Mediterranean tour,"[68] the cruise lines that emerged following the beginning of the jet age were specialized carriers, operating symbiotically with the expansion in global tourism that air travel had facilitated. The full-time cruise ships that emerged during the 1970s were almost exclusively operated under "flags of convenience" from jurisdictions such as Liberia, Panama and the Bahamas to avoid the high taxes and wages existing in most American and European jurisdictions and thus dramatically reduce operating costs. A 1973 inventory of passenger ships noted that since the jet age arrived on the North Atlantic there had been a 75 percent decrease in the number of passenger vessels flying the American and British flags and a 40 percent reduction in French-flagged passenger ships. Conversely, the number of passenger ships flying the Liberian flag had increased by 270 percent, Panamanian registry had grown 190 percent and Greek-flagged ships were up 260 percent.[69]

Cruise lines soon began to work cooperatively with airlines, booking many of their passengers on air packages that included flights to and from their cruise. After 1974, only Cunard's *Queen Elizabeth 2* was making regular crossings between Europe and North America, offering transatlantic passage mostly during the summer months.[70] Today's cruise ship industry is growing and profitable, but it offers a very different kind of mobility from what the ocean liners once provided. For most of today's long-distance ocean travelers, their ship has become a destination in itself whose transport function is only incidental to the "fun" of cruising. Only a handful of travelers now use ships for travel between continents. Nearly all people disembarking today's great passenger ships head directly to the airport to complete their journey.

This great transformation highlighted the ways in which military conflict can generate technology that becomes available for civilian transport. The same forces that triggered a sudden suppression of civilian travel by automobile during the Second World War produced the infrastructure and technology that would vastly expand mobility once hostilities were over. The longer runways and larger aircraft that were put to civilian use immediately following the war instantly turned pre-war flying boats into museum pieces. While the jet engine took a decade to become widely used in civil aviation, its arrival accelerated the speed and extended the scope of this transport revolution. Once the aircraft manufacturers had been cajoled into building civilian jets, the means to make intercontinental flights a widely used transport option were at hand.

As in the railway revolution in the UK more than a century before, the ascendancy of a faster and more powerful transport alternative did not doom its predecessor to an industrial extinction. Ocean liners largely disappeared, but cruise ships emerged as a thriving and profitable niche for marine transport. Once the technology breakthroughs attained in pursuit of military objectives had diffused through the civilian transport sphere, both flying and sailing were barely recognizable from their previous manifestations, yet both have remained a vital element among the world's transport options.

The High-Speed Rail Revolution of 1960 to 1985

By the late 1950s, growth in aviation and car ownership was drawing travelers away from long-distance trains. For some major railway companies these trends would lead to bankruptcy. That of the Penn Central Corporation was the most spectacular of these downfalls, the largest corporate collapse of its time.[71] In the UK, an investigation into railways' future that came to be known as the "Beeching Report" proposed major reductions in British Rail's network. It started from the premise that "...the industry must be of a size and pattern suited to modern conditions and prospects."[72] The modern conditions were seen to require shrinking Britain's rail network to eliminate capacity that was not being used as travelers chose driving and flying alternatives. In 1958, a special inquiry conducted by the US Interstate Commerce Commission also emphasized the fact that travelers were leaving the rails for other mobility modes. The Commission noted,

> the inescapable fact...seems to be that in a decade or so [the passenger train] may take its place in a museum along with the stagecoach, the sidewheeler and the steam locomotive... [T]his outcome will be due to the fact that the American public now is doing about ninety percent of its travelling by private automobile and prefers to do so.[73]

But that is not what happened. The railway's anticipated slide into obsolescence was decisively reversed. This first occurred through exceptional projects, notably Japan's pioneering Shinkansen between Tokyo and Osaka and France's innovative Train à Grand Vitesse (TGV) between Paris and Lyon. But the passenger train's second arrival as a renewed means of passenger travel has spread across much of Europe, and also beyond Japan to Korea and Taiwan.

The railway's anticipated slide into obsolescence was decisively reversed by high-speed rail.

The catalyst for this rail transport revolution was the introduction of technology that enabled trains to become much more effective at meeting an important subset of travel needs: trips between

200 and 1,000 kilometers along densely populated corridors. This required reconceptualizing the rail mode's historic role as a universal mobility provider. Banister and Hall suggest that the "strangest point of all" in this exceptional transport comeback is that high-speed trains "...are literally the fastest things on earth...represent an incremental technology, evolving out of the basic steel wheel on steel rail...system that George Stephenson borrowed...from the Northumbrian colliery tramways of his apprenticeship."[74] The high-speed rail revolution thus involved creating a new relationship. It was between the train technology that had evolved over more than a century in delivering universal mobility, on the one hand, and the specialized technology that had been developed to move people rapidly over the busiest travel corridors, on the other hand.

While Europe and North America were extending the frontiers of flying and driving, it was the Japanese who pioneered an organizational and technical redesign of passenger trains that yielded a breakthrough in the rail mode's performance. The attributes of the renewed passenger train were first brought together in the Japanese Shinkansen, which means "new trunk line."

First, the railway's role in providing mobility was no longer conceived of as being universal. Instead of being developed around the premise that early railways had embraced, namely that trains could move everything everywhere, the Shinkansen was designed to add new capacity on Japan's busiest transport route, the corridor between Tokyo and Osaka.[75] Plans to build a new rail line between Japan's two largest cities had been approved in 1939, with construction actually starting in 1941.[76] Had this initial development not been destroyed by bombing in 1944, Japan's post-war passenger rail trajectory might have been quite different. For one thing, the pre-war railroad had been run as part of a public bureaucracy that had changed little since the 19th century. In 1949, the Japanese National Railway (JNR) emerged as a public corporation, a new form of public enterprise. This attenuation from the civil service was intended to curb the power of militant public employee unions that appeared threatening to the American occupying government.[77] JNR would enable a more dynamic approach to rail management.

The second ingredient of this transport revolution was significant government support for rail technology research. The breakthroughs in rail vehicles, infrastructure and propulsion systems that made the Shinkansen concept a success were achieved only after the war, when Japan's industrial trajectory was intentionally redirected away from aerospace and military production. During this formative period of the high-speed rail revolution, the UK, Canada, the USSR and the US were devoting much of their research capacity to building new aerospace and atomic energy capacities.[78]

Instead of working for military suppliers and aerospace industries in the early 1950s, some of Japan's top electrical, mechanical and civil engineers were drawn to the Railway Technical Research Institute (RTRI), the scientific arm of the old rail bureaucracy that had become part of JNR. Their goal of developing high-speed intercity passenger trains was something that had dropped off the agendas of other nations' technology policies. In May 1957, RTRI held a conference marking the organization's 50th anniversary. Engineers presented a paper entitled *High Speed Railway in the Future* that outlined how newly designed infrastructure and rolling stock could enable a passenger train to cover the 550-kilometer distance from Tokyo to Osaka in three hours. Key design elements were a purpose-built track and signal system, and the use of lightweight electric multiple-unit trains that could attain unprecedented speeds. The presentation was seen as an "epoch-making event" in modern railroad development because it was the first time a new technology was put forward to gainsay the conventional wisdom that trains were becoming obsolete.[79]

At the time the RTRI went public with its proposed innovation, Japan's parliament had already approved a five-year plan for incremental upgrading of the narrow-gauge line between Tokyo and Osaka. Such an approach would not have yielded the performance breakthrough that the Shinkansen's designers eventually achieved. JNR's president, Shinji Sogo, created a way to move beyond such incrementalism by launching a special commission of investigation into Japan's railway future in May 1956. Directed by JNR's vice-president of engineering, Hideo Shima, the commission explored the need for an entirely new rail line between Tokyo and Osaka.[80] In a late-life memoir, Shima emphasized the revolutionary spirit that animated his work on railway reinvention:

> At the time, air and car traffic were showing remarkable growth. I thought that building a line that would soon fall behind the advancing transport world would be regrettable for the future of JNR and in meeting social expectations. I decided to build a railway that would be useful and rational for a long time into the future.[81]

The commission released its findings at about the time that the RTRI presented its breakthrough in high-speed train technology. Taken together, these findings showed a promising fit between the dramatic growth in travel demand that was projected to require expansion of all modes and the high-speed train's ability to offer a major increase in capacity. The commission's most conservative forecast predicted that travel between Tokyo and Osaka would more than double between 1957 and 1975, while freight volume would more than triple. Even taking construction of a proposed superhighway between Tokyo and Osaka into account, JNR's analysis projected that roads could handle just 10 percent of the growth in passenger travel and only 5 percent of the increase in freight traffic.[82]

Within a week of the RTRI symposium, JNR president Sogo formally requested the Ministry of Transport to authorize "improving" the line between Tokyo and Osaka, known as the Tokaido ("east sea route") line. His request cleverly sought to gain approval in principle for rail development that went beyond the incremental upgrading that had been called for in JNR's five-year plan.[83] Detailed cost estimates of building an entirely new high-speed line or pursuing more extensive upgrading of the existing narrow-gauge line were left to be developed after the government had given a green light to proceed.

Sogo's ability to gain this approval in principle gave JNR momentum in developing more detailed plans for deploying new rail technology. In August 1957, the JNR's Tokyo–Osaka line investigation commission began work and quickly focused on adapting the new high-speed technology to rail's most promising market segment in Japan. What would become the Shinkansen service concept was refined during this phase of analysis. The major innovation was to separate this new rail infrastructure from existing rail tracks and stations. As for most other railways, JNR's infrastructure had been designed to serve many purposes by accommodating local, intercity and long-distance passenger services along with freight trains on the same tracks. The Shinkansen would break from the universal mobility model that had predominated since the Liverpool and Manchester Railway.

Instead of trying to serve all mobility needs at the same time, the Shinkansen would focus on a niche that could take full advantage of the train's improved speed capabilities. Just as aviation innovators were doing during the 1950s, JNR's leadership proved adept at creating a mobility option that could offer the traveling public something new compared with what railways had previously provided and what other modes could deliver. Shinkansen was designed to be faster than a car, while also more frequent, cheaper and more convenient than a plane. This design proved to be a winner, with its first success being a formal approval from the Japanese Cabinet for the project's launch on December 19, 1958.[84]

Gaining a policy commitment to realize the Shinkansen concept was neither straightforward nor inevitable. Shinkansen skeptics could be found both within Japan's rail industry and abroad. These critics saw no need for the major expenditure in new technology and infrastructure because they viewed rail technology as being of diminishing value compared to the advances in aviation and road transport. A nation that was determined to catch up with, and even overtake, US and European economic development had made a costly error by investing so heavily in rail.[85] One critic wrote,

In less than 25 years the Shinkansen became the backbone of Japan's intercity rail system.

> Unfortunately, motor vehicle and air transport capacity was 10 years too late to save the country from the expense of the Shinkansen... Up until the mid-1950s, JNR did not have the funds to build the Tokaido Shinkansen and if it had waited until the end of the 1960s before deciding whether to launch the project, it would have questioned the need for constructing the new line. Instead, it probably would have added more double-tracked narrow-gauge sections to the high traffic-density areas.[86]

Financing the Shinkansen tested the resolve and ingenuity of railway innovators, who had to articulate their vision to those who were less inclined to see trains as having a bright future. When Japan's cabinet had authorized construction of the new line in 1958, a budget of ¥194.8 billion was approved, the equivalent of $541 million in 1958 and $3.84 billion in 2007.[87] Japan's fiscally constrained government approached the World Bank for a loan of $140 million for the project ($995 million in 2007 dollars).[88]

After detailed evaluation of the project, the Bank agreed to loan $80 million on April 24, 1961 (about $600 million in 2009 dollars). This was the third largest loan that the Bank had made and its largest transaction to date with Japan.[89] *Business Week* reported that the World Bank's support for Shinkansen "...enrages many an American railroad man who would love to have access to similar sources of financing."[90] But what railway executives outside Japan had not yet grasped, and would take decades to figure out, was that access to a major infusion of capital was dependent on providing something more than just the same kind of mobility that railways had been delivering for decades. In securing the World Bank loan, JNR leadership convinced a major international organization operating at "arm's length" that a reinvention of the passenger train would succeed. In short, JNR took its transport revolution to the bank.

World Bank funding raised the stakes of the Shinkansen project because the Japanese government's international credit rating now depended on JNR turning this initiative into a commercial success. JNR's strategy was illustrated by an executive, interviewed shortly after the inauguration, who stated, "We just can't lose money with so much business waiting for us."[91]

Shinkansen has produced a steady stream of profits on the Tokaido line, offsetting to some extent the losses on additional lines that were developed in less populated and prosperous corridors.[92] One report noted in 2005, "Income from the Shinkansen lines totals about $19.2 billion per year, which is 47 percent of the JR group's rail operations income."[93] When JNR was restructured and partially privatized in 1987, the Shinkansen network was valued at ¥8.5 trillion (equivalent to just over $100 billion in 2005).[94] These assets comprised the Tokaido, Sanyo, Tohoku and Joetsu Shinkansen lines, totaling 1,833 route kilometers. There have since been 619 additional route kilometers of Shinkansen lines built, making for a current network of 2,452 route kilometers; 590 route kilometers are under construction.[95]

The Shinkansen concept has proven itself and created a transport revolution in so doing. In less than 25 years the Shinkansen became the backbone of Japan's intercity rail system, absorbing much of the forecast growth in travel that accompanied Japan's rapid economic expansion in the 1970s and 1980s. Japan's internal air services grew much more slowly than would have been the case without Shinkansen. Between Tokyo and Osaka, Japan Airlines' load factor (share of seats occupied) dropped below 50 percent during the first year of Shinkansen operation. All Nippon Airways, the other major carrier between Tokyo and Osaka, reported an eight percent drop in passengers, in a market that was reported to be growing at seven percent annually.[96] By the 1990s, Shinkansen began to show signs of maturity in that annual travel volumes stabilized in the range of 70–75 billion passenger-kilometers a year, and now account for about 30 percent of intercity rail travel.

Ordinary trains still play a vital transport role, but there is no question that the Shinkansen has become established as the core intercity rail service. As Japan's achievements in reinventing the railway became apparent, the lesson of this dramatic departure from traditional passenger services was not lost on railway entrepreneurs in other countries.

It was France that first succeeded in transplanting the Shinkansen's revolutionary potential into a society with much higher levels of motorization and domestic air travel. By adapting Japan's high-speed rail formula to work in western Europe, the French established this innovation as a transport revolution with a much greater relevance than many experts had judged possible.

France's TGV initiative was spurred by a decline in the rail sector, which was losing traffic to air and road competition.

France had a well developed air and highway network that was expanding to meet post-war intercity travel growth. France thus had a less compelling rationale than Japan for investing in high-speed trains. To deliver a long-term pay-off by reducing the need for highway and airport expansion, the French TGV would need to attract travelers away from cars and planes. This was a different proposition from providing new mobility to people who had not yet become accustomed to the speed of air travel or the convenience of the car. Thus transport experts in the 1970s were equivocal about the transferability of the Shinkansen's success to Europe and North America. But France had a particular impetus to innovate in the rail sector that had not been at work in Japan.

Whereas the management of JNR had conceived of the Shinkansen as a means to build upon an existing growth trend, France's TGV initiative was spurred by a decline in the rail sector, which was losing traffic to air and road competition. To reverse these declining fortunes, French railway management, government officials and rail equipment manufacturers developed a new approach to designing, deploying and operating passenger trains. This required recasting the relationships between government, public enterprise and private industry to deliver a new transport option.

Expending the financial and political resources needed to create a successful future for trains was a higher priority for the French government than for its counterparts in the UK and North America. French railways had been nationalized just before the Second World War when, in return for taking control of insolvent carriers, the government assumed responsibility for both past rail debts and future rail financing. The Société Nationale des Chemins de fer Français (SNCF)—French National Railways—was created as a "mixed enterprise" akin to a public-private partnership in today's parlance. The investors who had been bought out in 1938 remained silent partners, holding 49 percent of the joint venture's equity. Profits from a consolidated and "modernized" SNCF were to pay dividends to these investors until 1982, when their shares would be repurchased.[97] But the profits never materialized, necessitating that dividend payments be made from the public treasury during SNCF's first 43 years of operation.

SNCF's financing arrangements caused French government officials to encourage private- and public-sector collaboration in an effort to bolster the rail mode's commercial fortunes. This approach contrasted with their US and UK counterparts, who emphasized either public or private efforts to address the rail mode's challenges. British Rail was a fully public enterprise, for which the primary remedy for loss-making services was seen to be the retrenchment proposed in the Beeching report. In the US, private railways took the lead in economic restructuring by petitioning regulatory bodies—notably the Interstate Commerce Commission—for permission to abandon unprofitable services. But in France, the passenger train's incipient decline stimulated a broader consideration of policy options that blended public and private innovation.

In France's public sector, administrative reorganization had been identified as the key to economic renewal in an influential 1967 policy review known as the "Nora Report."[98] It presented a set of commercial principles for future operation of publicly owned enterprises, putting "...emphasis upon the need for them to operate as much like private enterprises as possible, putting commercial requirements before public service."[99] Reluctance to embrace such a paradigm was dispelled by the oil-shock-induced economic crisis of the 1970s, when reducing public enterprise deficits became imperative.

The Nora Report proposed differentiating potentially competitive market services from the social and public interest functions pursued by public enterprise. Where public enterprise was to produce a good, or provide a service, below its cost, a contract would be called for to specify quantities and prices and the subsidy that government would allocate in the public interest. Rather than providing indirect subsidies through accumulating public enterprise deficits, debt write-offs or other forms of creative accounting, the Nora Report called for explicit valuation of what a public service was worth. Conversely, commercially competitive services were given a "green light" to make money.

SNCF management embraced the Nora Report's recommendations as a validation of their goal to reinvent rail service from within and became enthusiastic adopters of its approach.[100] Application of these principles in 1971 led to creation of a five-year business planning model, contracts with government for non-commercial services, and other aids to give management more effective conditions for commercial development. But in order to reverse the train's commercial fortunes, these tools would need to be deployed in a strategy that focused on meeting new mobility needs.

The high-speed train revolution reveals the capacity of government and industry to collaborate on both the technical and the organizational redesign of a mature mobility mode.

During the 1970s, SNCF's management began planning to break away from the traditional railway model of offering a universal mobility mode, as the Japanese had done a decade earlier. But SNCF would have to do even better in creating a mobility alternative because the train would now be competing for travelers against established air and road networks. As in Japan, France's point of departure for this transport revolution was its most heavily traveled corridor, between the Ile-de-France, the region centered on Paris, and the Rhône-Alpes region centered on Lyons, whose respective populations in 1982 were just over 10 million and just over 5 million.[101]

As had occurred in the Tokyo–Osaka travel corridor, the existing rail infrastructure between Paris and Lyon had reached the point of overload. Adding track capacity was judged essential, but the decision to develop an entirely separate line for high-speed passenger trains and devote existing tracks to freight and local passenger train movements provided the essence of France's rail transport revolution. SNCF planners set a goal of linking Paris and Lyon by train in two hours to make rail the preferred mobility mode between these cities. A two-hour train ride would thus prove especially attractive to those making a same-day return trip between Paris and Lyon.

The question facing SNCF's leadership was what kind of infrastructure and propulsion technology could yield a two-hour travel time at a cost acceptable to officials who held the French treasury's purse strings? A key factor in reaching an answer was for SNCF to remain integrally involved in developing what would come to be known as TGV. Like JNR, SNCF became a leader in developing high-speed trains by nurturing its own design and engineering capacity through close partnerships with private manufacturers, rather than by leaving such efforts mostly in the hands of external designers. This approach gave SNCF executives the ability to fend off competing transport innovations, which would have diluted, and quite possibly diverted, their efforts to launch this transport revolution.

The competing French technology that had begun to draw resources and initiative away from TGV development was the Aérotrain, an innovative vehicle suspended on an air cushion generated by huge turbofan engines, while pulled along a track by a linear induction motor (an unusual type in which elements of the electric motor are in both the track and the vehicle rather than entirely in the vehicle). Research support for this innovation had been provided by France's Regional and Economic Development Ministry, with no participation by the Transport Ministry. During the late 1960s and early 1970s, development of TGV and the Aérotrain project proceeded in parallel.

This competitive development of new high-speed ground transport resembled German efforts during the 1970s and 1980s to develop both the Inter-City Express, based on high-speed rail technology, and the superfast Transrapid, which utilized magnetic levitation technology.[102] Unlike the German Federal Railway, SNCF was able to convince its political leaders that deploying TGV would meet both commercial and technical needs, and that proceeding with the Aérotrain would squander resources and inhibit a turnabout of the railway's fortunes. As a result, the TGV project moved into advanced development during the 1970s and the Aérotrain project was canceled.

An important effect of SNCF's partnerships with the railway supply industry was to align the capacity for introducing new rail technology with the pay-offs from deploying it. This linkage between a public railway operator and major private manufacturers made it possible to add an industrial development dividend into the investment calculus applied in making the commitment to high-speed rail. In addition to the revenues anticipated from bringing travelers back to the rails, there would be profits and jobs in an industrial sector that was seen to be shaky.

When considering the trade-offs around a major investment, public officials often tend to value maintaining existing jobs and profits more highly than developing new jobs and future profits. This is especially true if the existing sector appears threatened with industrial decline. French rail manufacturing was sufficiently at risk from a decline in passenger train travel, while also being a beneficiary from the gains of rail revitalization, for this to weigh in favor of government's investing in the TGV. Investing in the less certain pay-off from the Aérotrain's new technology also appeared likely to come at the expense of existing jobs and profits in the rail supply sector.

The results of TGV's development and implementation paralleled Japan's experience. It demonstrated that a high-speed rail-based transport revolution could work in a "western" setting. France's high-speed rail initiative provided the clearest evidence of passenger trains' potential to meet a modern mobility need better than any alternative, and in so doing generated profits in a competitive travel market. Figure 1.4 documents the rapid growth of TGV ridership in the first decade of its operation between Paris and Lyon. Patronage climbed from 6.1 million in 1982, the TGV's first full year of service, to 37 million in 1991.

During the late 1990s, TGV evolved from a train primarily serving the Paris to Lyon travel market into a high-speed ground transport network radiating from Paris to most corners of France. Unlike the Shinkansen, whose standard track gauge prevented interoperability on Japan's narrow-gauge rail network, TGV could make full use of SNCF's conventional electrified lines to extend service beyond the high-speed infrastructure. Through such incremental expansion, as well as additional high-speed tracks to Tours, the Belgian border and Channel Tunnel, Marseille and most recently Strasbourg, TGV has become an important element of the French railroad system.

Figure 1.5 shows how travel by TGV ridership grew in the 1990s and beyond, in comparison with travel by other French railways (SNCF).

TGV's share of travel grew from just over a quarter to just over a half of the total. Travel by TGV grew continuously, and appears to have reversed a decline in rail travel in France in the mid-1990s. The TGV has become the mainstay of intercity travel by rail.

This high-speed rail transport revolution has spread beyond France's borders and is evolving into a trans-European high-speed train network.

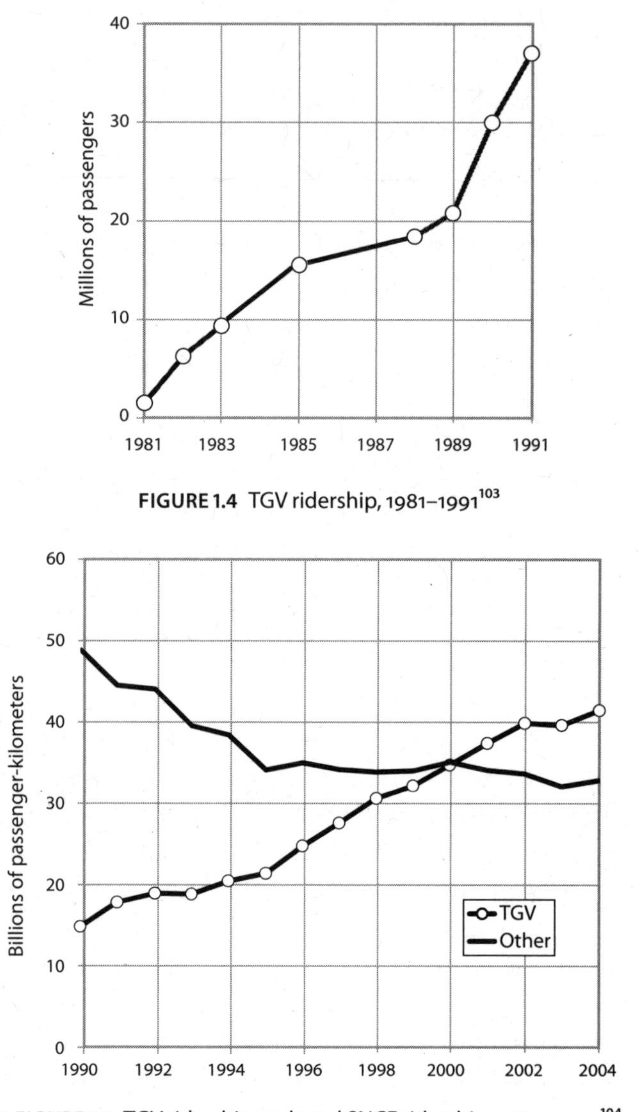

FIGURE 1.4 TGV ridership, 1981–1991[103]

FIGURE 1.5 TGV ridership and total SNCF ridership, 1990–2004[104]

From Stockholm to Rome and London to Berlin, high-speed trains have become the mainstay of Europe's intercity train service. They contribute important diversity to the mix of intercity travel options that is much less developed in North America, but has seen adaptation in Taiwan and Korea.

The high-speed train revolution reveals the capacity of government and industry to collaborate on both the technical and the organizational redesign of a mature mobility mode. The breakthroughs that launched both the Shinkansen and TGV required looking beyond both the existing arrangements by which railways had long established their role as a universal mobility provider, and the breakthroughs in aerospace and road transportation. When available evidence suggested to many in the UK and the US that the train was reaching the end of the line in its industrial development, innovators in Japan and France identified a promising combination of new technology and technique, and then succeeded in introducing it to their venerable railway institutions so as to renew their roles as mobility providers.

The Air Freight Revolution of 1980 to the Present

People began moving freight aboard aircraft, particularly mail, almost as soon as the first powered aircraft flew. The first recorded carriage of mail by plane was in India in 1911: some 6,000 letters and postcards were flown from Allahabad across the river Ganges to Naini, a distance of eight kilometers. They were postmarked with a special "First Aerial Post" cancellation, with the proceeds going toward the Oxford and Cambridge Hostel for Indian students started by the Holy Trinity Church in Allahabad. The US postal service established regular airmail delivery between New York and Washington DC in May 1918.[105] Since then, the most urgent freight shipments, such as emergency medical supplies, have been moved by air. There was considerable growth in goods movement by air in the decades after the Second World War, but the way in which air transport supported commerce and trade changed considerably during the 1980s.

This change followed the launch by Federal Express of a hub-and-spoke network for dedicated air cargo flights and the integration of this new air freight network with local trucking to offer overnight door-to-door delivery across North America, which became known as air express. The resulting transport revolution then expanded to cover and connect all continents.[106] It has enabled and accelerated structural changes in manufacturing, retailing and distribution that are often considered an integral part of contemporary "globalization." According to Rodrigue et al., "the fundamental question [regarding transport's role in globalization] does not necessarily reside in the nature, origins and destinations of freight movements, but in *how this freight is moving*" [emphasis in original].[107]

Before this revolution, air freight was largely overshadowed as an aviation industry concern by airline passenger operations. Most cargo that moved by air went in the bellies of passenger planes, on schedules that had been developed for the movement of people.[108] Shippers had to deliver their own cargo to the airport or pay a freight forwarder or local courier company to collect and deliver it.[109] As well, postal services offered "special delivery," but this did not guarantee a specific arrival time. These air cargo arrangements could move freight faster than ordinary mail, railway express or trucking, but they did not provide the guaranteed expedited delivery that people now take for granted and rely on.

In the US, overnight delivery actually became less available when airlines acquired wide-bodied jet aircraft in the 1960s, consolidated passenger schedules and cut back their overnight flights. Fewer overnight flights meant less overnight air cargo capacity. Moreover, unlike railways in the US, which had integrated local delivery into their "Railway Express" service, airlines were slow to match their considerable speed between airports with seamless ground connections. US airlines appeared satisfied to leave the work of collecting, assembling and delivering small shipments to freight forwarders, who received 40 percent of the industry's cargo revenue in the late 1960s and early 1970s.[110]

The arrangements for a transport revolution that would transform the relationship between aviation and freight were first applied by Fred Smith, a pilot and owner of Arkansas Aviation Sales. Smith believed that moving air cargo on dedicated planes that were not providing passenger service, and integrating it with a door-to-door delivery operation, would yield a faster and more efficient freight transport option that could unleash an immense demand. Smith's ideas were first mooted in a Yale University economics course in 1965, where he identified the hub-and-spoke concept as the key to overnight delivery, effected chiefly by air. The first iteration of this idea was not particularly well developed, nor was it well received by his professors, but Smith was determined to explore its potential. He had two independent studies completed in 1972 indicating an untapped market for express delivery of small packages in the order $1 billion, about $5 billion in 2009 dollars.[111]

Smith initially focused on the movement of checks for clearance by the US Federal Reserve Bank as a promising niche to enter the overnight delivery market. This led him to name the fledgling courier company Federal Express. In the 1970s, clearing a check drawn on a bank in one Federal Reserve district and deposited in another could take up to four days, because of the time it took to move the checks between the banks. In 1970, Smith estimated the float on such transactions at $3 million daily, about $14 million in 2009 dollars. This money could be saved by Smith's proposed delivery service, which would dispatch small "Federal Express" jets to each Reserve Bank district in the late afternoon, fly the checks to a central sorting station by the middle of the night, and then deliver checks to each Reserve district the next morning. Checks could easily be moved on specially configured Falcon executive jets, which were just below the aircraft size that would require route and schedule approval from the US Civil Aeronautics Board.[112]

By combining the hub-and-spoke aviation network with a ground delivery system, FedEx took the lead in American, then global, delivery of urgent shipments, which it has never relinquished.

Although the Federal Reserve Bank did not become a customer of Smith's new company, the Federal Express name stuck. FedEx began operations on April 17, 1973, flying 14 planes to 25 cities. That night, 186 packages were collected directly from shippers, driven to the nearest airport where a FedEx jet was waiting, flown to the carrier's Memphis hub and sent on to destination airports where trucks completed their delivery the next morning. The logistics worked, but FedEx lost $29 million developing the business over the next two years. By 1975, the model had blossomed. By combining the hub-and-spoke aviation network with a ground delivery system, FedEx took the lead in American, then global, delivery of urgent shipments, which it has never relinquished. FedEx now serves over 375 airports in 220 countries with 672 planes, handles over 6.5 million shipments daily (3.3 million of them express), with Fiscal Year 2006 revenue of $32.3 billion ($21.4 billion express).[113] The International Air Transport Association has regularly ranked FedEx first in scheduled freight tkm, noting that the carrier flew 15.1 million tkm globally in 2006.[114]

As both its freight volumes and markets served grew, FedEx became a technology pioneer. It developed leading-edge information systems and use of the Internet that opened the door to real-time logistics management on a global scale. In 1979, FedEx launched "COSMOS (Customers, Operations and Services Master Online System), a centralized computer system to manage people, packages, vehicles and weather scenarios in real time." The carrier had installed an electronic communication system in all its delivery vehicles by 1980, and introduced a computer-based shipping system for its customers in 1984. Instead of trying to operate a growing global logistics network using paper copies of waybills and sorting packages by hand, FedEx collaborated with suppliers to develop bar code labels, scanners and automatic sorters. Their bar codes for shipments went far beyond retailing applications so that each package's bar code contained all the information needed for effective logistics. By 1982, FedEx had machines to print bar-code labels, scanners to read them and a computer system to keep track of the information. In 1986, a hand-held device was deployed—the SuperTracker—that could digitize signatures. Online tracking of shipments was introduced in 1994.[115]

The explosive growth of air freight that followed FedEx's launch of the transport revolution soon attracted three major competitors: United Parcel Service (now UPS); Dalsey, Hillblom and Lynn (now DHL); and Thomas Nationwide Transport (now TNT). Each adapted their existing organizations and technology to provide integrated pickup, air freight and delivery across expanding, eventually global networks.[116] UPS, which began as a local delivery service for department stores in Seattle in 1907, was operating a national package delivery service, mostly by truck, at the time FedEx started up. In 1982, UPS began offering overnight express delivery, and by 1988 the company launched its own airline to fly express shipments across America and beyond. UPS's massive sorting facility and air hub are described in Box 2.4 (Chapter 2).

In 1969, DHL began flying customs paperwork for ships en route to the US west coast across the Pacific so that cargos could be cleared ahead of the ship's arrival. It entered the overnight parcel delivery business in 1979, established an air freight hub in Cincinnati in 1983 and added a European hub in Brussels in 1985. DHL has been particularly successful in Europe, and today is a wholly owned subsidiary of Deutsche Post. Volumes at DHL's Brussels hub have grown from 60 tonnes nightly in 1988 to 1,000 tonnes nightly in 2000. TNT was an Australian trucking company dating from 1946 that established a British subsidiary in 1978 and launched its own air operation in 1987. TNT, now based in the Netherlands, created a European "super hub" in Liège, Belgium in 1998 notable for its location on the Trans-European high-speed rail network.[117]

Behind the innovation of FedEx and its competitors, and in response to intensive lobbying by the carrier, governments liberalized policy frameworks that had limited the reconfiguration of airline routes into hub-and-spoke networks. Commercial liberalization in the airline industry began with air cargo, when the US Civil Aeronautics Board raised the weight limit on air taxi (unscheduled) services in 1972, thereby allowing FedEx to launch its hub-and-spoke system using Falcon jets. The removal of all restrictions on cargo aircraft size in 1977 enabled FedEx to expand its capacity.[118]

> The transport revolution in air cargo has changed both the way that many products are created and how they are distributed.

The US Airline Deregulation Act of 1978 pioneered the elimination of restrictions on which routes an air carrier could fly and what prices could be charged. The European Union emulated this deregulation of aviation at a slower pace, stretching the process from 1984 to 2001 with a series of three legislative packages that progressively deregulated flying between member states. Air cargo was usually at the vanguard of airline deregulation.

While the elimination of regulations stimulated air cargo innovation, the implementation of international trade agreements—including the North American Free Trade Agreement and Europe's single market—and the World Trade Organization's increasingly liberal trade regime have encouraged growth in the demand for fast, reliably scheduled movement of cargo over long distances.[119]

Air transport effectively shortened the time required to keep production lines moving across continents. Integrators such as FedEx offer a "just in time" delivery of both manufacturing inputs and finished products that reduce the need to maintain inventory and open the door to flexible production arrangements in which different parts of the production process exploit lower costs in varied locations. E-commerce has connected this global logistics system directly to the individual consumer, with integrators fulfilling orders taken over the Internet in as little as 24 hours.[120] In a highly publicized sales phenomenon, FedEx teamed with Amazon.com to deliver Harry Potter books directly to purchasers on the day of their release, moving 250,000 copies in 2000 and 400,000 in 2003.[121] For the final book in the series, Amazon is said to have shipped 2.2 million copies for delivery in the US on July 21, 2007, using several companies.[122]

The transport revolution in air cargo has changed both the way that many products are created and how they are distributed. This revolution introduced new information technology to the movement of goods, but such technical innovation followed organizational change. Air transport networks were established around cargo hubs and surface transport was integrated to create a door-to-door delivery system that could make the most of aircraft performance characteristics.

Air cargo's transformation—from an adjunct to passenger flights into an everyday link between shippers and recipients around the globe—also highlights the role that a visionary entrepreneur can play in launching new mobility options. Fred Smith forged unprecedented relationships among aviation, surface transport and information systems to deliver a service that had not previously existed. Although the intended launch customer did not turn out to be an early adopter of overnight delivery, many more individuals and organizations found a need for this service once it emerged. Transport revolutions can thus depend on the capability of an entrepreneur to bring a new mobility option into being, so that others may make use of it.

Reflections on Past Revolutions

Gaining Perspective on Major Mobility Changes

The five examples of transport revolutions we have explored can help prepare readers for the major changes we see coming. A prerequisite for making the most of a major change is learning to distinguish it from the ongoing adaptation of existing arrangements. This is not as easy as it might seem. Media and advertising constantly hype very modest mobility changes as big and important breakthroughs, often with an emphasis on the promise of particular technological features. As we spelled out at the beginning of this chapter, we believe the term transport revolution should be reserved for major changes in the movement of people and freight.

In describing the above five examples, we noted several additional features—beyond the magnitude of the change in transport activity—that may be useful in identifying transport revolutions:

1. From the launch of railway service, we see that *revolutions can have a high degree of unpredictability in their immediate outcomes.* Entrepreneurs in the UK expected to create a major change in how freight moved, but the first railways turned out to yield a major change in how people moved. This unpredictability extended to the adaptation of horse-drawn coaches and barges, which quickly reconfigured to provide alternative services (e.g. local feeder service by horse coaches). Anticipating the effects of a revolution may be quite challenging. The divergence between "conventional wisdom" and unexpected results of change offers a signal that the change was indeed revolutionary.
2. The US pause in expanding motorization shows that *revolutionary change can move quickly.* What may seem to be entrenched mobility patterns can change dramatically in response to government interventions such as regulation of vehicle production and fuel distribution. Striking results emerged from this brief period when the US made enhancing the efficiency of its transport system a top priority. Individual sacrifice was justified in the name of national security, a recipe for legitimating sudden and dramatic behavioral change.
3. From the revolutionary shift in transatlantic travel, we see that *moving faster exerts a powerful attraction, especially when the cost of such speed is no more than that of moving more slowly.* We also see how the dynamics of a transport system play out as the established modes adapt to new circumstances and find a role where their technology can meet new needs. Because moving faster confers great advantage to armed forces, military research and development is a prime source of new technology that can revolutionize civilian transport.
4. From the high-speed rail revolution, we see that major change does not always reflect the linear notion of progress in which newer mobility modes eventually displace older ways of moving about. The experts who suggested that passenger trains were obsolete, and that they would join the stage coach in transport museums by the end of the 1960s were (or should have been) surprised by the Shinkansen and TGV. *When mature organizations and established technologies are redesigned to yield higher performance, the resulting revolution can unlock significant amounts of hidden value within the transport system.* Major assets such as central city rail terminals that would have been

otherwise written off gain a new life at a fraction of the cost of replicating their functions for another mode (e.g., adding a new airport).

5. From the revolution in air freight, we see that *changing the relationship between transport modes can yield just as big a difference as introducing organizational and technical innovations to a single mode.* The door-to-door speed of freight movement was significantly increased without any increase in the speed of the aircraft and trucks that moved the shipments. This was done by creating a network of aircraft and delivery vehicles that could deliver a new form of mobility—guaranteed overnight delivery—which has changed the ways in which people do business.

Characterizing Profound Mobility Change Along Four Dimensions

Assessing the significance of a transport revolution using four dimensions of change facilitates understanding of what the revolution means for human society. The dimensions are *scale* (people and freight move farther), *speed, efficiency* (usually fuel efficiency but sometimes other kinds) and *accessibility* (best thought of in terms of who cannot use the system, whether for physical, financial or other reasons). Over the course of history these key attributes have been valued differently. The attraction of doing better along one or more of these dimensions has inspired visionaries, entrepreneurs and leaders to develop new transport options that have persuaded large numbers of people to change the ways in which they have traveled or shipped goods.

Change that expands the scale and speed of transport is easily noted. The expressway, the airport and the cargo hub have a physical presence that embody the expansion of trade and travel across much of the globe. The desire to be well connected to other places appears to be strong, as does the desire to obtain products from distant places. The five revolutions we highlighted above provide considerable evidence that people will pay to move farther and faster, but there is also evidence that such a willingness to pay has its limits. Supersonic air travel existed between 1976 and 2003, but could not be sustained at fare levels that would pay for its long-run costs.[123]

The efficiency and accessibility dimensions of a transport revolution strongly influence economic and social opportunities. However, the resulting benefits and burdens are unevenly distributed, unless there are compensatory mechanisms. Some revolutionary changes can trigger decline, and even collapse, as seen in the abandoned UK shipyards in Glasgow and Birkenhead following the transatlantic jet revolution. The same forces of change unleashed by jet aircraft made the Hawaiian Islands into a thriving economy fueled by tourism. But, as the distance between the Clyde and Mersey rivers and the beaches of Waikiki makes clear, the winners and losers from a transport revolution are often quite disconnected.

The evidence of a particular transport revolution may be most apparent in just one of these dimensions, but the effects of change could eventually be revealed along all four dimensions. The changes that result from a transport revolution can be so large as to create barriers to further change. When new forms of technology or new organizational arrangements prove successful and attract huge investments in major infrastructure, their existence creates resistance to further change. The infrastructure "locks in" a particular combination of technology and organization. The "lock in" provides resistance to further change. The value of existing structures, both physical and organizational, is perceived to be so high that further innovation is seen to be too costly.[124] This is why transport revolutions are very much the exception in the history of human mobility, and are largely unfamiliar to those who work in transport and those who use it.

The rest of this book focuses on identifying the nature of the interconnected problems that could motivate dreamers and doers to put forth innovative solutions and then move them into the mobility mainstream. We turn next to situating the performance of contemporary transport systems, the first step in gauging their potential for such change.

2

Transport Today

Introduction

This chapter describes transport today, touching on history, trends and causes. It deals with the movement of people and freight by land, water and air. It sets the scene for the discussions in Chapter 3 about transport and energy, in Chapter 4 about transport's adverse impact and in Chapters 5 and 6 as to where transport could or should be heading.

Travel and the movement of freight have been part of human experience since the migrations of our distant ancestors out of Africa, first to Europe and Asia and then to Australasia and the Pacific islands. Among the most remarkable journeys were those to the Americas: from Asia in the millennia before written history—across what is now the Bering Strait to as far south as Tierra del Fuego—and from Europe and Africa during the last millennium and perhaps before.

Societies across the world have progressed in military, economic and social matters—not always at the same time—to the extent they have mastered and improved upon the movement of people and freight. Over the years, effective transport brought advantage to numerous peoples: the Phoenicians, Romans, Mongols, Venetians, Incas, Dutch, British and Americans, among others. During the last 200 years, the links between transport and economic development have become increasingly tight.

Until the 19th century, travel everywhere was uncomfortable, dangerous and enormously time-consuming. Freight movement posed even greater challenges. The barriers of distance were overcome where feasible by use of inland waterways, including canals, but mechanized rail transport made the real difference in accelerating the scope and volume of transport. The linking of two earlier inventions—wheels on smooth iron rails and the steam engine—allowed widespread motorized transport across land, and the beginning of a new era in the mobility of people and goods. Also important was the linking of the steam engine to the paddle wheel and propeller to provide motorized transport over water.

Rail transport began to give way to road transport in the first part of the 20th century, although the main expansion in the use of road vehicles has occurred since 1945. Air transport arrived soon after motorized road transport, allowing high-speed travel over great distances and ready access to remote places. Ocean freight still dominates the carriage of products and raw materials. Transport's evolution since the mid-19th century is shown in Figure 2.1. Developments since 1990 are summarized later in Figure 2.14.

Until the 19th century, travel everywhere was uncomfortable, dangerous and enormously time-consuming.

Motorized transport has facilitated and even stimulated just about everything now regarded as progress. It has helped expand intellectual horizons and deter starvation. Comfort in travel is now commonplace, at a level hardly dreamed of in former years even

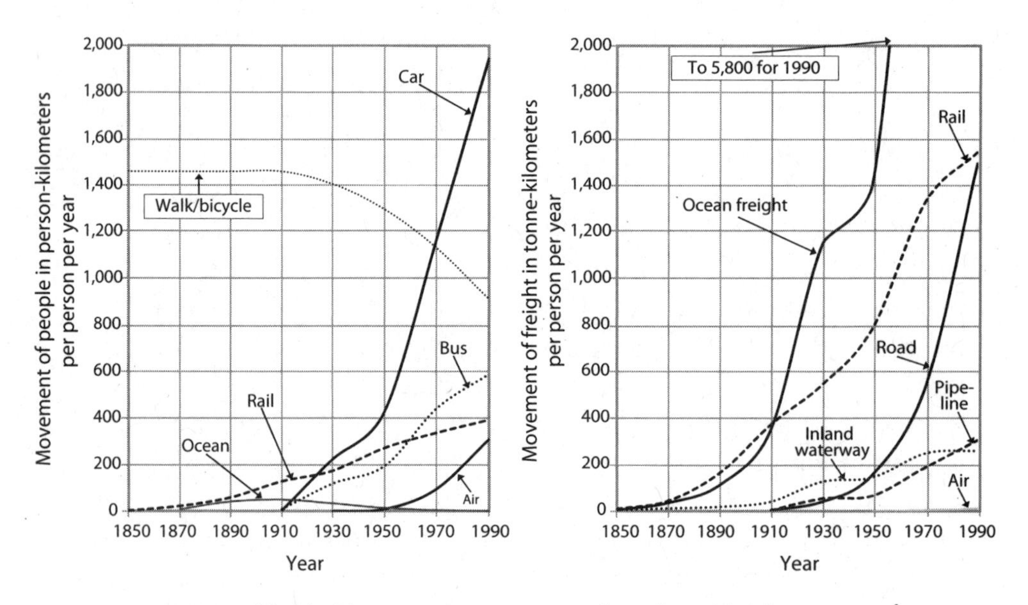

FIGURE 2.1 World-wide per-capita movement of people and freight, 1850–1990[1]

by royalty, as is ready access to the products of distant places. Motorized transport has also facilitated some of the low points of recent human history, including the Holocaust and the Soviet gulags.

The growth of personal road transport—chiefly cars—has been closely associated with two of the major phenomena of the 20th century: growth in material well-being and expansion of democratic institutions. Ownership of a car—usually the most expensive of consumer purchases—has assumed in rich countries the status of a democratic right. As a token of passage into adulthood, qualifying to drive a car can be more important than qualifying to vote.[2] In central and Eastern European countries, relaxation of prohibitions on car ownership often preceded enfranchisement,[3] and may have contributed to it.

The unusual case of Hong Kong shows that the desire for car ownership is not universal. There, according to a survey, only one in a hundred university students owns a car and less than one in five would want one. The author of the survey report concluded, "...if public transport is generally perceived to be good and cheap, it can suppress demand for cars."[4] Other factors are relevant in Hong Kong's case, including the high cost of car ownership and use, and the limited opportunities for driving.[5]

By 2007, world totals for the *motorized* movement of people and freight had grown to truly remarkable levels. Estimates of these, and the fuel used, are in Box 2.1. Most often, the totals are in *trillions* (millions of millions), numbers almost beyond comprehension. The present chapter provides a fine-grain analysis of this movement.

The next section concerns the *local* movement of people—their everyday travel, to work places and other places, by motorized and non-motorized means. Sometimes we make a loose distinction between higher-income and lower-income countries and urban regions. "Higher-income"—as well as "affluent"—refers to countries or regions that had a per capita Gross Domestic Product (GDP) of more than about $10,000 in 1995. Also, because of the way in which some data are available, we instead occasionally distinguish between members of the Organization for Economic Cooperation and Development (OECD),[7] which are mostly higher-income countries, and other countries.

There are major differences in local travel, not only between higher- and lower-income places, but also within the same country and even

BOX 2.1 World motorized transport in 2007[6]

- Some 900 million road vehicles traveled 13 trillion (13×10^{12}) km, on 10 million km of paved roads and 20 million km of unpaved roads.
- About two thirds of these vehicles were passenger cars, including SUVs, vans and other personal vehicles. Most of the remainder were trucks, but a substantial number were buses and — especially in lower-income countries — two-wheeled vehicles.
- Most motorized movement of people was by *road*, in cars or buses, a total by road of some 25 trillion person-kilometers (pkm), where one pkm is a person moving through one km. *Rail* and *air* travel totaled about 0.5 trillion and 2.0 trillion pkm, respectively.
- Most motorized movement of freight was on water. This totaled about 45 trillion tonne-kilometers (tkm), where one tkm is a tonne of freight moving through one km. As well, 8 trillion tkm moved by road, 7.5 trillion tkm by rail and 0.2 trillion tkm by air.
- The *value* of what was moved by air was about a third of the value of all freight moved, even though, in terms of tkm, air freight was less than 0.5 percent of total freight movement.
- Perhaps another 2 trillion tkm of freight movement was by pipeline; there are few good data on this mode.
- About a third of freight transport was movement of oil and oil products, just over half of which was used to fuel the movement of people and freight. Oil products were also used to heat buildings and make electricity and as a feedstock in the manufacture of plastics, fertilizers, pesticides, pharmaceuticals and other products.
- Oil products totaling some 2 trillion liters fueled more than 95 percent of motorized transport. The remainder was fueled by electricity, natural gas, propane and coal.

within the same urban region. We lay out some of the reasons for these differences.

The following section focuses on how people move over longer distances, within their countries and between countries. This section mainly concerns people in richer countries. Poorer people engage in relatively little longer-distance travel. Air travel is a major focus of this section, and we spend some time describing a major phenomenon in this industry: the rise of low-cost air carriers.

The final section, before a brief conclusion, presents data on freight movement, a much neglected topic for which there are relatively few data, especially about local freight movement. There is some focus on freight transport by water, which is the main way freight travels over long distances (see Figure 2.1 and Box 2.1).

The brief conclusion—entitled "Transport tomorrow?"—sets out one of many projections about the future of transport based on a continuation of recent trends. We explain why such a future is unlikely to happen.

How People Move Locally

Differences Among Urban Regions

When the topic of transport is raised, people often think first about local travel, particularly the journey to and from work. More people travel during the morning and afternoon "rush hours" than at other times, and that is when there are the most concerns about congestion and other transport matters. Horror stories about local travel usually involve rush-hour travel. They include reference to people-pushers hired to cram passengers on to trains in Tokyo, three-hour commuting trips by car in the Los Angeles and Chicago areas, and two-hour walks to work during a strike by employees of Paris's suburban rail system.

Transport planners focus on rush-hour journeys because they determine capacity requirements: for example, how many lanes of roadway are required or how many suburban trains must be run. As a result of this focus, there is more information about trips to and from work than about other trips. However, commuting trips appear to be slowly declining as a share of all local travel, even on weekdays.[8] If you factor in weekend trips—there can be more local travel on Saturdays than on some weekdays[9]—commuting trips represent only about a quarter of all local travel.

Transport planners focus on rush-hour journeys because they determine capacity requirements.

In this section we focus where we can on *all* local trips, not just the journey to and from work. By "local," we mean a trip of less than about 50 kilometers.[10] "Trip" or "journey" refers to travel from one place to another. Visiting a friend requires two trips, one there and one back.[11]

There is enormous variation in how people move locally. Table 2.1 shows 2001 data on the shares of journeys made by car and other modes in 48 mostly affluent urban regions, ranked according to car's shares. Most of these urban regions are in Europe, because that is the focus of the organization that collected the data. The urban region with the lowest share of trips made by car (Hong Kong) and the three with the highest shares (Chicago, Dubai and Melbourne) are not in Europe. Nor are two other urban regions in Table 2.1: Singapore and São Paulo, the latter having the lowest car share of the group apart from Hong Kong and several Eastern European urban regions.

Even among the Western European urban regions in Table 2.1 there is large variation. They range from Amsterdam, where only a third of trips are made by car and more than a quarter by each of cycling and walking, to Manchester, where two thirds of trips are made by car and there is little use of public transport and almost no bicycling.

We prefer the term "urban region" to city because a city can refer to only a small part of an urban region. An extreme case is the City of London (UK), which has only about 0.1 percent of the residential population of the urban region (Greater London), and about 7 percent of its employment. When there is no ambiguity we sometimes use "city" as a synonym for urban region. We do it to provide variety, and because it's shorter.

In the cities in Table 2.1, low car use tends to be associated with high levels of transit use and, to a lesser degree, with high levels of journeys by foot. Causal relationships are not clear. Does Hong Kong have such low car use because public transport is so good and because it is so compact? Or is public transport good and the city compact because car use is low? The truth is likely a mix of both. Singapore has much higher car use than Hong Kong, even though car ownership is rationed in Singapore and much more expensive. The higher car use in Singapore could be because this island city and country, while smaller than Hong Kong in overall territory, has been less compactly developed.

Other factors are at play, including availability of parking, levels of disposable income and fuel prices. The data for these and other factors are often questionable. Strong interpretations about the causes of the use of particular transport modes should perhaps be resisted until we have better information than we have today about how people move locally.

More extensive data on mode shares are available for 1995. They are summarized in Figure 2.2, which groups urban regions by country, or by sets of countries defined geographically and economically. Figure 2.2 shows how local travel in North America, Australasia and the Middle East was dominated by travel by private car, which was also the most used mode in affluent Asian urban regions.

TABLE 2.1 Percentage shares of local trips by car or motorcycle, public transport, walking and bicycling, 48 cities, 2001[12]

	Car or motorcycle	Public transport	Walking	Bicycling
Hong Kong	16	46	38	0
Moscow	26	49	22	2
Krakow	28	40	32	1
Warsaw	29	52	19	0
Budapest	33	44	22	1
São Paulo	34	29	37	0
Amsterdam	34	15	26	26
Bilbao	35	16	48	1
Prague	36	43	20	1
Vienna	36	34	27	3
Berlin	39	25	26	10
Bern	40	21	30	9
Munich	41	22	28	9
Valencia	41	12	46	0
Helsinki	44	27	22	7
Singapore	45	41	10	4
Zürich	46	18	20	15
Paris	46	18	35	1
Graz	46	23	25	5
Stockholm	47	19	34	0
Barcelona	47	22	25	6
Hamburg	47	16	25	12
Lisbon	48	28	25	0
Seville	48	10	41	1
Rotterdam	48	10	23	19
Copenhagen	49	12	19	20
London	50	19	30	1
Madrid	51	15	30	4
Geneva	51	22	26	0
Turin	54	21	24	1
Lyons	54	11	34	1
Marseilles	54	13	32	1
Rome	56	20	23	0
Newcastle	57	14	25	4
Bologna	57	16	25	1
Oslo	59	14	26	1
Brussels	59	11	23	8
Stuttgart	59	15	21	4
Clermont-Ferrand	61	6	32	1
Lille	63	6	29	2
Athens	64	28	7	1
Nantes	64	13	21	2
Gent	65	5	16	14
Glasgow	66	11	23	1
Manchester	68	9	21	2
Melbourne	76	6	17	1
Dubai	77	7	16	0
Chicago	88	6	5	1

Figure 2.2 also represents data for many urban regions in lower-income countries. In lower-income Asian cities the car was the most used mode but was responsible for only a minority of trips, as was true in affluent Asian cities. In the remaining places, another mode was the most used: public transport (Latin America and Eastern Europe), walking (Africa) and bicycling (China). We describe below how car ownership and use in lower-income countries are rising rapidly. For the moment, we note from Figure 2.2 that car use's share was a little higher in the ten "Other Asian Cities"[14] than in the five "Affluent Asian Cities."[15] This suggests that the level of car use may not be simply a matter of wealth, but may also be the result of availability of public transport, urban density and perhaps other factors.

The level of car use may not be simply a matter of wealth, but may also be the result of availability of public transport, density and other factors.

One key difference between higher- and lower-income urban regions is not evident in Figure 2.2. It concerns how much of public transport travel is by rail. In higher-income regions and those of Eastern Europe, rail trips are typically about a third of all public transport trips, considerably more if distance traveled is the measure. Lower-income urban regions usually have high shares of travel by public transport, but fewer

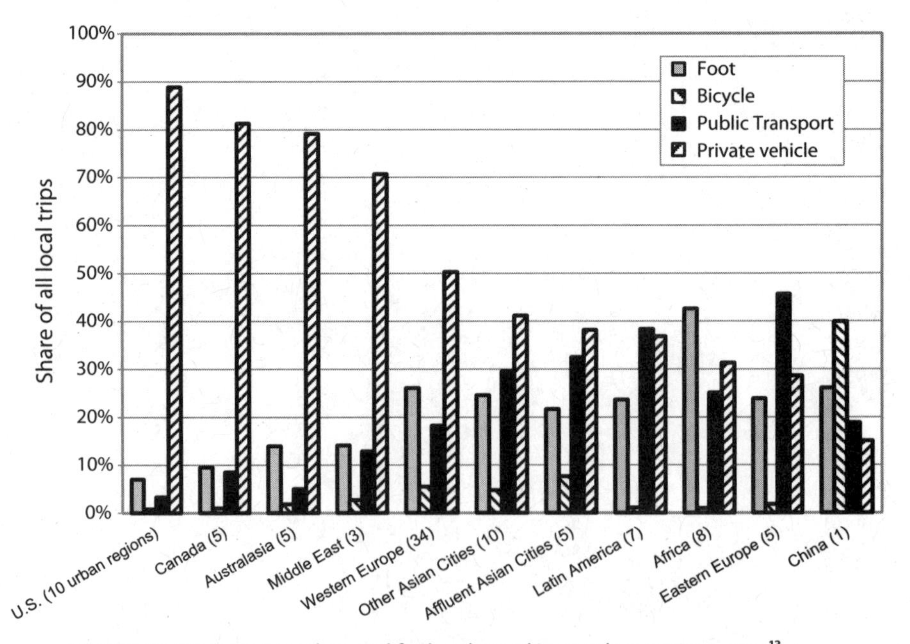

FIGURE 2.2 Transport modes used for local travel in 93 urban regions, 1995[13]

of these trips are by rail, usually less than a sixth.[16] In both higher- and lower-income countries, most public transport trips are by diesel-fueled buses. This is more the case in lower-income countries, where the buses are usually old and often retired from the fleets of higher-income cities.

Differences in Travel Within a Region

Local travel can also vary considerably *within* an urban region, where there can be quite different patterns in the central city and the suburbs. Figure 2.3 shows this for concentric parts of Canada's Toronto region—population about five million in 2001—which has its downtown business district close to Lake Ontario. As you move away from the downtown, from the inner core to the outer core, and then from the inner suburbs to the outer suburbs, the shares of trips by foot, bicycle and public transport fall, and the share by car increases. The travel pattern of residents of Toronto's inner core is similar to that of an affluent Asian city (see Figure 2.2). The travel pattern of residents of the outer suburbs is typical of North American urban regions.

Figure 2.3 provides details for the four parts of the Toronto region. Residential densities decline as you move away from the core, and car ownership and the amount of travel both increase. The differences in car ownership are not simply a matter of income, because per-capita income is roughly the same in each of the four parts of the region.

The pie charts in Figure 2.3 (and the bars in Figure 2.2) represent shares of *trips*, with a 1 km trip on foot counting the same as a 10 km trip by public transport or a 30 km trip by car. Trips by foot are usually shorter than trips by other modes. Trips by suburban train are often the longest. If person-kilometers (pkm) are used as the measure of travel,[18] rather than trips, the shares for walking and bicycling as shown in Figure 2.3 will be much smaller. When examining travel as human behavior, it often makes more sense to discuss it in terms of trips. When examining travel's impacts or energy use, comparisons in terms of pkm can be more meaningful.

Another measure of travel is time spent traveling, important because of suggestions that people are inclined to spend a total of about an hour a day moving from one place to another. The notion of a constant travel-time budget can be a powerful explanatory principle for local travel and will be returned to below.

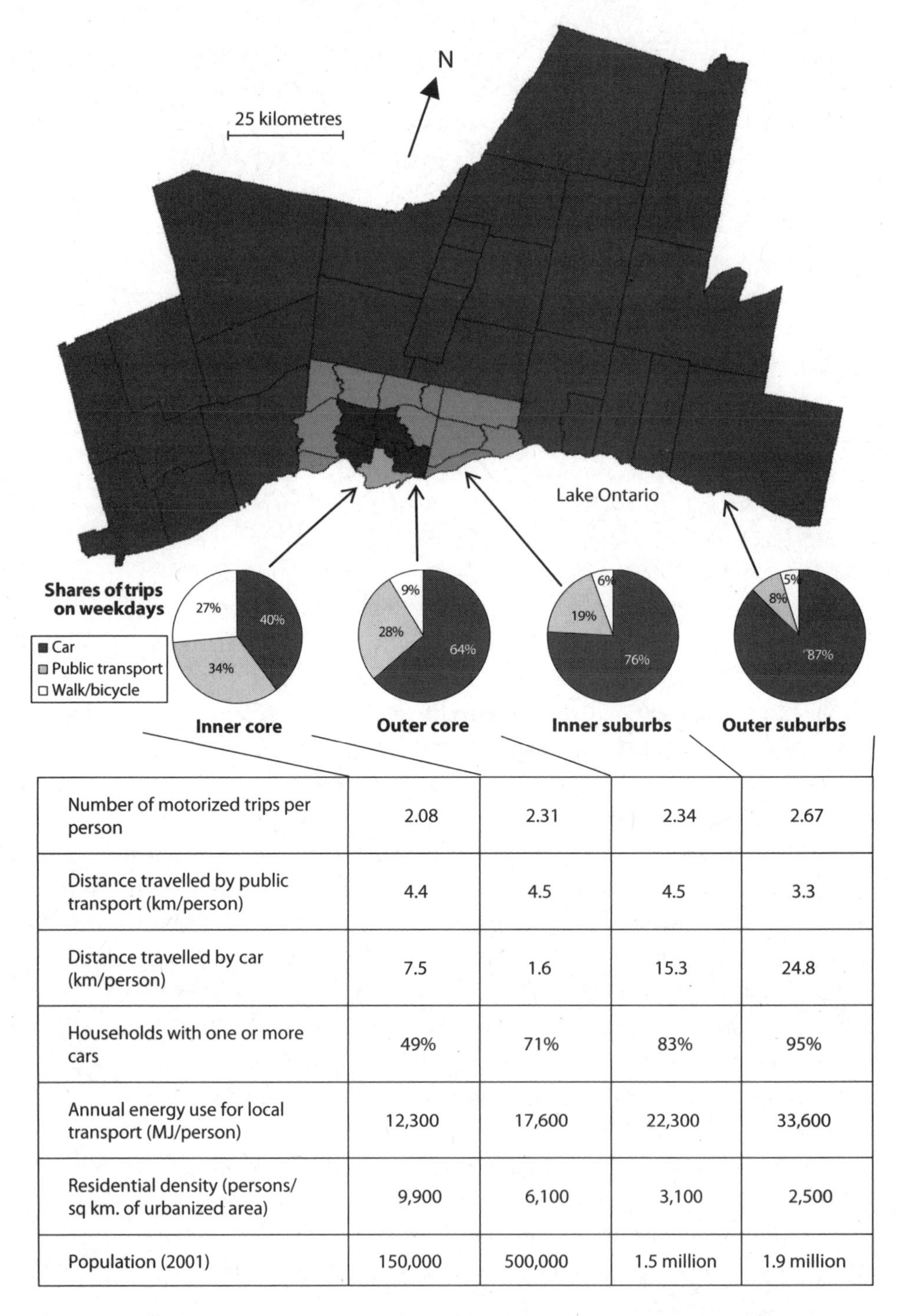

	Inner core	Outer core	Inner suburbs	Outer suburbs
Number of motorized trips per person	2.08	2.31	2.34	2.67
Distance travelled by public transport (km/person)	4.4	4.5	4.5	3.3
Distance travelled by car (km/person)	7.5	1.6	15.3	24.8
Households with one or more cars	49%	71%	83%	95%
Annual energy use for local transport (MJ/person)	12,300	17,600	22,300	33,600
Residential density (persons/ sq km. of urbanized area)	9,900	6,100	3,100	2,500
Population (2001)	150,000	500,000	1.5 million	1.9 million

FIGURE 2.3 Travel features of residents of different parts of the Toronto region, 2001[17]

Car Ownership in Europe and the US

The most important differences among countries concern car ownership. Figure 2.4 shows data until 2000 for 16 lower-income Eastern European countries, including Russia, and 15 higher-income Western European countries, together with US data. The US seems to have been moving toward a possible plateau as an ownership rate of one car per person is approached. There's less indication in Figure 2.4 of a slowing down in the growth of car ownership in Western Europe, where in 2000 the number of cars per 1,000 residents was only 60 percent of the US rate. In Eastern Europe, where the rate was just a little over half of the Western European rate, there seems to be no suggestion at all of a slowing down.

The relative growth in car ownership in Eastern Europe has been extraordinary. In 1980, there were about 20 million cars in Eastern Europe, compared with about 120 million in Western Europe. Between 1980 and 2000, both Eastern and Western Europe each added 60 million cars, with

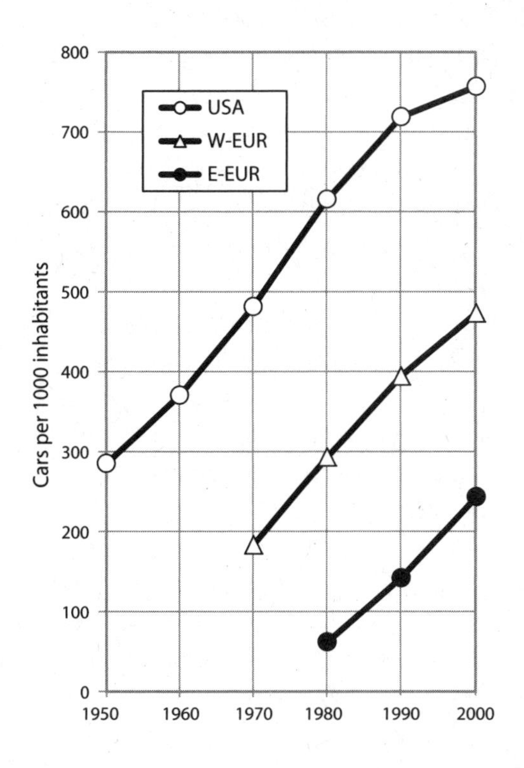

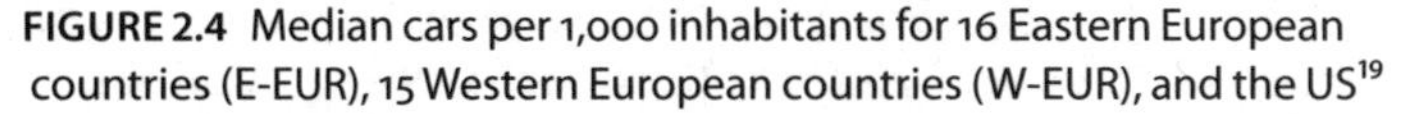
FIGURE 2.4 Median cars per 1,000 inhabitants for 16 Eastern European countries (E-EUR), 15 Western European countries (W-EUR), and the US[19]

no population increase in Eastern Europe and an increase of about eight percent in Western Europe. In the US over the same years, the increase was just over 40 million, to a total there of about 225 million automobiles, a rate slightly below the increase in population during this period, which was about 24 percent.

Car Ownership and Use in China, Particularly Beijing and Hong Kong

In China, the rate of growth has been even more remarkable than in Eastern Europe. There were less than a million cars on the road in 1985 and more than 17 million in 2004, an average growth rate of 18 percent per year. Even more striking was the growth in privately owned cars. They comprised only 2 percent of all cars in China in 1985 and 65 percent in 2005.[20] Data for 2006 suggest that new car production in China was about 35 percent above the 2005 level, similar to the average rate of increase during the previous five years.[21]

Even with these extraordinary increases, the number of cars per 1,000 inhabitants remains low in China. It was 13 per 1,000 in 2004, compared with 280 in Eastern Europe, 470 in Western Europe and 760 in the US. Nevertheless, if recent rates of growth continue, China will pass the US in total number of cars on the road in about 18 years. Then, China's ownership rate would be about 200 cars per 1,000 inhabitants, the same as it was in the US in 1940, in Western Europe in 1970 and in Eastern Europe in 1995.

There are large differences in car ownership among the different parts of China, as in all lower-income countries; much more so than in higher-income countries. In Beijing, there were 108 cars per 1,000 residents in 2004, 72 percent of which were privately owned. In Gansu, Guizhou and Jiangxi provinces, there were fewer than 5 per 1,000, and only 42 percent of these cars were privately owned.

It's instructive to compare Beijing's relatively high number of cars on the road with that for another Chinese city, Hong Kong, which had 79 cars per 1,000 inhabitants in 2004.[22] Beijing residents are on average much less wealthy than Hong Kong residents (about $4,000 vs. $24,000 of GDP per capita). They nevertheless own more cars. More than two thirds of motorized trips in Hong Kong are made on its remarkably efficient and comprehensive public transport system. Public transport's share of trips in Beijing is not known, but it is likely much lower. Beijing's 15 million people make 14 million public transit trips a day while Hong Kong's seven million make 11 million trips a day, almost 70 percent more per person.

Bicycle Use in China

Until recently, the most widely used transport mode in Beijing was bicycling. In 2001, there were 10 million bicycles and hardly more than a million cars in the city. The balance is changing rapidly in response to official promotion of cars over bicycles. This includes removal of wide and extensive bicycle lanes to add to capacity for motorized traffic, notably cars.[23] According to one report, 60 percent of Beijing's work force cycled to work in 1998, but only 20 percent cycled in 2002.[24]

The favoring of motorized traffic over bicycles and pedestrians in Beijing and some other cities in China has been criticized by the Chinese government. In June 2006, Qiu Baoxing, Vice Minister of Construction, said China should "maintain its large bicyclists" population rather than diminish it, as it will help ease energy shortages and accelerate urban development."[25] His ministry had already ordered restoration of some of the bicycle lanes.

The effect of such central government interest is uncertain. The editor of *China Economic Quarterly* has been reported as saying—in respect of unsafe coal mining—"The clash is between the central government's desires and the local government's pressing economic needs, and in 99 cases out of 100, local government wins out."[26]

Shanghai, China's most populous city, appears to have reversed a movement to rid major roads of bicycles and now encourages bicycle use again. Public transport is the priority, however, so much so that car ownership is limited by a system of auctioned entitlements to purchase cars, similar to Singapore's system,[27] that keeps Shanghai's car ownership rate far below Beijing's (37 vs. 108 passenger vehicles per 1,000 persons).[28]

Bicycle use, increasingly with some electrical powering, remains popular in most parts of China. Some 5–10 times as many bicycles are sold in China each year as cars.[29]

Cars in China Are Small

An important consideration when comparing car use in China—and in many other lower-income countries—with car use in higher-income countries is the average size of the vehicle. In China, almost 90 percent of

BOX 2.2 Road vehicles in use in higher- and lower-income countries worldwide[30]

The table on the right shows how road vehicles (four or more wheels) are unevenly distributed between higher- and lower-income countries. Higher-income countries have less than a fifth of the world's population but three quarters of the world's cars and more than half of the larger road vehicles.

(Data in this table are for 2001)	Higher income	Lower income
Number of countries for which data are available	39	97
Total population of these countries (millions)	991	4,465
Cars in use in these countries (millions)	503	131
Trucks and buses in use in these countries (millions)	63	56
Cars/1,000 inhabitants	507	29
Road vehicles (≥4 wheels) /1,000 inhabitants	571	42

The chart below shows the considerable growth in these vehicles worldwide: by 80 percent between 1981 and 2001. The growth was greater in lower-income than in higher-income countries (225 percent vs. 57 percent). However, because so many more vehicles are in higher-income countries, they had a much higher share of the growth (61 percent vs. 39 percent).

Missing from these data are not-well-documented numbers of two- and three-wheel motorized vehicles, which are used mostly in the urban areas of lower-income countries, particularly in Asia. One report suggests that worldwide there were more than 100 million such vehicles in 1999, growing at the rate of 7 percent/year. Another report indicates that 8.9 million motorcycles were sold in China in 1998 (down 10 percent from 1997), compared with sales of 1.6 million vehicles with four or more wheels. See also the data on India in Box 2.3.

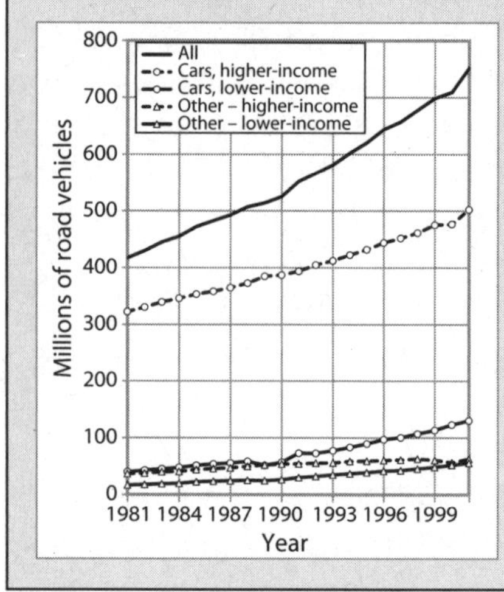

Even less well documented is the number of non-motorized bicycles on the road. One source suggests that the number manufactured worldwide in 2003 greatly exceeded car production (105 vs. 42 million units). China produced 70 percent of the bicycles, and exported 70 percent of its production. Some 20 million went to the US, where, as everywhere else, bicycle sales exceeded car sales.

cars on the road are classified as small (i.e., with an unladen weight of no more than about one tonne).[31] In Canada, by contrast, only about 40 percent of cars on the road weigh less than 1.2 tonnes. Another 30 percent are larger cars, and the remaining 30 percent are mostly sport utility vehicles (SUVs), pick-ups and passenger vans.[32] In the US, less than 5 percent of model year 2004 light-duty vehicles (cars, SUVs, vans, pick-ups) weighed less than 1.25 tonne.[33]

In lower-income countries, rural residents likely travel much less than urban residents because there is little opportunity for motorized travel.

Car sizes and weights in Europe are between those in Canada and China. In 1997, 63 percent of cars sold in Europe weighed less than 1.1 tonne. Almost all of the remainder were larger cars; some were light trucks, vans and SUVs used as personal vehicles.[34]

Vehicle weight is usually the most important factor in fuel use, particularly when accelerating and hill-climbing. Other factors include engine and drive-train efficiency and streamlining.

Travel by Urban and Rural Residents

So far, we have been describing local travel in urban areas. In lower-income countries, rural residents likely travel much less than urban residents because there is little opportunity for motorized travel. These rural dwellers comprise almost half the world's population, and most of them rarely travel other than by foot or bicycle. Although the world may seem highly motorized, with nearly a billion motor vehicles in 2007 for its 6.5 billion people, most of these vehicles are in higher-income countries and in the urban parts of lower-income countries. Box 2.2 shows world trends in vehicles on the road. Box 2.3 provides a focus on India, another "waking giant" in transport and many other matters.

By contrast, people who live in rural parts of affluent countries travel considerably more than urban residents. In these rural areas, activities away from home can require extensive travel: to a job or physician 50 km away or a supermarket that may be farther. In the US in 2001, the average trip by a rural resident was 12.2 km compared with 9.3 km for the average non-rural resident. Over the year, rural residents traveled an average of 26,900 km; urban residents traveled 21,600 km.[36]

There was a similar difference between rural and urban residents of the UK in 2002–2003, although overall amounts of travel were much

BOX 2.3 Motorized road vehicles and their use in India[35]

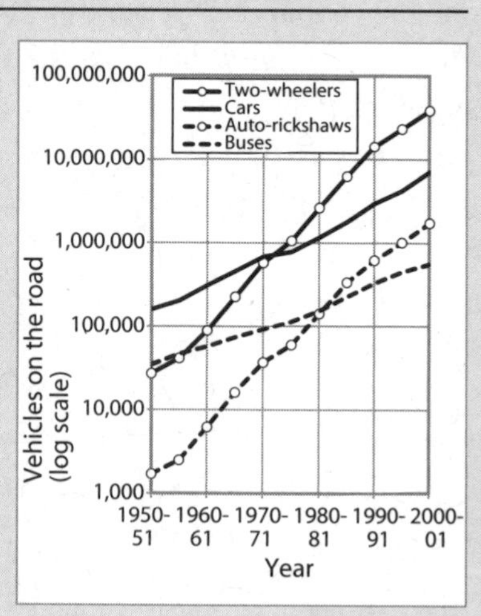

The upper chart shows the growth in four kinds of motorized vehicle on the road in India. Auto-rickshaws are light, three-wheel vehicles usually used for carrying passengers for hire, but also used for carrying freight. Their number has passed the number of buses (1.7 vs. 0.6 million in 2000–2001) as the number of two-wheelers has passed the number of cars (39 vs. 7 million in 2000–2001).

However, as the lower chart shows, most person-kilometers in India are performed by bus. The annual pkm *per bus* is extraordinary: it was 4.2 million in 2000–2001. By comparison, the pkm per *car,* two-wheeler and auto-rickshaw was 40, 9 and 59 *thousand,* respectively.

Nevertheless, the growth in two-wheelers and auto-rickshaws has also been extraordinary: 15 and 12 times, respectively, between 1981 and 2001, compared with 6 times and 4 times for cars and buses. Meanwhile, India's population grew from 680 to 1,020 million.

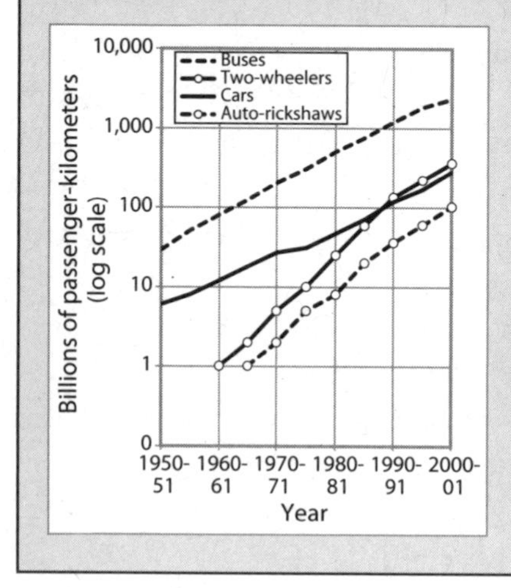

The overall ownership rates remain low: 7 cars per 1,000 inhabitants in 2000–2001 and 38 two-wheelers per 1,000. The car ownership rate is similar to China's overall rate (see text), but the annual rate of growth during the 1980s and 1990s was lower in India (9 vs. 18 percent).

Note that the two charts have logarithmic vertical scales.

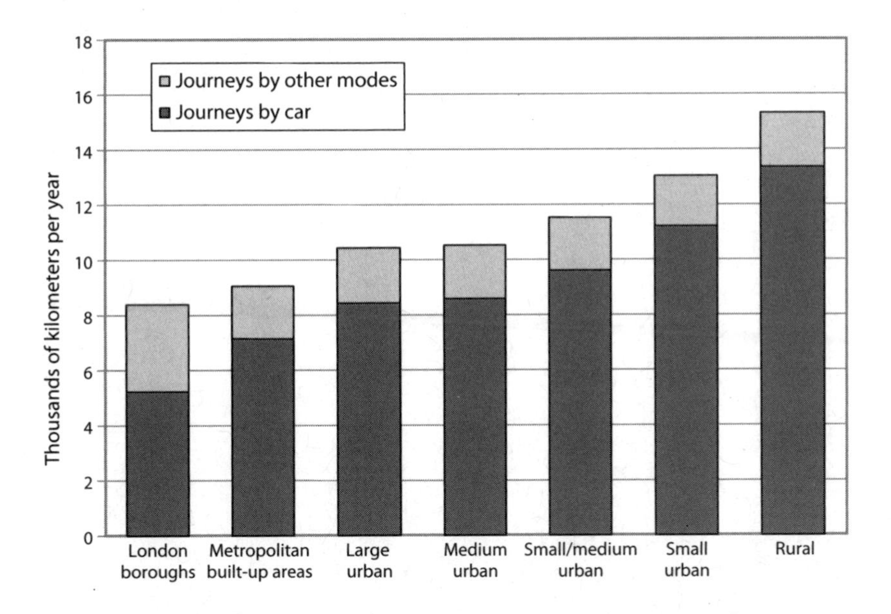

FIGURE 2.5 Annual travel by residents of different types of area in the UK, 2002–2003[37]

lower. Rural residents traveled 15,300 km a year compared with an average of 11,000 km for all residents. Available UK data reveal a finely scaled gradation according to the size of the community, illustrated in Figure 2.5. Except for London, the differences among types of settlement lay almost entirely in the amount of travel by car; the amount of travel other than by car was close to being the same in all places. In London there was more than twice as much travel by public transport as in the rest of the UK.

A factor contributing to the large difference in public transport patronage between London and the rest of the UK may have been the nature of the privatization of local bus services implemented in the 1980s. The system introduced in London was "competition for the road": franchised routes were—and are—tendered in batches. Outside London there was—and mostly still is—"competition on the road": any operator meeting safety standards and giving due notice can operate anywhere. The result was a sharp decline in bus use outside London and an increase in bus use in the capital.[38] Moreover, operation of the London metro (Underground) system was not privatized; it is responsible for many more passenger-kilometers than London buses.[39]

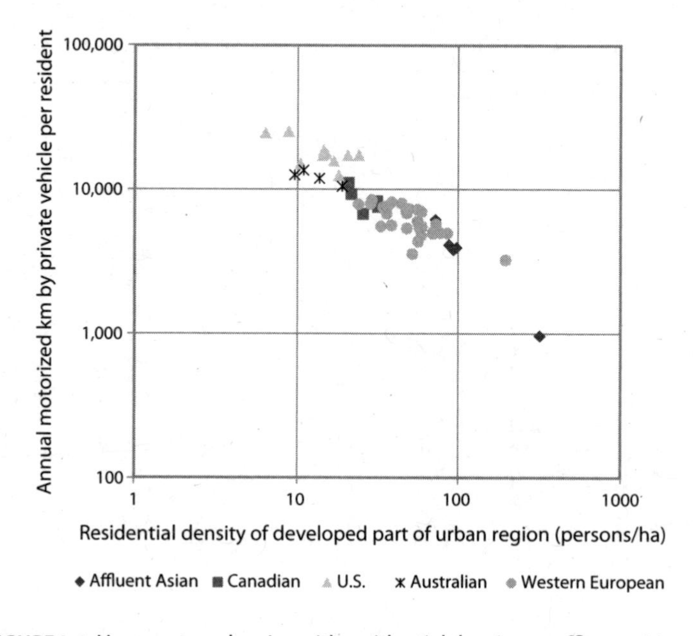

FIGURE 2.6 How car travel varies with residential density, 52 affluent cities, 1995[40]

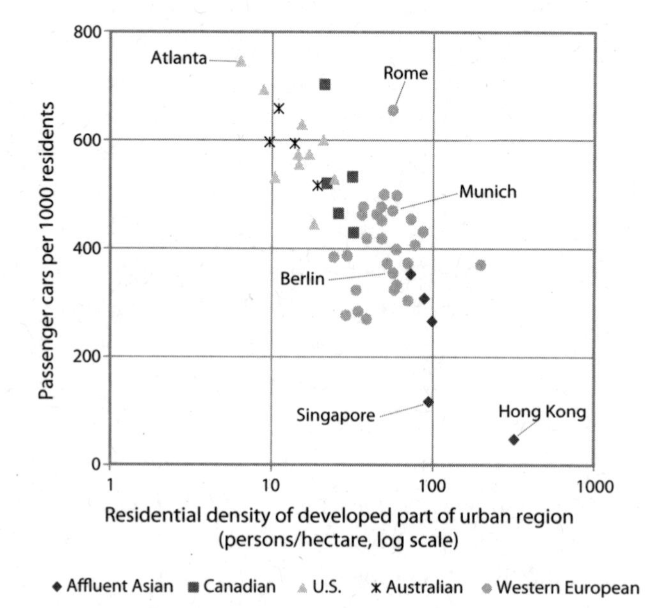

FIGURE 2.7 How car ownership varies with residential density, 52 affluent cities, 1995[41]

Factors in Local Travel

Figure 2.5 suggests that in higher-income places community size is one factor in amounts traveled, particularly amounts traveled by car. Figure 2.3 shows that settlement density—the number of people and employment places in a given land area—is another factor. People travel less overall and less by car when they live in denser rather than less dense areas. Likewise, there is more travel by car in spread-out cities, as shown in Figure 2.6.

Car ownership is the linking factor between residential density and amount of travel by car. Figure 2.7 shows how these two factors are linked, and the enormous variation in both factors, even among affluent cities. Hong Kong, at one extreme, had a residential density of 320 persons per hectare (pp/ha) in 1995. Atlanta's density, at the other extreme, was 6.4 pp/ha. These two cities' rates of car ownership in that year were 46 and 746 per 1,000 residents, respectively.

Car ownership is the linking factor between residential density and amount of travel by car.

Other factors than density are involved in car ownership. Figure 2.7 shows that Rome, Munich and Berlin all have residential densities of close to 58 pp/ha, but their car ownership rates are very different. Public transport use is not evidently a factor in car ownership because residents of these three cities make similar numbers of public transport trips. Trips by walking and cycling appear to be more related.[42] However, it's not clear whether people in Rome walk and cycle less because they own more cars or they own more cars because they walk and cycle less.

The other factors could include the priority given to pedestrians and cyclists (high in Berlin) and the availability of legal or illegal parking spaces (high in Rome). Cultural factors could also be involved. Owning and using a car in the city could be more acceptable in Rome than in Berlin. However, the ways in which such factors might work to produce such large differences in car ownership remain unclear.

Considering all places—not just urban regions in richer countries—the main factor in car ownership, and thus in urban travel, is personal or household income. Figure 2.8 shows how road motor vehicle ownership varies with an indicator that is usually related to average income: GDP.[43] When GDP per capita is more than about $5,000, car ownership appears to rise steeply with income.

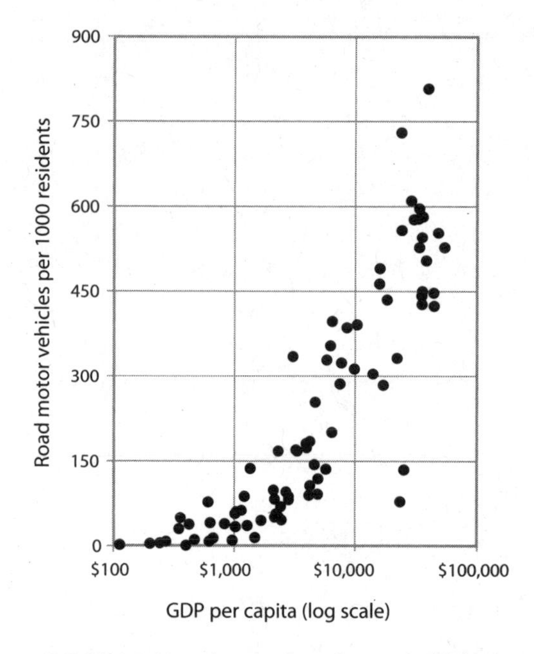

FIGURE 2.8 How the number of motor vehicles in use varies with income, 85 jurisdictions, 2003[45]

Four outliers are labeled in Figure 2.8. The highest per-capita rate of vehicles on the road is that of the US, which may not be surprising. The next highest rate is that of New Zealand, which surprised the authors enough to induce a specific check. New Zealand government data show that in 2003 there were 2.93 million motor vehicles on the road for 4.01 million residents.[44] Two other outliers are affluent jurisdictions with unusually low numbers of vehicles on the road per capita, Hong Kong and Singapore, both of which have featured already in this chapter: Hong Kong for its extraordinarily high density and high public transport use and Singapore for its explicit rationing of car ownership.

More on the Costs of Car Use

For new cars, ownership and other fixed costs are usually the major costs in car use. This is illustrated for the US and the UK in Table 2.2, which shows fixed costs (costs of finance, depreciation, insurance, licensing) and variable costs (fuel, maintenance) for late-model cars in 2005. In each case, fixed costs were 71 percent of total costs, which were about 12 percent higher in the UK. Costs per kilometer differed more. They were 47 percent higher in the UK, chiefly because fewer kilometers were driven.

TABLE 2.2 Costs of new car ownership and operation, UK and US, 2005[46]

	UK	US
Fixed costs ($)	6,269	5,569
Variable costs ($)	2,591	2,265
Total costs ($)	8,859	7,834
Fixed costs as share of all costs	71%	71%
Kilometers driven	18,670	24,150
Cost per kilometer ($)	0.47	0.32

The higher fuel costs in the UK were almost exactly offset by the lower amounts of travel.

A car loses very roughly 40 percent of its value during the first four years. It is usually driven less when it is older but has higher maintenance costs. These various factors together produce the result that fixed costs are roughly 70 percent of total costs during the first and the second four years of a vehicle's life.

The share of fixed costs increases with the price of the vehicle. For the vehicles on which the UK data in Table 2.2 are based, the range was from below 60 percent for a lower-cost vehicle to above 80 percent for a higher-cost vehicle. The share has also changed over time as cars have become relatively more expensive, but also on account of the cost of fuel. In the US, the share was 55 percent in 1975 and rose steadily to 78 percent in 2004 before falling to 71 percent in 2005.

Insurance costs are among the fixed costs in Table 2.2, which is usually how they are charged. Arguments have been made that they should be a cost that varies with distance traveled, to provide increased actuarial accuracy and an incentive to drive less. On the last point, "variabilization" of insurance costs has been estimated to reduce distance traveled by 10–12 percent.[47] However, this estimate does not take into account the loss of the deterrent value of insurance costs on car ownership. If car ownership is the main factor in car use, as is argued below, the deterrent value of upfront insurance costs could have a stronger effect on distance traveled than an additional cost that varies with distance traveled.[48]

Key Constancies: Travel Time per Day and Annual Distance Traveled per Car

A concept used to help explain the variation in local travel among different places and at different times is the *constant travel-time budget.* This

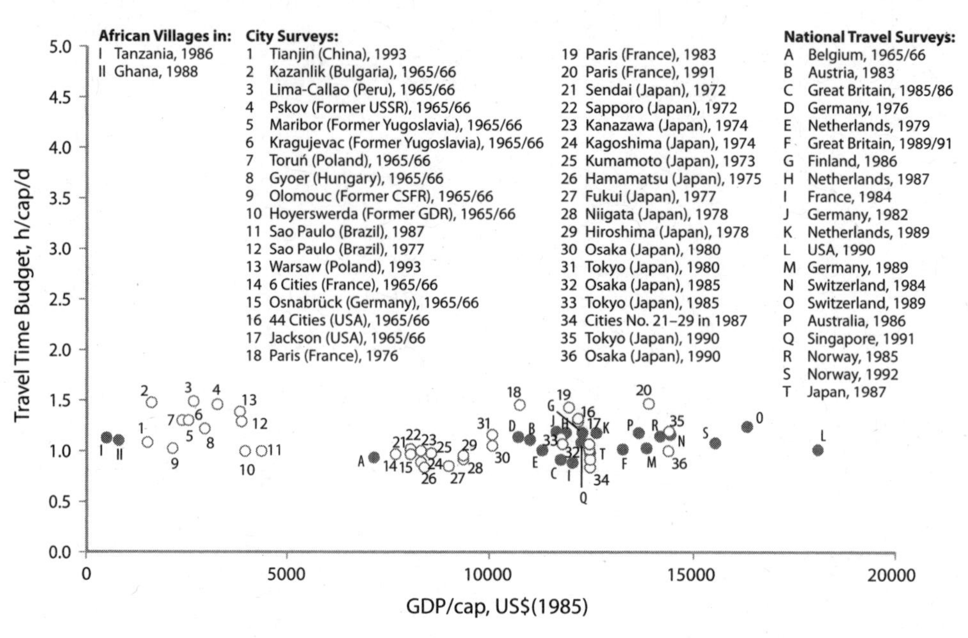

FIGURE 2.9 Travel time and GDP[49]

term refers to the disposition of people to travel for no more than about an hour a day, and to organize where they live, work, shop and socialize accordingly. Available data from a variety of situations, shown in Figure 2.9, suggest that this is indeed what happens, whether in African villages, where most trips are on foot, or in the US, where nearly all local trips are by car.

Peter Newman and Jeffrey Kenworthy have used this constancy to explain why cities before motorization were 5–8 km across.[50] This is a distance that can be traveled by foot in an hour. Similarly, cities where public transit prevails are 20–30 km across, and cities where cars prevail can be 50–60 km across. Other factors are needed to explain the availability of cars and the particular locations of homes and employment. Nevertheless, the notion of the travel-time constant provides a framework for understanding the spatial configuration of urban regions.

If a car is owned it will be used.

Another key constant in travel is annual distance traveled per vehicle. Figure 2.10 shows that for a particular country this distance is remarkably stable across time.[51] The number of cars on the road, and thus the total distances traveled, increased greatly between 1970 and 1995 for all

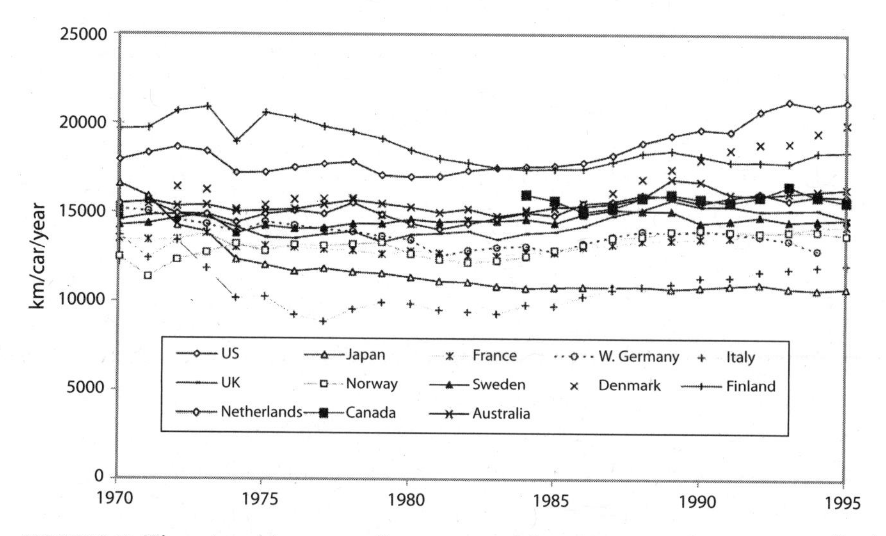

FIGURE 2.10 Kilometers driven annually per automobile, various countries, 1970–1995[52]

the countries represented in Figure 2.10. Because of the constancy of distance traveled per car, it was as if in each case the only factor determining the amount of vehicular movement was the number of vehicles, that is, car ownership provides a sufficient cause of car use. Put another way, if a car is owned it will be used, with the amount being determined by the particular circumstances in which it is owned.

There are consistent variations within countries according to where the car is owned and other circumstances. Figure 2.5 has already provided an example of how, in the UK, the amount of car driving varies with where the owner lives.[53] Another factor is the age of the vehicle. Table 2.3 shows that in the US a new vehicle is driven about twice as much as a vehicle that is ten years old or older.

Table 2.3 reinforces the overall constancy of annual kilometers per car from year to year by showing little change in this average between 1969 and 2001. This constancy occurred in spite of the variation with vehicle age and aging of the vehicle fleet across this period. In 1977, 28 percent of person vehicles on the road were one or two years old. In 2001, only 15 percent of vehicles were so new.[55] Also, between 1969 and 2001, vehicles per household increased from 1.2 to 1.9 and household size fell from 3.2 to 2.6.[56] The relative constancy of kilometers driven per car notwithstanding these many changes provides compelling evidence of the importance of car ownership as a factor in car use.

TABLE 2.3 Kilometers moved annually by vehicle age, US household vehicles 1969–2001[54]

	Year of survey					
Vehicle Age	**1969**	**1977**	**1983**	**1990**	**1995**	**2001**
0 to 2 years	25,267	23,271	24,610	27,055	25,898	23,966
3 to 5 years	18,025	17,822	19,154	22,058	22,537	21,292
6 to 9 years	15,611	14,804	14,891	20,204	20,291	18,673
10 or more years	10,461	10,871	11,302	14,767	14,095	12,654
ALL	18,668	17,186	16,600	20,049	19,676	17,828

It follows that limiting car ownership would be an effective factor in limiting car use. However, most strategies for limiting car use do not address ownership. On the contrary, as John Adams has noted, politicians of many stripes have advocated growth in car ownership, tempered by the occasional wish that the cars be left in the garage more of the time.[57] Indeed, the implicit objective of much policy making in respect of cars is that they should be owned but not used. This flies in the face of the above evidence that if people have cars they will use them.

Most strategies for limiting car use do not address ownership.

Apart from a few places, notably Singapore,[58] little has been done to restrain car ownership. Thus the effect of measures directed at restraining car ownership, in comparison with measures directed at restraining car use, has been little studied. Some support for the greater potential of restricting car ownership rather than use comes from the large literature on how price changes affect fuel use, amounts driven, car ownership and other variables. In general, it suggests that changes in variable costs have less impact than changes in fixed costs, where both kinds of cost are applicable.[59] (Where transport is a purchased service as in public transport and for-hire freight transport, there is no fixed cost from the user's perspective. In these cases, the use of the service seems relatively sensitive to changes in its cost.)

Causes of Local Motorized Travel by Car

In higher-income places, most local travel is by car. Why people travel by car, if one is available, hardly needs explanation. For most local journeys in most places car travel provides unmatched comfort, speed, privacy and convenience in travel. The greater challenge concerning local travel may be that of explaining why people who can afford to travel by car would not travel by car.

Equally unnecessary is the need to explain why people travel locally. The basic requirements for work, provisioning, socializing and recreation are universal. There can be differences in the amounts of each of these, and the associated amounts of motorized travel. These differences are sometimes of interest—for example, the shopping habits of inner-city and outer suburban residents—but the basic needs for travel can be almost taken for granted. It is entirely possible to live without leaving the home, and even more to live in a manner that requires little or no motorized travel, but the former is rare except when there is ill-health or disability, and the latter is rare in higher-income places.[60] Almost everyone travels locally. In higher-income places this travel is motorized and mostly by car.

To the extent that car use is determined almost completely by car ownership—if people have cars they will use them—the challenge becomes that of explaining why people might forgo car ownership. The critical factors appear to be relative costs of ownership and use, and settlement density. The cost constraints are self evident; although the relative importance of ownership and operating costs in deterring ownership is not well understood. Ownership costs in particular can be higher in denser places, chiefly because of the cost of parking. There are other factors associated with higher densities that may have more effect in restraining ownership. In denser places destinations are more often a walk or a bicycle ride away, and public transport is usually more available. Car travel can be slower or less convenient than these alternatives, or insufficiently better to justify a commitment to car ownership.

This account of the factors in car use offers land use and transport planners two kinds of recourse when they seek to reduce car use. One is to raise the relative cost of car ownership. The other is to increase settlement densities. There is a theoretical third alternative. It is to provide surroundings that induce people to forgo car ownership without increasing settlement densities.

The planning principle that speaks to inducing avoidance of car ownership has been dubbed the *EANO principle* (Equal Advantage for Non-Ownership).[61] The objective is to create a milieu in which it is at least as advantageous to live without a car as with a car. This involves providing alternative ways of moving people and goods, or reducing the need for this movement. It can also involve offsetting the car's enormous convenience, perhaps by ensuring that car-owners cannot park near their homes.[62] In practice, providing such surroundings when densities are low is impracticable because there are not enough people to justify public transport service or the location of shopping or recreational facilities.

In considering what makes people travel locally, particularly by car, and thus how this travel may be changed, we favor explanations and strategies such as the foregoing that focus on the context of the travel rather than those that emphasize the decision-making processes of travelers. We can't dispute assertions such as, "travel, in general, and driving, in particular, are about making choices,"[63] but we do feel that too much effort may be expended on figuring out what determines these choices and how the choices might be changed rather than what determines actual travel behavior and how that can be changed. We believe that travel behavior is appropriate to the contexts in which it occurs. If an aspect of a context changes—for example, the price of fuel—transport behavior will change accordingly. Appeals to personal choices and decision-making processes seem mostly irrelevant.

This perspective gives us optimism that as we approach a period of high prices or scarcity of current transport fuels—detailed in Chapter 3—there will be appropriate adjustments in human behavior. Transport activity changed dramatically during previous transport revolutions—as illustrated in Chapter 1—and similar adjustments can occur again, as we will discuss in Chapter 5.

How People Move Across Distances

Longer-Distance Travel in Context

Even in higher-income countries, only a relatively small amount of travel involves longer-distance trips. Figure 2.11 shows that for US residents in 2001, trips of more than 80 km comprised only 2.1 percent of all trips, although 29 percent of total person-kilometers. These shares are lower in other higher-income countries, as is detailed below. In lower-income countries, they are likely very much lower, with almost all residents of these countries engaging in no longer-distance traveling at all. Most travel is a local matter.

Longer-distance travel is nevertheless of importance for two reasons.

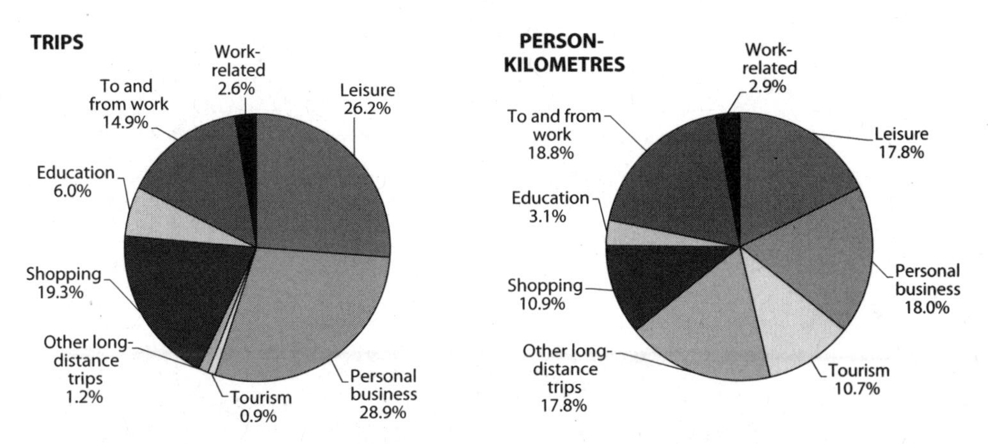

FIGURE 2.11 Trips and person-kilometers by trip purpose, US, 2001[64]

One is that it may have been growing faster than local travel, as discussed below. The other is that much of it is by air, in some ways the most environmentally damaging and energy intensive of modes, as explained in Chapter 4. There is a third reason, discussed in Chapter 5. During the next few decades, long-distance travel, particularly travel between continents, may have to change more than local travel, chiefly because the fuel will not be available to support substantial amounts of air travel.

How Longer-Distance Travel Is Performed, and Why

Before moving to trends in longer-distance travel, it may be useful to discuss how longer-distance travel is performed. Some of this depends on where the traveler lives. New Zealanders, for example, are more likely than residents of most other countries to make long-distance trips by air. Elsewhere, there are usually more alternatives. Table 2.4 shows how people in the US and in Western European countries make longer-distance trips.

Most longer-distance trips are made by car, more so in the US than in Europe. An evident difference is the number of trips made by bus and train: about a quarter of all longer-distance trips in Western Europe in 2001–2002 were made by these modes, but only 3 percent in the US. Table 2.5 shows, for the US only, that the share of trips by car declines sharply with trip distance in favor of traveling by air, which dominated return trips longer than 1,200 kilometers.

One thing that stands out in Table 2.4 is the much larger number of longer-distance trips performed by Americans compared with Europeans.

TABLE 2.4 Modes of longer-distance trips, US, 2001, and EU15, 2001–2002[65]

	EU15		US			
Mode	Trips/ person	Share of trips	Trips/ person	Share of trips	Distance/ trip (km)	Share of distances
Car	1.7	65%	8.2	89%	524	56%
Air	0.2	6%	0.7	7%	4,641	41%
Bus	0.3	12%	0.2	2%	787	2%
Train	0.4	14%	0.1	1%	804	1%
Totals:	2.7	97%	9.1	100%		100%

Note: EU15 longer-distance trips are all two-way trips, for any purpose, of more than 200 kilometers. Other modes (e.g. ship, motorcycle) are not represented; hence the total of shares can be less than 100%. US longer-distance trips are all two-way trips, for any purpose, of more than 160 kilometers.

TABLE 2.5 Modes of long-distance travel in the US by distance, 2001[66]

	Trip distance in kilometers				
	80–800	800–1,200	1,200–1,600	1,600–2,400	>2,400
Personal vehicle	95.4%	61.8%	42.3%	31.5%	14.8%
Air	1.6%	33.7%	55.2%	65.6%	82.1%
Other	3.1%	4.4%	2.5%	2.9%	3.2%
This distance's share of all longer-distance trips	89.8%	3.1%	2.0%	2.3%	2.8%

Overall, there were more than three times as many, although the differences would not be quite so large if the criteria for longer-distance trips were the same (see the note at the bottom of Table 2.4). Americans made more than four times as many longer trips by car and more than three times as many by air.

Even though Americans made many more longer-distance trips, the purposes of the trips were about the same. In both Europe and the US about 70 percent of longer-distance trips were made for tourism or personal matters, just under 20 percent of trips were for business and just over 10 percent were commuting trips.[67]

Tourism

Tourism is among the most highly organized parts of the transport sector. In 2004, the World Tourism Organization (WTO), headquartered in Madrid, became the 15th specialized agency of the United Nations, with the status of the World Health Organization. It joined two other transport-focused specialized agencies of the UN, the International Civil Aviation Organization (ICAO) and the International Maritime Organization, whose headquarters are respectively in Montreal and London. Tourism involves more than transport, but transport availability provides the framework within which most tourism occurs.

Data provided by the WTO give the impression that Europe is the focus of tourist activity. For example, a chart at the WTO's website suggests that in 2003 Europe had almost 60 percent of the world's total.[68] What is not entirely clear is that these data refer to "international tourist arrivals" only. Short trips across international borders in Europe count as tourist activity, but long trips in North America are for the most part not counted because they occur within one country. Only 2 percent of longer-distance trips beginning in the US are to a destination outside the US.

Tourism is among the most highly organized parts of the transport sector.

We have already noted that US residents make more than three times as many longer-distance trips as Europeans for tourism and personal business. Data on domestic tourism in, say, Russia, Brazil or China are not available to allow a proper analysis, but a reasonable conclusion for the moment may be that most of the world's tourism takes place within the US.

Trends in Air Travel

The major incident in recent years to impact longer-distance travel was the series of terrorist attacks on the US in September 2001. The effects were chiefly in the US and mostly on air travel, evident in Figure 2.12, which shows travel by plane and car in the US from 1990 to mid 2004. In October–December 2001, air travel was down 19 percent compared with the previous year, a decline that continued into 2002. By 2005, air travel had returned to 2000 levels.

International travel of all kinds and non-business travel declined the most after September 2001. Business travel was relatively unchanged. The number of shorter long-distance trips (80–160 km) increased. These trips are almost all made by car, but the increase hardly shows in Figure 2.12 because it was small in relation to the tides of local travel that ebb and flow in the US each day.

The impact of the events of September 2001 in the US is evident in the data shown in Figure 2.13 on domestic and international travel throughout the world. By 2005, air travel had returned to the trend of mid and

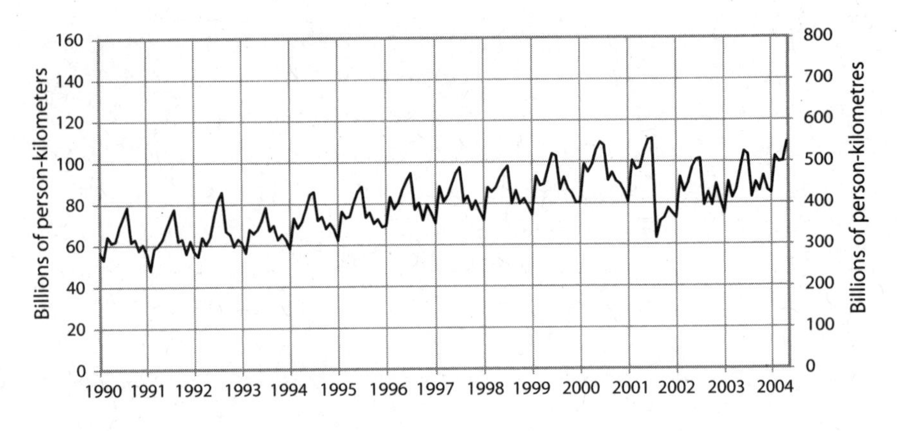

FIGURE 2.12 Monthly road and air person-kilometers in the US, 1990–2004[69]

late 1990s. The projections for 2006–2008 in Figure 2.13 are for an even steeper rate of growth than in the 1990s.

Low-Cost Carriers

The projections in Figure 2.13 take into account what may presently be the most important contributor to growth in air travel: the proliferation and growing success of low-cost air carriers, also called discount carriers, non-traditional carriers and, aptly, no-frills carriers. These airlines became a feature of the air industry when air transport was partially deregulated in the 1980s in North America and in the 1990s in Europe. They reduce costs by doing some or all of the following:

- Use one type of aircraft only, to reduce maintenance and employee training costs.
- Use a simplified route structure, to raise aircraft utilization.
- Minimize time at airports, to maximize revenue-earning use of aircraft.
- Use secondary airports, to reduce airport costs and turnaround times.
- Have unreserved seating, to speed up boarding.
- Pay lower or performance-related wages, to reduce personnel costs.
- Require employees to perform several roles, to reduce personnel costs.
- Provide no "frills," for example, in-flight meals and pillows, except for payment.
- Sell tickets mostly from company websites, to reduce transaction costs.

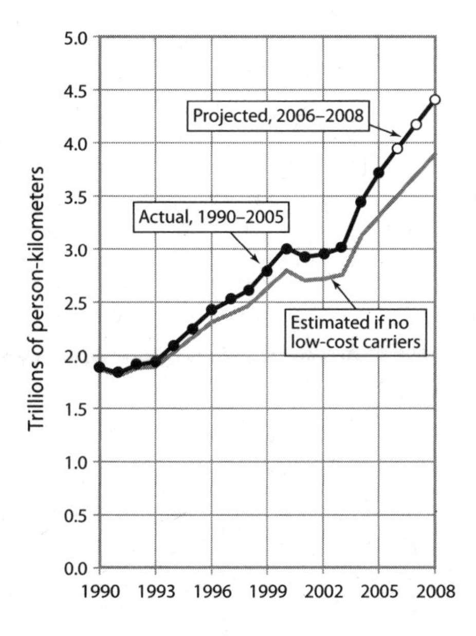

FIGURE 2.13 Travel by scheduled airlines, world, 1990–2005 (actual) and 2006–2008 (projected), and estimated travel without emergence of low-cost airlines[70]

- Charge very low fares for advance booking, to fill planes and market a low-cost image.

Their lower costs allow these carriers to charge lower fares overall than traditional carriers, also known as legacy carriers because they bear the burdens—notably underfunded pension arrangements—of a less competitive past. Some fares can seem absurdly low. In November 2004, one of the authors flew by Ryanair from London (Stanstead) to Berlin (Schönefeld) and back. The fare for the outgoing flight was £0.99 ($1.85). For the return flight it was £4.99 ($9.25). With taxes and fees, etc., the return flight cost £33.70 ($62.70). The fuel cost alone per passenger for each 912 km flight would have been in the order of $12.45—more than the higher of the two fares,[71] leaving nothing for crew, maintenance, aircraft purchase and investors. Yet, Ryanair is among the most profitable airlines in the world. How can this be?

One point is that the very low fares are usually available only several weeks in advance of the flight, and then only for some flights, often at inconvenient times. As the flight date approaches, fares increase according to how many seats remain available. Just before the flight date, and even at other times, the fare for a seat on a low-cost carrier can be higher than the comparable fare charged by a legacy carrier.[72] Overall, however, the fares charged by low-cost carriers are lower.

Perhaps the most important reason why low-cost airlines can charge lower fares and be profitable is their low employee costs. Ryanair and legacy carrier British Airways (BA) provide an extreme comparison. In 2004, Ryanair carried 27 million passengers and required 2,300 employees to do this. BA carried 35 million passengers and had 51,900 employees: 17 times as many employees per passenger. Some of this is because BA has a more extensive route system, but most of it is because Ryanair uses staff more efficiently—and it pays them less too. Worse, BA has a long-standing, underfunded pension scheme and a legacy of other entitlements by former employees. A friend of one of the authors, who held a senior position with BA some decades ago, still flies annually on BA for virtually nothing.

The low-cost movement is now more evident in Europe and increasingly in Asia, but it started in the US. There, Southwest, the main pioneer, remains almost consistently profitable while legacy carriers are often close to, or in, bankruptcy. Nevertheless, Southwest is more like, say, American Airlines than Ryanair is like BA. For example, Southwest has 36 employees per 100,000 passengers, four times as many as Ryanair, which has nine. American has 82 per 100,000, just over half as many as BA, which has 148. In North America, low-cost and legacy carriers have converged, chiefly in that the service levels of legacy carriers—and the fares they charge—have plummeted often below those of low-cost carriers.

Worldwide during the last few years, 15 or more low-cost carriers have been starting up each year, although not all have survived. The result has been real decline in air fares, in part because legacy carriers have charged less to meet the prices of the low-cost carriers. For example, US domestic air fares fell by 18 percent in real terms between early 1995 and late 2006,[73] and the fall in average fares paid may have been even steeper in Europe and elsewhere.

Much of the decline in fares occurred before 2002, when the price of jet fuel began to rise steeply. By the end of 2005, airlines' fuel costs in the US had for the first time become larger than their labor costs.[74] In the third quarter of 2006, they were reported as 27.4 percent of "operating expenses," the highest share since 1982.[75]

The cost of a liter of jet fuel increased threefold between 2002 and 2006, stayed about the same throughout 2007, doubled further by mid 2008, fell by more than half late in 2008, and by mid 2009 was at the 2006–2007 level. The respite from high fuel prices in late 2008 and early 2009 was not enough to overcome a new challenge to the aviation industry: the global economic recession and associated loss of revenue. Passenger and freight kilometers both fell in 2008 and are expected to fall much more in 2009, by 8 and 17 percent, respectively.[76] The worldwide industry recorded a net loss of over $10 billion in 2008, chiefly on account of high fuel costs. A similar loss is expected for 2009, even though fuel costs are projected to be much lower and services have been reduced dramatically. The aviation industry's revolution may already be under way.

By the end of 2005, airlines' fuel costs in the US had for the first time become larger than their labor costs.

According to the European Low Fares Airline Association, about 40 percent of the business of low-cost carriers has been diverted from legacy carriers and 60 percent are trips that would not otherwise have been taken. Low-cost carriers have about 25 percent of the market in Europe and North America, and less elsewhere. These estimates have been used to develop an assessment of the impact of the low-cost carrier phenomenon, which is a feature of Figure 2.13.[77]

Another factor of note is the development of high-speed rail, which in some places has diverted passengers from air travel.[78] Operators of high-speed trains—and other trains—are adopting some of the methods of low-cost airlines, chiefly to compete with these airlines.[79] Fares for travel by rail between some cities in Europe have fallen dramatically, especially for off-peak trips booked far in advance. The main result is to have well-occupied trains, which can be even more advantageous to operators than well-occupied planes, because additional passengers make no practical difference to a train's fuel use, but can sharply increase the fuel required for a plane's take-off.

The overall result of fare wars is to increase the amount of traveling. The actual traveling may be more fuel efficient—because trains and planes are better occupied and fuel use per person-kilometer falls—but the net result can be more fuel use, especially if trains or planes make additional trips.

Determinants of Longer-Distance Travel

For shorter-distance journeys, we highlighted car ownership as the main factor, and settlement density and costs of owning and operating a car as the main factors in car ownership. We also noted time spent traveling as a possible limiting factor. We said that the need to travel at all requires little explanation. No such simple account of longer journeys can be as plausible. Longer-distance travel is relatively rare. For the most part it is a luxury, although some longer-distance travel is impelled by personal or commercial business requirements. Characterizing tourist travel in particular as a luxury highlights its sensitivity to price.[80]

The potential demand for this particular luxury seems high. University students and retirees, in particular, appear to value longer-distance travel and engage in it when it is possible and affordable. The main reasons students give for international travel include: to escape from stress and responsibilities; to see and learn; and to enjoy a different climate.[81]

Thus, the important determinants of growth in longer-distance travel are perhaps mostly in the availability of affordable means. Other factors influence the choices of destination and time, and occasionally—as in major terrorist incidents—whether travel occurs at all.

If there is a desire to reduce longer-distance travel, the obvious remedy would be to raise its price. If prices increase, say because of increases in the cost of fuel, longer-distance travel will likely decline.

How Freight Moves

Freight Transport Is Mostly Ignored

Worldwide, about half of transport energy use is for the movement of freight, and the share is slowly growing; it's up from about 38 percent in 1971.[82] Freight movement may be responsible for an even higher share of emissions from transport. Yet, compared with the movement of people, the movement of freight is neglected. For example, a standard US university text on transport, *The Geography of Urban Transportation,*[83] devoted 14 of the 400 pages of its text to freight movement. Freight transport may be neglected in policy development too. For example, in Canada the movement of freight contributed over half of the *increase* in greenhouse gas emissions from transport between 1990 and 2002. Nevertheless, in the Government of Canada's 2002 climate change plan, the movement of freight was to provide only one quarter of transport's reduction in these emissions.[84]

We exemplify this relative lack of attention to freight transport in that here we devote many more pages to moving people than to moving freight. Our regrettable excuse is that there is less to say about how freight moves.

Freight movement may be responsible for an even higher share of emissions from transport.

Nevertheless, we do know that modern freight transport is an extraordinary phenomenon. An illustration is the distribution of live lobsters harvested off the coast of Nova Scotia. They are stored in a comatose state for up to six months in a million individual cells in a massive building in the port of Arichat before shipping to Europe, Japan and, above all, the US. Most of the US-bound lobsters go by road to the UPS air hub in Louisville, Kentucky—30,000 per truck, a 30-hour trip (see Box 2.4)—and then by air to hundreds of destinations across the US. They arrive at restaurants and retailers alive and snapping, soon to be boiled. In Chapter 1 we described the development of the "hub-and-spoke" distribution system by the Federal Express Co. (FedEx), on which the UPS system appears to be based.

Freight Movement by Mode

Figure 2.1 has already illustrated what may be a surprising feature of freight transport: on a tonne-kilometer basis, two thirds of freight moves by water. If only international trade is considered, the share moved by water is over 95 percent. However, on a value basis, water's share of international trade is only about 50 percent, with aviation carrying most of the remaining value, as shown in Table 2.6.

Underlying Table 2.6 are large differences in the value per tonne of goods shipped. These differences are illustrated in Table 2.7 for four of the five modes by which freight is shipped between the US and Canada or Mexico. Note that it cost 12 times as much to ship by road as by rail, and three times as much to ship by air as by road. However, of the modes shown, air freight has the lowest cost per dollar value of the item shipped. As a percentage of freight value, shipping by road appears to be much more expensive than other modes.

Water modes are not shown in Table 2.7 because of their wide variation in cargo value and shipping cost. The latter varies from the extraordinarily low value of about $0.60 per tkm, to ship iron ore from Australia to Europe, to about $5.00 per tkm, to move goods by barge along US rivers and canals.[88] Most of what moves by water is low-value bulk goods, but

BOX 2.4 Lobsters' wild ride[85]

UPS once leased old gas stations, furnished them with sawhorses under four-by-eight plywood sheets, and used the old gas stations as centers for sorting packages. Now they have the Worldport, as they call it—a sorting facility that requires four million square feet [370,000 m^2] of floor space and is under one roof. Its location is more than near the Louisville International Airport; it is between the airport's parallel runways on five hundred and fifty acres [223 ha] that are owned not by the county, state or city but by UPS. The hub is half a mile [0.8 km] south of the passenger terminal, which it dwarfs. If you were to walk all the way around the hub's exterior, along the white walls, you would hike five miles. You would walk under the noses of 727s, 747s, 757s, 767s, DC-8s, MD-11s, A-300s—the fleet of heavies that UPS refers to as "browntails." Basically, the hub is a large rectangle with three long concourses slanting out from one side to dock airplanes. The walls are white because there is no practical way to air-condition so much cavernous space. The hub sorts about a million packages a day, for the most part between 11 PM and 4 AM. Your living lobster, checked in, goes off for a wild uphill and downhill looping circuitous ride and in eight or ten minutes comes out at the right plane. It has traveled at least two miles inside the hub. The building is about seventy-five feet [22.9 m] high, and essentially windowless. Its vast interior spaces are supported by forests of columns. It could bring to mind, among other things, the seemingly endless colonnades of the Great Mosque of Córdoba, but the Great Mosque of UPS is fifteen times the size of the Great Mosque of Córdoba.

some relatively high-value items also move by water, including cars, agricultural equipment and other large manufactures.

Figure 2.14, which represents all transport, domestic and international, shows how the amount of movement of freight has been rising at a faster rate than the amount of movement of people. This is evidently true overall, and it is true for each mode. For example, movement by road of freight and people increased between 1990 and 2003 by 123 and 32 percent, respectively. Figure 2.14 also shows, along with Figure 2.1, how water modes dominate freight movement and road modes dominate the movement of people.

Perhaps the most striking feature of Figure 2.14 is the huge surge in movement of goods by ship, more than 80 percent of which is intercontinental trade. More than a third of this freight movement—more than half until the 1980s—involves movement of oil and oil products.[90]

TABLE 2.6 Mode shares of international trade's transport activity[86]

Mode	Tonne-kilometers	Shipment values
Water	96.7%	49%
Road	1.5%	11%
Rail	1.0%	3%
Pipeline	0.5%	2%
Air	0.3%	35%

TABLE 2.7 Value of freight and transport cost, by mode, US[87]

	Pipeline	Rail	Road	Air
Freight value per tonne	$666	$911	$2,839	$86,816
Transport cost per 1,000tkm	$10	$15	$180	$551
Transport cost as percentage of goods' value	1.5%	1.7%	6.3%	0.6%

Countries differ greatly as to the extent and mode of their domestic freight movement. Five rather different activity patterns are represented in Figure 2.15, exhibiting relatively high and low use of each of road, rail and water modes. Canada has the most even distribution among the three main modes, but is at the upper extreme in amount of freight activity per capita, reflecting her dispersed, relatively small population.

The lowest levels of freight movement by rail shown in Figure 2.15 are in Japan and Europe, where movement of people by rail is at a relatively high level. Conversely, the highest levels of use of rail for freight are in North America, where travel by rail is relatively rare.

Figure 2.15 does not reflect freight movement by pipeline because data on this mode are readily available for only three of the countries and regions. If included for China, Europe and the US, pipeline transport would have accounted for 1, 3 and 20 percent, respectively, with commensurate reductions in the shares of the other modes. Pipelines usually carry fossil fuels, notably oil and natural gas. They also carry other liquids and gases, including the carbon dioxide moved from Beulah, North Dakota, some 330 kilometers to the world's largest sequestration project in Weyburn, Saskatchewan, where it is used for enhanced oil recovery.[92] (CO_2 sequestration and enhanced oil recovery are discussed here in Chapter 3.)

Slurries are also transported by pipeline, notably slurried coal, that is, ground coal mixed with water. The world's longest slurried coal pipeline runs for 440 kilometers between an open-pit mine in Arizona and

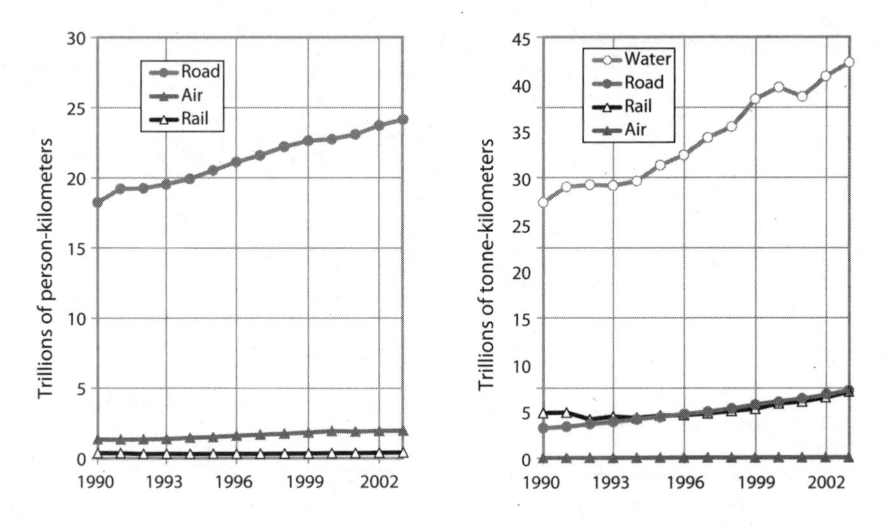

FIGURE 2.14 Movement of people and freight worldwide, 1990–2003[89]

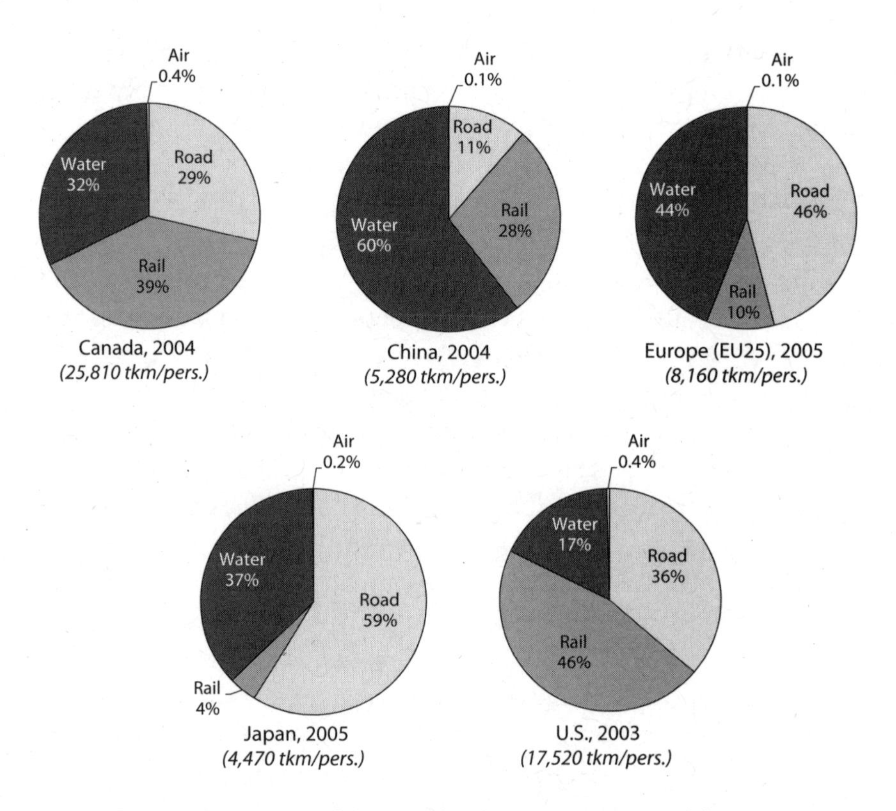

FIGURE 2.15 Mode shares of domestic freight movement (tonne-kilometers) for the indicated years, excluding movement by pipeline[91]

an electricity generating station in Nevada. It carries about 4.5 million tonnes of coal annually, representing about two billion tkm of freight movement.[93]

Relatively little is known about freight movement overall.[94] Even less is known about how freight moves within urban regions. This is the converse of the movement of people, where often more is known about movement within urban regions than about movement over longer distances.

Freight movement within urban regions appears to be different from movement between urban regions and between continents.

The character of freight movement *within* urban regions appears to be different from movement between urban regions and between continents. Nearly all local freight movement is by truck, often in small vehicles carrying loads well below their capacity. Some of the urban traffic includes heavy trucks at the beginning or end of long-distance trips, or passing through the urban region. Most freight movement within urban regions seems to be local. For example, a study of traffic in Edmonton, Canada, a city of some 850,000 people, showed that 93 percent of commercial traffic trips began and ended within the urban region, 5 percent began or ended elsewhere, and 2 percent involved vehicles that were passing through.[95]

Data for Canada suggest that about as much was spent in the late 1990s on moving goods within urban regions as was spent on moving goods between them. However, whereas most of the *intra*city movement was by road vehicles owned by businesses with a stake in what is being moved, most of the *inter*city movement by road was by vehicles operated by companies in the business of carrying other companies' goods.[96] Canada may have lower-than-average shares of freight movement within urban regions because of her long intercity distances. In other places, freight movement within urban regions may be even more important. Data on this matter do not appear to be available.

Shipping Containers

A major feature of goods movement is growth in the use of shipping containers—standardized metal boxes used for moving just about anything that can be packed into them. Among containers' virtues is the ease with which they can be moved between modes, from ship to rail or road vehicles and vice versa, thereby reducing handling costs. They are chiefly

FIGURE 2.16 Stern view of the *Emma Mæsk,* near fully laden with over 5,000 two-TEU containers[97]

used for moving intermediate (partially finished) products and finished products.

When shipping containers were first introduced (see Box 1.1) they were much smaller than current containers. Then they became a standard 6.1 m (20 feet) in length, the other dimensions being 2.4 m (width) and 2.6 m (height). Such a container is said to be one TEU (twenty-foot equivalent unit). Most containers today are twice as long (12.2 m), but with the same width and height, and are thus two TEU. The largest container ship, launched in August 2006, carries some 11,000 TEU (see Figure 2.16), although most carry no more than half this number. A train 1.5 km in length, carrying double-stacked two-TEU containers, can have a total load of close to 330 TEU.

Container activity is difficult to compare with other freight movement because the basic data are usually in terms of TEUs handled rather than tkm.[98] TEUs handled at the world's ports increased from 128 million to 337 million TEUs between 1994 and 2004.[99] This is a rate of increase in excess of ten percent annually, considerably more than the overall increase in freight transport activity illustrated in Figure 2.14.

BOX 2.5 Modern inland waterway freight movement[101]

In the thousand feet (305 meters) in front of Mel are fifteen barges wired together in three five-barge strings. Variously, the barges contain pig iron, structural iron, steel coils, furnace coke, and fertilizer. Each barge is two hundred feet long. Those with the pig iron seem empty, because the minimum river channel is nine feet deep and the iron is so heavy it can use no more than 10 percent of the volume of a barge. The barges are lashed in seventy-six places in various configurations with hundreds of feet of steel cable an inch (2.5 cm) thick... The *Billy Joe Boling,* at the stern, is no less tightly wired to the barges than the barges are to one another, so that the vessel is an essentially rigid unit with the plan view of a rat-tail file. In the upside-down and inside-out terminology of this trade, the *Billy Joe Boling* is a towboat. Its bow is blunt and as wide as its beam. It looks like a ship cut in half. Snug up against the rear barge in the center string, it is also wired tight to the rear barges in the port and starboard strings. It pushes the entire aggregation, reaching forward a fifth of a mile (0.3 km), its wake of white water thundering astern.

More specific comparison of container and other maritime traffic suggest that container movement may comprise about 12 percent of total marine tkm.[100] As noted above, about 40 percent of the total comprises the movement of oil and oil products. Another 30 percent consists of bulk goods carried in bulk carriers, notably iron ore, coal and grain. The remaining 18 percent of marine tkm are performed by general cargo ships, chemical tankers, liquefied gas carriers and miscellaneous vessels.

Not included in these shares are relatively small amounts of freight moved along rivers and canals, other than waterways plied by ocean-going vessels such as the Panama and Suez canals. Movement along inland waterways can take some unusual forms. One is illustrated in Box 2.5.

In 2005, it cost perhaps $2500 to move a two-TEU container across the Pacific,[102] or about $33 per cubic meter of usable space. If five flat-screen televisions and their packaging take up one of a container's 75 cubic meters, this means than the ocean shipping cost for an item that may retail for $700 would be about $9. Shore-side transport and handling costs could triple this amount, but this possible total of about $27 is still less than four percent of the retail price. Intercontinental transport prices would have to rise considerably to put a damper on international trade.

Factors in Freight Transport Activity

As with the movement of people, the challenge is to explain the extent and growth in freight transport activity rather than why it occurs at all. The recent growth in freight movement has been impressive, far more than the growth in the movement of people. For the latter, there can be some fairly simple explanations. The extent of local travel is largely determined by car ownership, which in turn is determined by the costs of car ownership and operation and by settlement density. The extent of longer-distance travel is largely determined by its relative cost.

Intercontinental transport prices would have to rise considerably to put a damper on international trade.

One simple explanation of freight transport activity is that it is the result of economic activity. When economic activity increases, so does freight movement. However, the case can be made that the main causal relationship is in the opposite direction: that is, freight transport activity drives economic activity.[103] The likely truth is that both occur. Efficient freight movement facilitates economic activity that in turn stimulates more freight movement.

The growth in the movement of goods by sea (see Figure 2.14)—which comprises most freight movement (see Table 2.6)—has not been evidently accompanied by increases in the average distance traveled by each tonne of freight or in the average real value of each tonne. These factors have been more or less constant, at least since about 1980, as is illustrated in Figure 2.17. Also, even though there has been a large increase in the amount of air freight, the average distance moved by each tonne has, if anything, declined.[104]

The relative constancies in Figure 2.17 do not give ready comfort to explanations of increased freight activity in terms of processes of globalization. These processes in part comprise the manufacture of increasingly more valuable goods at increasingly greater distances from where they are used. If such globalization were the main cause of increased freight movement, both distance per tonne and value per tonne might have been expected to increase.

More local factors could be relevant, including the manufacturing practices known as outsourcing and just-in-time delivery. Outsourcing involves the sub-contracting of elements of a manufacturing process to a plant that may be in a distant place. Increased outsourcing is undoubt-

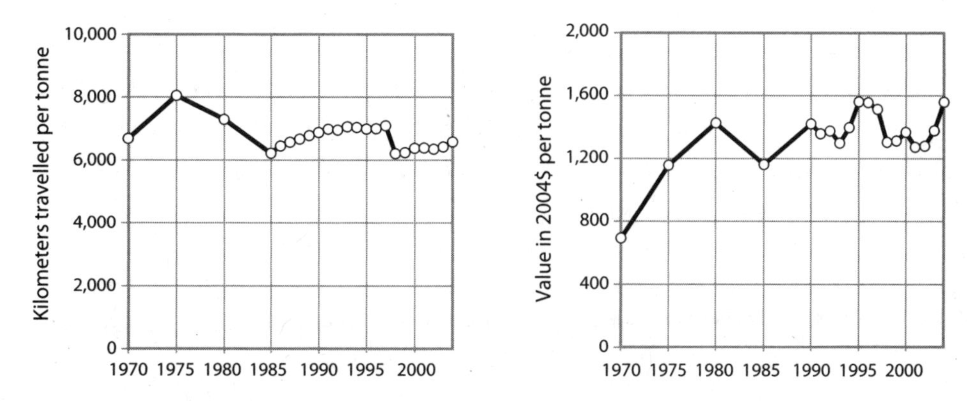

FIGURE 2.17 Average distance traveled by marine freight (left panel) and value per tonne of international trade (right panel), 1970–2004[105]

edly a modern business practice. However, in several industries, notably car manufacture, it has been accompanied by tighter process integration, even to the point of having the firm to which work is outsourced actually located in the plant of the outsourcing entity.

Just-in-time (JIT) delivery refers to the substitution of warehousing by precisely scheduled deliveries of required materials, which in practice can mean more frequent deliveries, possibly involving less-than-full vehicles. JIT requirements are unquestionably prevalent, although they may have been moderated by the high and growing costs of road shipment—which is usually essential for JIT—and the challenges posed by growing road congestion.

Analysis of US data suggests that two factors contributed more or less equally to the growth in truck vehicle-kilometers in the 1990s. Between 1990 and 1999, average trip distance increased by 17 percent and the number of trips increased by 18 percent.[106]

Transport Tomorrow?

In this chapter we have focused on recent and current transport activity, at least until 2007. Until 2007, there was widespread agreement that unless checked by government policies or perhaps by unavailability of fuel, transport activity would continue to increase, particularly in poorer countries. The most authoritative projections of "business-as-usual"[107] transport activity were those of the International Energy Agency, developed as part of the Sustainable Mobility Project (SMP) of the Geneva-based World Business Council for Sustainable Development.[108]

These business-as-usual projections are set out in Figure 2.18. Note that waterborne modes are not included, nor freight movement by air or pipeline. As illustrated in Figures 2.1 and 2.14, waterborne modes are particularly important for freight transport.

The upper panel of Figure 2.18 represents a projected increase in total person-kilometers worldwide by 50 percent between 2000 (SMP's base year) and 2025, and by 129 percent between 2000 and 2050. Person-kilometers in China were projected to increase at the highest rate among the countries and regions: by 110 percent between 2000 and 2025 and by 339 percent between 2000 and 2050. Available data suggest that SMP greatly underestimated the annual rate of growth in pkm in China. During 2000–2005, it was 7.4 percent rather than the 2.9 percent projected by the model.[110] North American pkm during 2000–2005 were also greatly underestimated. SMP projected an annual rate of growth of 0.8 percent, but the actual rate in the US was 2.3 percent.[111] The actual rates for both China and the US were 2.5 and 2.7 *times* the projected rates. Freight activity was also underestimated, but not as much as the movement of people.

These projections were, and may even still be regarded as, the most authoritative projections of transport activity. SMP's major underestimation of increases in the motorized movement of people in particular highlight the extraordinary nature of recent trends in transport activity. They also point to how unprepared countries such as China and the US may have been for changes in the supply and cost of transport energy.

We should stress that SMP's authors did not propose that the above "business-as-usual" projections will or should happen. Indeed, they suggested that the projections do not meet seven goals required for transport sustainability:

1. Reduce conventional emissions from transport so that they do not constitute a significant public health concern anywhere in the world.
2. Limit greenhouse gas (GHG) emissions from transport to sustainable levels.
3. Reduce significantly the number of transport-related deaths and injuries worldwide.
4. Reduce transport-related noise.
5. Mitigate traffic congestion.

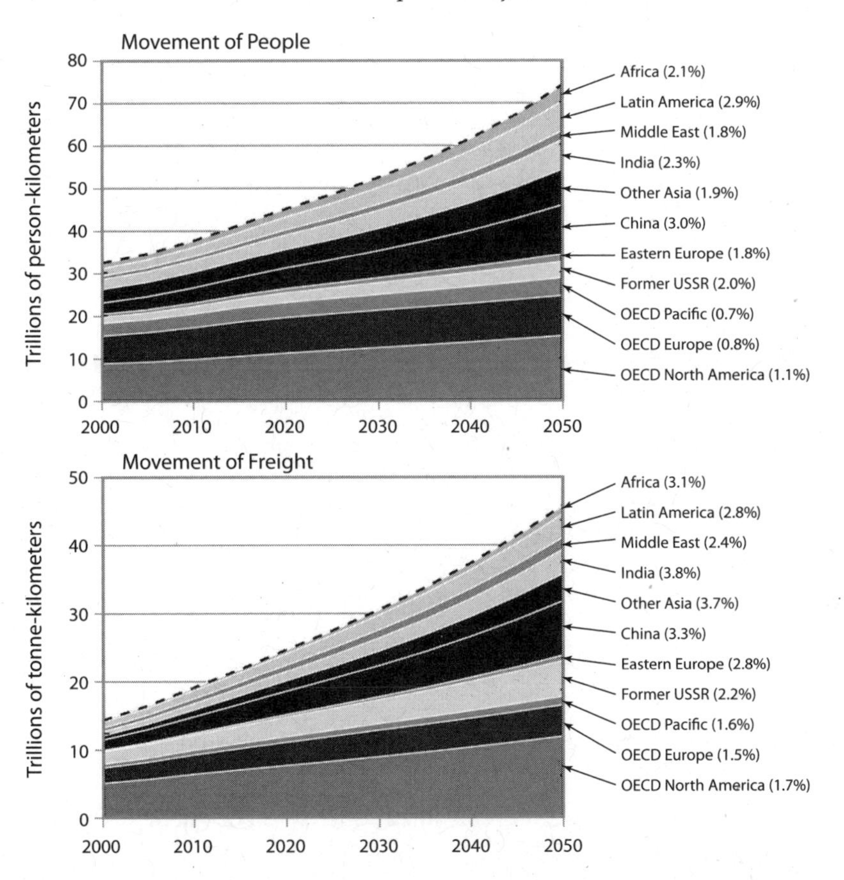

FIGURE 2.18 "Business-as-usual" projections until 2050 of motorized movement of people and freight (percentages are projected annual rates of increase, 2000–2050)[109]

6. Narrow "mobility divides" that exist within all countries and between the richest and poorest countries.
7. Improve mobility opportunities for the general population in developed and developing societies.

The world changed in 2008, when there appears to have been the sharpest fall in travel for decades. The evidence for this is mostly anecdotal except for the US, which produces data on vehicle movement—and an extraordinary range of other features of US society—with exemplary timeliness and accessibility. Figure 2.19 shows per-capita movement of light-duty vehicles (i.e., personal vehicles such as cars and SUVs) in the US for each year since 1970 and illustrates the steepness of the decline in travel in 2008.

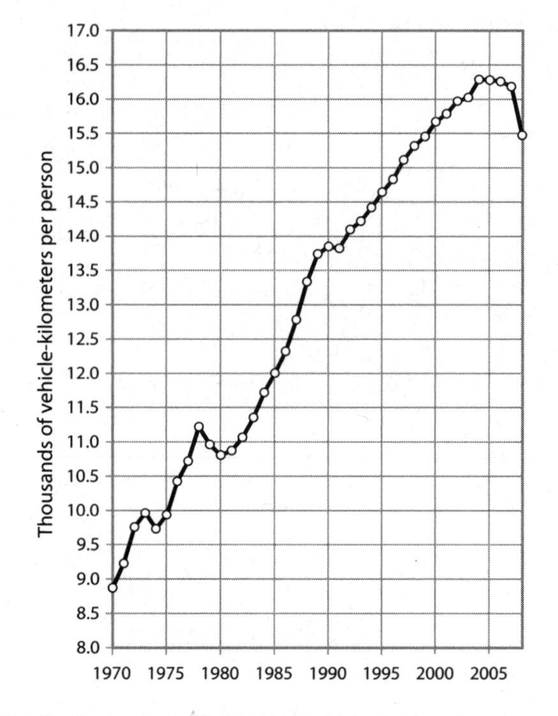

FIGURE 2.19 Movement of light-duty vehicles in the US, 1970–2008[112]

Figure 2.20 provides a more detailed description of the movement of light-duty vehicles for 2000–2008. It shows gasoline prices too.

The figure suggests that the period from 2000 to 2008 can be divided into four parts:

- During the first part, from 2000–2003, gasoline prices remained more or less constant. Vehicle movement rose steeply.
- During the second part, from 2004–2006, gasoline prices rose by more than 50 percent. Vehicle movement continued to increase, but at a slower rate than during 2000–2003.
- During the third phase, from 2007 until the third quarter of 2008, gasoline prices were firmly above what may have been the alarming price of $3.00 per gallon and still increasing. On some days and in some places they went above what may have been the even more alarming price of $4.00 per gallon. (The price of diesel fuel went above $4.70 per gallon.) Vehicle movement began to fall, at first slowly and then, during 2008, steeply.

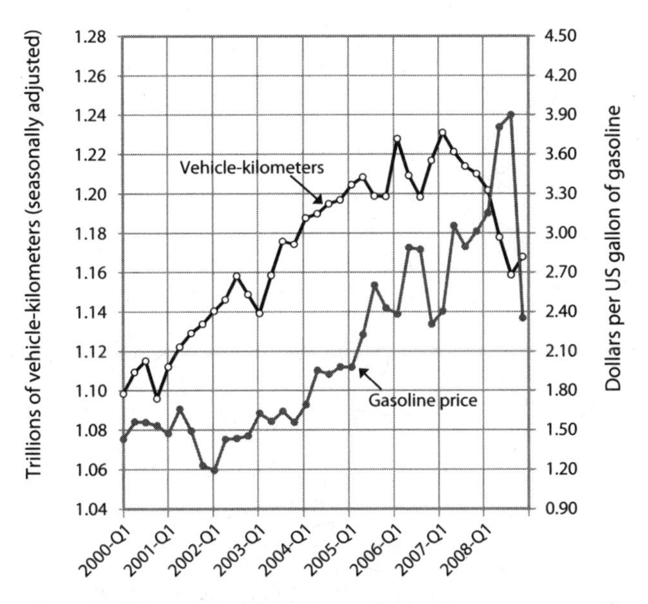

FIGURE 2.20 Movement of light-duty vehicles (quarters, seasonally adjusted) and average retail gasoline price, US, 2000–2008[113]

- In the fourth phase, which began with the fourth quarter of 2008, or perhaps just before, gasoline prices fell steeply and vehicle movement began to increase again. Incomplete data for the first quarter of 2009 suggest that this increase in vehicle movement is continuing.

Thus, a rational account of the last few years could be this. When gasoline prices are low, transport activity increases. Increases in gasoline prices slow down the rate of growth in activity. When a particular price level is passed, transport activity begins to fall, eventually reducing demand for gasoline. The reduced demand causes prices to fall, at which point transport activity begins to increase again.

Missing from this account are the causes of the rise in gasoline prices from 2004–2008 and the roles of the US and global economic recessions that became evident during 2008. These matters are addressed in Chapter 3 and again in Chapter 6.

We are in evidently in a period of considerable turbulence in matters related to transport activity, particularly energy matters. This turbulence could be signaling the start of an era of transport revolutions.

3

Transport and Energy

Introduction

Products of petroleum oil fuel almost all today's transport. Much of this chapter concerns prospects for continued sufficient supply of oil to meet anticipated transport activity and for replacing oil as the main transport fuel. Our conclusion is that world oil production will soon begin to fall and that electricity is the most likely replacement, with much of it eventually produced from renewable resources. We discuss electric vehicles, as well as vehicles that use oil products and other fuels. Finally, we consider how enough electricity might be produced to replace oil as the main transport fuel.

Today's extensive oil use is a recent phenomenon.

Today's extensive oil use is a recent phenomenon. Figure 3.1 shows that more than 50 percent of the oil ever used has been consumed since 1986 and more than 95 percent of the world's total oil consumption has occurred since the beginning of the Second World War. The cumulative total consumption of 1.16 trillion (10^{12}) barrels at the end of 2008 appears to be approaching about half the oil that could ever be extracted.[1] We believe this milestone heralds the beginning of an essentially unavoidable decline in the amount that can be produced—and thus consumed—in any year, as we discuss below.

According to conventional economic notions, when an item is abundant its price is relatively low and when it becomes scarce its price rises. The higher price suppresses consumption to the level of availability of the

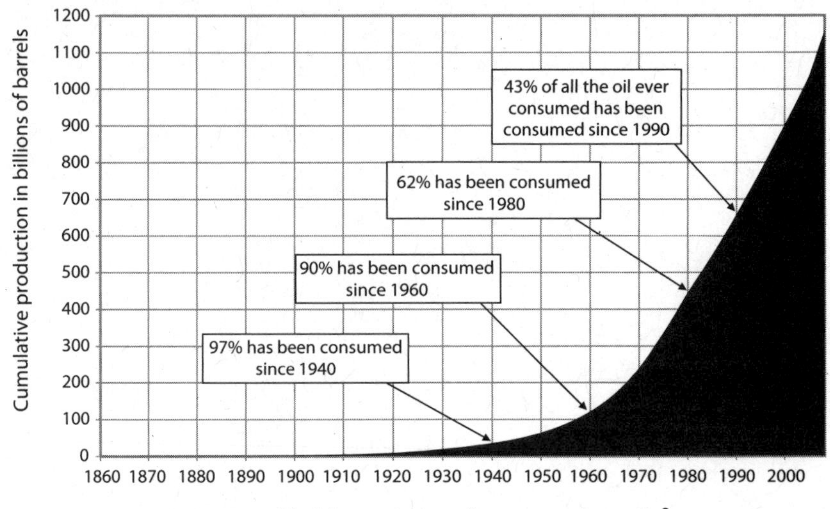

FIGURE 3.1 World cumulative oil consumption, 1860[2]

item. Scarcity thus produces a new equilibrium of price and consumption. The higher price can also increase supply in the form of more expensive alternatives to the item. The alternatives become feasible to produce, and their price may decline because of the quantities produced.[3] In the case of oil, the alternative could be difficult-to-reach oil whose production was not profitable at the lower price. The alternative could also be another fuel such as ethanol that can be intrinsically more costly to produce than oil. Such additions to supply reset the equilibrium again, this time toward a lower price and more consumption, but perhaps not to such a low price as before and with consumption still below the initial level. In this chapter we explore how the price of oil can change, especially in the light of the extraordinary events of 2008, when the price of a barrel of crude oil reached its highest ever level of $145.29 on July 3 but by December 19 had fallen to $33.87, a decline of 77 percent.[4]

Oil and Its Future

Production of Transport Fuels

Some 95 percent of the fuel used for transport is a liquid petroleum product made from crude oil.[5] Cars run mostly on gasoline (petrol), although increasingly in Europe they run on diesel fuel.[6] Diesel fuel is denser, less volatile and contains more usable energy per unit volume than gasoline. Trucks run mostly on diesel fuel, although smaller ones use gasoline.

Small boats use gasoline. Large ships invariably use diesel fuel, or a dense, high-sulfur variant of diesel fuel known as bunker fuel. Non-electric railway locomotives mostly use diesel fuel, usually for generators powering electric motors that drive the wheels but also, in smaller locomotives, for engines that drive wheels directly. A few locomotives still use coal, chiefly to give tourists a sense of rail travel in earlier times. Jet aircraft use a form of kerosene, which is similar to diesel fuel. Propeller aircraft use aviation gasoline, also known as avgas. It is similar to what automobile gasoline used to be like in that it contains a lead compound to reduce uncontrolled ignition of the fuel ("knocking").

The three main fuels—road gasoline, diesel fuel and aviation gasoline—correspond to the three main types of internal combustion engine (ICE) that today propel almost all transport. In two of these types of ICE, fuel ignites within a closed cylinder, expands, and moves a piston whose action is converted into rotary motion. In one of these two types, which uses gasoline as a fuel, ignition is achieved by a carefully timed electric spark. In the other type, ignition occurs when diesel fuel is subjected to high pressure by the returning piston. These two kinds of ICE thus operate through making use of series of contained explosions of their fuels.

Some 95 percent of the fuel used for transport is a liquid petroleum product made from crude oil.

The third type of ICE, the gas turbine engine, burns fuel continuously, producing a high-velocity flow of exhaust gases. In a jet engine, which is a form of gas turbine engine, this flow of gases provides the thrust that results in propulsion. As well, a turbine, propelled by the exhaust gases, compresses fuel and air for ignition and powers other equipment. In a turbofan engine, used in most non-military jet aircraft, a fan driven by the turbine acts like an enclosed propeller and provides additional thrust. For other applications of gas turbine engines, including in some cars and locomotives, energy is recovered mainly from the turbine rather than from the direct thrust of the exhaust gases. The turbine's mechanical energy can be used directly or after conversion via a generator to electrical energy.

The three types of fuel are derived from crude oil in oil refineries, in what has been described as "one big fuming silo."[7] The crude oil is boiled at the bottom and its fumes rise into the column. The temperature of the column declines with increasing height, and different products condense

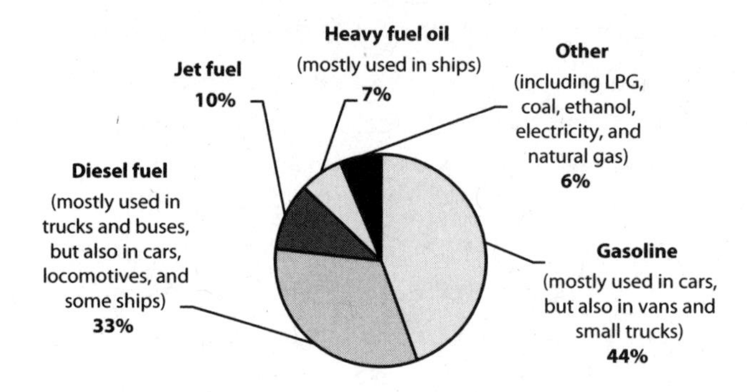

FIGURE 3.2 World end-use consumption of transport fuels, 2003[13]

out according to temperature: asphalt at 600°C, followed by lubricating oils and greases (400°C), heating oil, diesel oil and jet kerosene (200°C), naphtha (70°C), and propane and butane (20°C). There is much blending with lighter oils and gases, both of the crude oil before it is boiled and of the products of distillation. Notable among the products is naphtha, which is blended to form gasoline.

The refining process is moderately energy-intensive, consuming about 10 percent of the energy available in the materials produced. The remainder of the processes of extraction, conditioning and transport of conventional crude oil from well to vehicle require a further 5 percent.[8] What is meant by "conventional oil" is set out below.

World Consumption of Transport Fuels

World consumption of transport fuels in 2004 was 2.03 billion (10^9) tonnes or 14.9 billion barrels of oil or oil equivalent.[9] In energy terms, it was 91.0 exajoules, i.e., 91.0 billion billion (10^{18}) joules.[10] Figure 3.2 shows the shares for each fuel of total end-use transport energy consumption in 2003. Worldwide, transport energy was almost evenly shared between gasoline, on the one hand, and denser fuels including diesel fuel and jet fuel, on the other hand. There were strong regional differences. For example, in Europe, where most new cars have diesel engines,[11] diesel and other denser fuels comprised 71 percent of use of transport fuels. In Japan they comprised 53 percent, and in the US, where diesel-fueled cars are rare, they comprised only 43 percent.[12]

> Use of oil for transport has been rising at a higher rate than use of oil for other purposes.

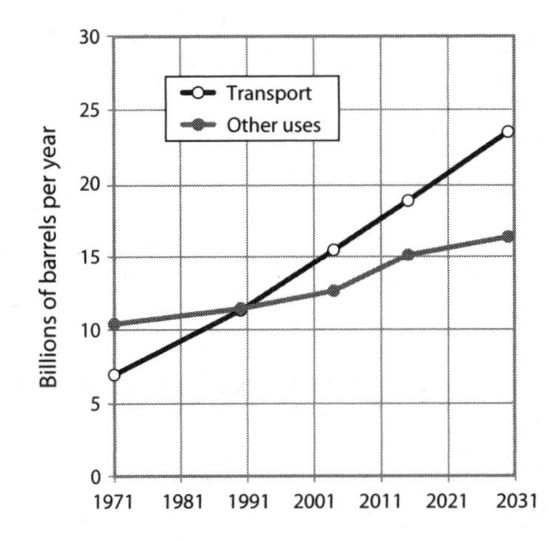

FIGURE 3.3 Actual and estimated oil consumption by purpose, world 1971–2030[17]

The oil used for transport represented about 58 percent of all end uses of oil products in 2004.[14] The remainder was used for making road surfaces (asphalt), heating buildings, generating electricity, and as a feedstock for plastics, pharmaceuticals, fertilizers and pesticides. Compared with this world average of 58 percent, transport comprised a higher share of oil use in North America (71 percent) and in Europe (61 percent) and a lower share in almost all of the rest of the world, including Japan (43 percent).[15]

Use of oil for transport has been rising at a higher rate than use of oil for other purposes. The International Energy Agency (IEA) has projected a continuation of this difference, as shown in Figure 3.3. Between 2004 and 2030, oil use for transport is to grow by 52 percent, from 15.4 to 23.4 billion barrels per year (bb/y), while oil use for other purposes is to grow by 29 percent, from 12.6 to 16.3 bb/y.[16]

The IEA expects more of the overall *growth* in annual oil use for transport between 2004 and 2030 to come from poorer rather than richer countries—5.3 vs. 2.5 bb/y—as shown in the upper chart in Figure 3.4, where richer means OECD member countries. Poorer countries have many more people (5.2 vs. 1.2 billion in 2004) and their populations are projected to grow more rapidly (1.1 vs. 0.4 percent per year). As a result, their increase in oil use *per capita* between 2004 and 2030 will be less than that in richer countries—1.1 vs. 5.3 barrels per person per year—as shown in the lower panel of Figure 3.4.

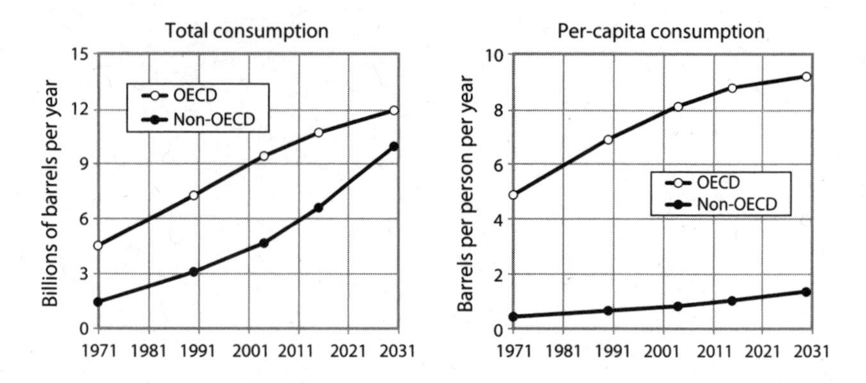

FIGURE 3.4 Actual and estimated total consumption (upper chart) and per capita consumption (lower chart) of oil for transport in OECD and other countries, 1971–2030[18]

Thus, the potential growth in oil use for transport can be seen in one or both of two ways. It can be seen as driven mostly by expansion of motorized transport in poorer countries or as continued, more intensive appropriation of resources by people in richer countries.

The key question, however, is not who will be responsible for the growth in oil use for transport but whether the projected increase can occur at all. Three factors could forestall the growth. One is unavailability of oil, discussed in this chapter. The second is action to curb oil use because of its environmental impacts, discussed in Chapter 4. The third is economic downturn or even collapse, perhaps the result of the first or the second factors, also considered in this chapter.

Oil Discoveries, Reserves and Extraction (Production)

The most important fact about oil availability is that the peak of discovery of previously undetected oil is long past and the rate of worldwide consumption is now three or more times the rate of discovery. This is shown in Figure 3.5. The worldwide rate of consumption has been close to 30 bb/y, and the rate of discovery has been below 10 bb/y.[19] A discovery of oil is any new quantity of underground oil identified through drilling or in other ways.

Discovered oil can be a *proven reserve,* meaning that it is estimated to have a 90 percent or higher probability of being extractable at current prices with current technology; a *probable reserve,* with 50–90 percent

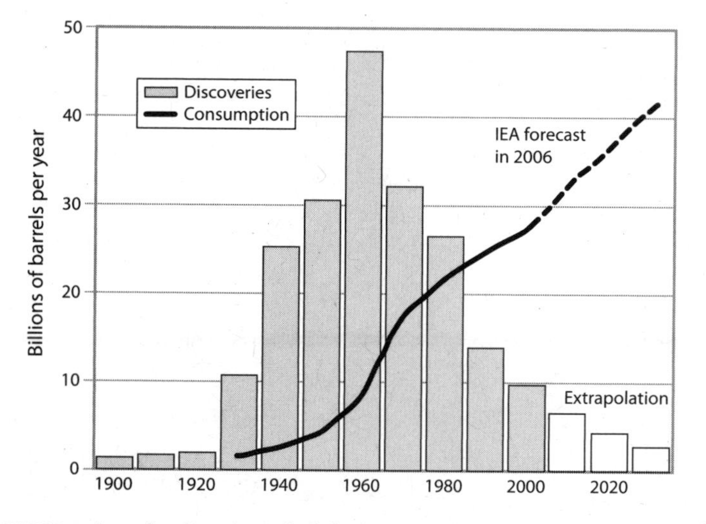

FIGURE 3.5 Actual and projected oil discovery and consumption, 1900–2030[20]

extractability; a *possible reserve,* with 10–50 percent extractability; or not qualify as a reserve.[21]

Oil reserves can translate into extraction of crude oil and then production of gasoline, diesel fuel and other oil products. The constraints on extraction have been a controversial matter, and in the next few paragraphs we give our view. The ultimate constraint is what is discovered—oil cannot be extracted if it is not there—but there are more severe constraints both as to how quickly and for how long oil can be removed from a given underground reservoir.

The most important factor appears to be the proportion of the oil that has been removed from an oil-producing area. When little has been removed, production is usually low because few wells have been drilled. As more wells are drilled, production increases until all the easily accessible oil has been removed, and then production begins to decline. Typically, the peak in production is reached when about half of the extractable oil in an oil-bearing sedimentary basin has been removed. The extractable oil is roughly the total of the identified proven and probable reserves,[22] that is, oil considered to have a better than even chance of being extracted under current conditions.

The process is illustrated in the following account by Michael Smith, which is organized around his Figure 19.1, reproduced here as Figure 3.6:

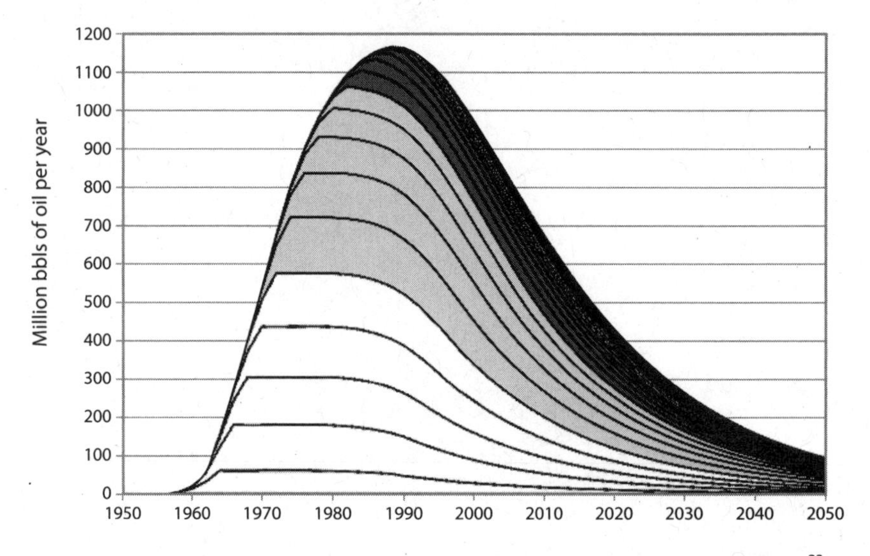

FIGURE 3.6 Ideal model of oil production from an oil-bearing sedimentary basin[23]

Consider the following ideal model of a sedimentary basin anywhere in the world. The first few fields are large and relatively easy to find. On average they come on stream within 3–4 years of discovery. Every field has a production profile determined by the following five factors:

1. reservoir characteristics (e.g., porosity, permeability, etc.),
2. fluid type (gas, oil, viscosity, etc.),
3. pressure and temperature,
4. production environment (e.g., onshore, offshore, etc.); and
5. level and timing of investment

Generally for each field, output rises to a plateau within 2–3 years, is stable for 2–3 years, and then declines at a rate of 5–15 percent per year, depending on the five factors listed above. As time passes and more fields are discovered, field size becomes progressively smaller. This process is illustrated for thirty fields in a hypothetical sedimentary basin in Figure 19.1.

It is clear from this simple model that the last half or so (last 15) of the thirty fields illustrated in Figure 19.1 do not affect the timing of the peak in any significant way. Nor does enhanced recovery affect the timing of the peak, since it is usually implemented after the peak has passed. Instead, later discoveries and enhanced recovery affect the rate of decline.[24]

What applies to an individual sedimentary basin also applies to a group of basins or to all the oil-bearing sedimentary basins in a country. In the 1950s, geologist M. King Hubbert applied this understanding to the oil resources of the lower 48 states of the US and concluded that the peak in production from them would occur between 1966 and 1972.[25] It occurred in 1970. Since then, peaks in production of *conventional oil* have been identified in 55 of the 64 countries that have produced significant amounts of this oil, and the peak in world production of conventional oil may have occurred as early as 2005.[26]

> The proposition that there will be a peak in oil production is mostly not controversial. Oil is a finite resource, and extraction of it cannot continue indefinitely.

Conventional oil is relatively easy to extract and process. It exists in readily accessible locations, often under pressure, and comprises more than 95 percent of the oil that has ever been extracted. *Non-conventional oil* is in remote places or requires much mining and processing, or both. Oil from polar regions and from Alberta's tar sands are examples of non-conventional oil.[27] Production of a refinable product from non-conventional oil can cost several times more than extraction of conventional oil.

Vigorous drilling and aggressive extraction techniques can push the peak back a little—so that it occurs, say, when about 55 percent of the reservoir has been depleted—but the post-peak decline in production will be steeper. According to the IEA, "The fall will be sharper than was the increase to peak if more than half of ultimate resources have already been produced when the peak is reached."[28]

Production of Petroleum Liquids Likely to Peak by 2012

Based on the kind of analysis developed by M. King Hubbert, with input from geologists in several countries, researchers at the University of Uppsala in Sweden have projected that the peak production of *all* petroleum liquids will occur by 2012.[29] This is represented in Figure 3.7, which shows production of conventional oil by region and production of non-conventional oil (heavy, deepwater, polar) by type. Heavy oil is chiefly produced from the tar sands (also known as oil sands) in northern Alberta, Canada. Deepwater oil is extracted through ocean depths of 500 meters and more,[30] chiefly in the Gulf of Mexico and the South Atlantic. Polar oil comes from Arctic regions of the US, Canada and Russia; it warrants classification as non-conventional because of its remoteness,

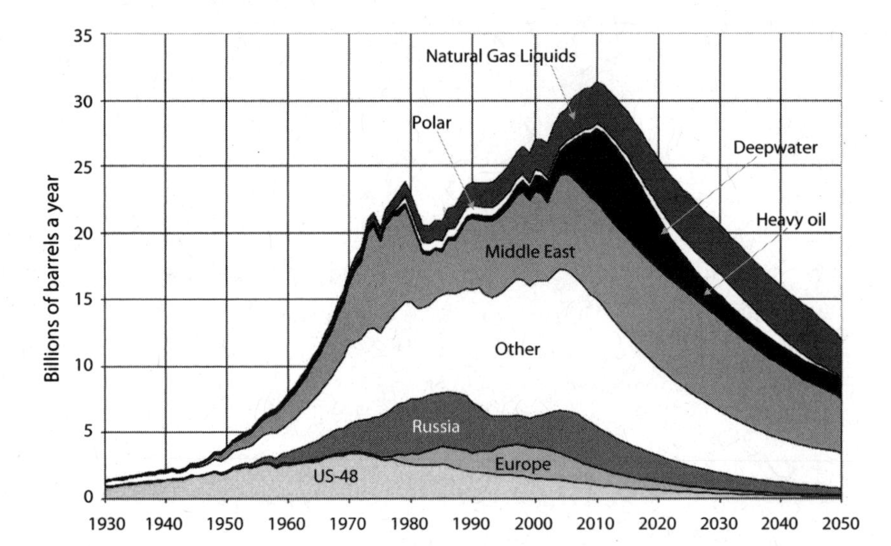

FIGURE 3.7 Actual and estimated production of petroleum liquids, by region or type, 1930–2050[31]

the difficulty of extraction and the heightened risk of environmental damage.

The top layer of the chart in Figure 3.7 refers to natural gas liquids. These are by-products of natural gas processing that are used to sweeten (i.e., make less dense) the outputs of oil refineries. They are mostly easily liquefiable gases, notably propane and butane. Their contribution to oil production is substantial, amounting to about 10 percent of the total in 2005, a share similar to that of non-conventional oil (i.e., deepwater and polar oil and oil from tar sands). The darkest layer refers to oil from Alberta's tar sands and similar sources. Oil from this source is a substantial and growing share of Canadian production (about 50 percent in 2009), but, as shown in Figure 3.7, less than 5 percent of world supply, rising to no more than 15 percent by 2050.

We believe the Uppsala projection in Figure 3.7 to be the most authoritative. Other predictions of peak production of all petroleum liquids should be noted, ranging from 2006 or earlier to beyond 2030.[32] The proposition that there will be a peak in oil production is mostly not controversial. Oil is a finite resource, and extraction of it cannot continue indefinitely. Usually, only the date of peak production is in question and, on occasion, the reason for the peak. However, some analysts suggest that oil consumption at something like the present level, and perhaps the present or even a lower price, can continue indefinitely, or at least for the foreseeable future.[33]

Just as it took time for a consensus to emerge that US oil production peaked in 1970, there is still controversy regarding the existence, let alone the actual date, of a peak in conventional oil production. One author has suggested that the consequences of a peak in global oil production are likely to be so severe that the debate over the exact timing of the peak "...will be forgotten in the turmoil of political events accompanying it."[34] By that measure, we have not yet passed the global production peak.

Note that Figure 3.7 shows actual oil production until 2007 and then estimated production *assuming that all that could be produced in any year would be produced.* This seemed a reasonable assumption in the years before 2008, when world oil consumption was expected to grow for many years to come, if sufficient supply would be available. The events of 2008 demonstrated that continued growth in demand for oil is not inevitable.

Events of 2008: Reality Checks

The left-hand panel of Figure 3.8 captures our thinking about the future of oil supply and demand at the time we completed the first edition of this book in July 2007. The solid line is actual, projected or maximum world production, based on Figure 3.7. The dotted line represents what in July 2007 was the most authoritative projection of demand for oil, that of the International Energy Agency (IEA), an organization described in *The Economist* magazine as "a think-tank funded by power-hungry countries."[35] This Paris-based OECD affiliate provides analysis and advice to its 28 members, which include most OECD countries.

IEA's demand projection as represented in the left-hand panel of Figure 3.8 was the business-as-usual (reference) projection in the 2006 version of its influential *World Energy Outlook* (*WEO*). IEA described this projection as "unsustainable," noting that sufficient supply would be unlikely to be available because of "under-investment, environmental catastrophe or sudden supply interruption."[37]

Whatever the particular uncertainties about supply in 2007, there were widespread expectations of continued increasing demand, or potential demand, and of impending supply difficulties. We believed—and still believe—that the primary cause of the steep rise in oil prices during

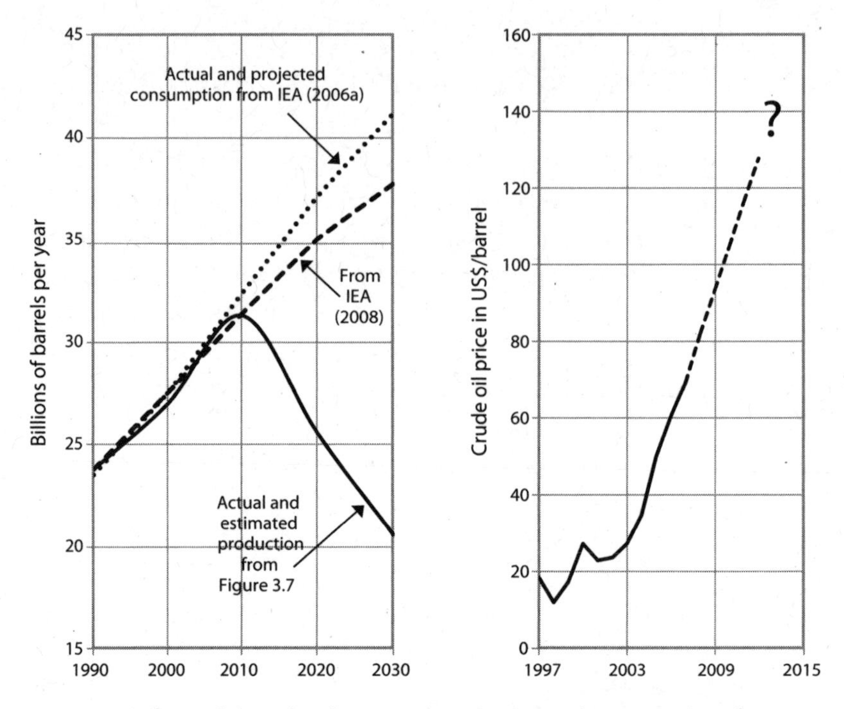

FIGURE 3.8 Left panel: Actual and expected production and consumption of petroleum liquids, 1990–2030. Right panel: Actual and expected price of crude oil, 1997–2012[36]

2002–2007 was actual or anticipated discrepancies between what people wanted to use and what was or could be available. We believed that the rise would continue at least until 2012 in something like the manner represented in the right-hand panel of Figure 3.8. There, the dashed line is a linear extrapolation of prices during the previous five years.

We expected too that not long after about 2012 the anticipated sharp divergence of supply and potential demand (see Figure 3.8) could well produce larger price increases. How large could such price increases be? According to the US National Commission on Energy Policy, "a roughly four-per-cent global shortfall in daily supply results in a 177 percent increase in the price of oil."[38] Another analysis suggested that a 15 percent shortfall could result in a 550 percent increase in the crude oil price.[39] Thus, prices in the years after 2012 could be much higher than those experienced during 2002–2007, when prices may have risen in anticipation of a peak in world oil production rather than as a consequence of it.

What actually happened in 2008 was surprising in two respects. Dur-

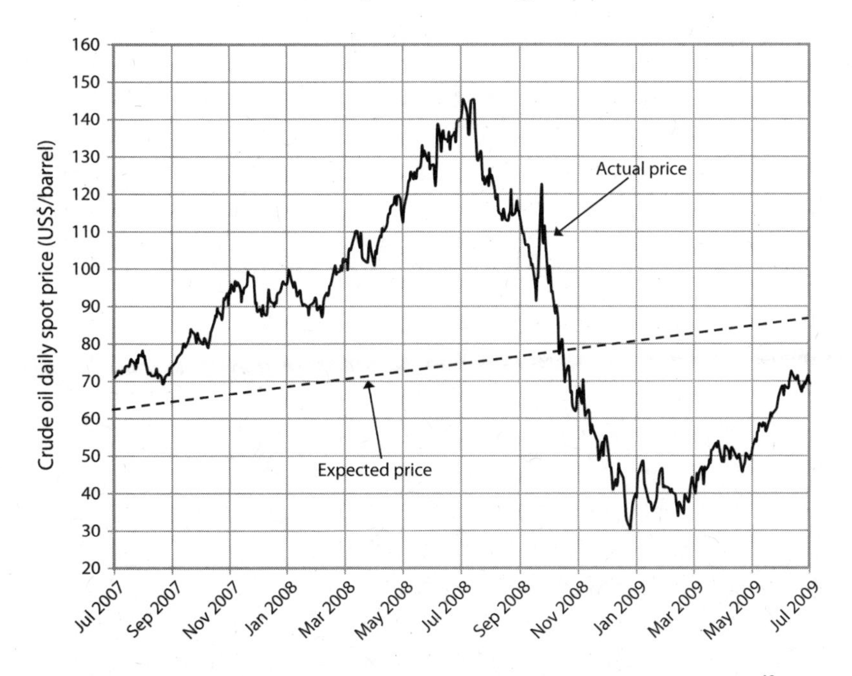

FIGURE 3.9 Actual and expected price of crude oil, July 2007 to July 2009[40]

ing the first half of the year, oil prices rose more steeply than we—and just about everyone else—expected. During the second half of the year, oil prices fell even more steeply than they had risen, an almost entirely unanticipated result.

Figure 3.9 shows what happened. The solid line shows the average daily spot price per barrel from July 2007 to April 2009. The dashed line shows our minimum expectation of a price increase; it reproduces the relevant part of the dashed line in the right-hand panel of Figure 3.8. Figure 3.9 shows the steep price rise between February and July 2008 to above $140/barrel. This was followed by an even steeper fall to just above $30/barrel in late December, interrupted by a brief rally in late September. By May 2009, the price had risen to above $60/barrel.

Oil is mostly traded in US dollars, and oil prices are almost always quoted in US dollars. The rise and fall in oil prices was accompanied by a fall and then a steeper rise in the value of the US dollar in relation to another major trading currency, the euro. This led to arguments that much of the price peak shown in Figure 3.9 was a currency phenomenon. However, the highest value of euro in relation to the dollar, in November, was only 27 percent above the lowest value, in July, and the oil price variation was many times that difference. Also, other major currencies varied differently with the dollar. The yen appreciated unevenly throughout 2008, by a total of 24 percent. The UK pound changed little in the first part of the year and fell sharply from August through December, falling in value by 28 percent.[41]

Moreover, at least one earlier analysis had found that currency volatility was a *result* of oil price shocks rather than a *cause* of them.[42] When the dust settles, a similar conclusion may emerge regarding the events of 2008.

Speculation in the markets for oil futures was another reason given for the unexpectedly steep increase in the price of oil during the first part of 2008. This would involve purchase by so-called "non-commercial traders" of commitments or options to buy or sell oil in the expectation or hope that the price would go up or down and a profit could be made.

Analyses of what might have been speculative activity are divided in their interpretation of whether speculation has driven up oil prices. Diametrically opposed views are offered, with evidence to support each of their claims. For example, the United States Senate's Permanent Subcommittee on Investigations concluded in 2006 that "speculators have, in effect, created an additional demand for oil, driving up the price of oil to be delivered in the future in the same manner that additional demand for the immediate delivery of a physical barrel of oil drives up the price on the spot market."[43] More recently, also favoring a strong role for speculation, an oil industry "insider" concluded that because several times more "paper barrels" are traded than there is actual production of oil, speculative activities could have exacerbated the extremes in the recent rise and decline of oil prices.[44]

At least one earlier analysis had found that currency volatility was a *result* of oil price shocks rather than a *cause* of them.

Yet, there appears to have been agreement among experts at a conference held by the US Energy Information Administration in April 2009 that speculators should not be blamed for record high oil prices in 2008, although there was no agreement as to what caused the high prices. The chief economist of the US Commodity Futures Trading Commission was reported as saying that an effort continues to determine why crude prices surged and then plunged in 2008.[45]

There had been an increase in the number of speculators operating in regulated oil futures markets since 2000, but they remained considerably outnumbered by "commercial traders," i.e., people who buy or sell oil based on their need for petroleum products.[46] However, only a small part of the world's oil is traded in futures markets such as New York Mercantile Exchange (NYMEX). Considerably more oil is bought and sold through "over-the-counter" private contracts, which have been estimated to account for several times the volume of trading on regulated futures markets.[47] Although limited credence is given by analysts in the energy business to notions that speculation is a prime cause of changes in oil prices,[48] the question nevertheless remains open to further investigation, and potential revelations.[49]

A basic problem with proposals that speculation drives changes in the price of oil is that trading activity is disciplined by the eventual need to buy and sell the oil that is traded. James Hamilton said this another way: "The key intellectual challenge...is to reconcile the proposed speculative price path with what is happening to the physical quantities of petroleum demanded and supplied."[50]

Hamilton's 2009 paper, published in the Brookings Institution's series on economic activity, provides one of the more thorough and plausible explanations offered so far by an economist seeking to understand the remarkable changes in oil price shown in Figure 3.9. He attributed the cause of the steep increase in the first part of 2008 to two factors. The main one was insufficient production of oil in relation to the demand for it. The other factor was the low price elasticity of demand for oil. By this he meant that, based on historical data, large price increases were required to change behavior in ways that would reduce oil consumption. For the run up in oil price in 2008, he estimated this elasticity to be −0.06, meaning that the price would have to rise by 17 percent to cause a 1 percent reduction in consumption.

The slump in oil prices in the second part of 2008, wrote Hamilton, was the result of a widespread fall in economic activity. This is not surprising. The tight link between changes in economic activity, as reflected in Gross Domestic Product (GDP), and changes in oil consumption is well known. It is illustrated in Figure 3.10, which shows these changes for the US on an annual basis across the period 1949–2007. The correlation between these changes is 0.67, which is highly significant.

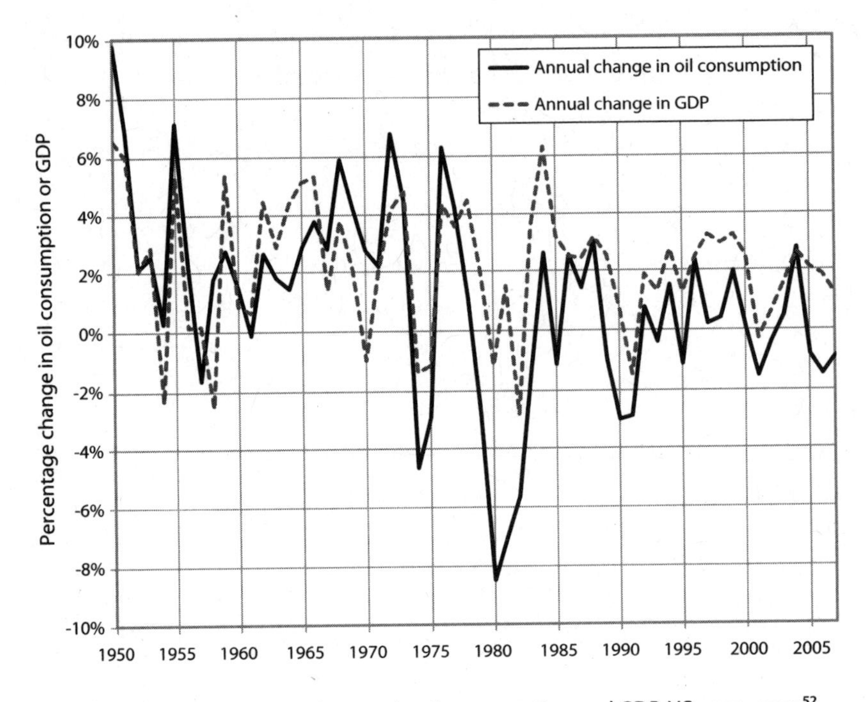

FIGURE 3.10 Year-on-year changes in oil consumption and GDP, US, 1950–2007[52]

It's usually taken that changes in GDP cause the changes in oil consumption, rather than vice versa. Oil consumption falls because there is less economic activity. The US GDP did indeed fall during the second half of 2008. The changes over the previous quarter for the four quarters of 2008 were 0.9, 2.8, −0.5, and −6.3 percent.[51]

What caused the widespread fall in economic activity? Some economists have said it was the prior surge in oil prices. One noted, "Like nearly every other recession of the postwar period, this one was triggered by a literally unbearable increase in the price of oil."[53] This economist noted too that the 2008 recession began earlier and became deeper in several other countries than the US, further implicating a more worldwide effect such a high oil prices than a US-specific effect such as problems in the housing sector. Hamilton had concluded that "almost all of the downturn of 2008 could be attributed to the oil shock," i.e., to the sharp oil price increase that ended in July 2008.[54] He allowed that the US economy was in stress because of problems in the housing sector and related and unrelated problems concerning financial institutions. These and other problems had continued for some years, he wrote, without causing a recession. It was the oil price rise that pushed the economy into recession, collapsed oil prices, and had a much broader impact domestically and internationally. Hamilton added that the oil price increases earlier in the decade, while large (see the right-hand panel of Figure 3.8), had been affordable. The increases in 2008, which put the average retail price of gasoline in the US above four dollars a gallon—almost four times what it had been in early 2002[55]—had gone beyond a threshold of affordability.

Rising oil prices may have contributed to problems in the US housing sector, in part through their impact on the far-flung suburbs of metropolitan regions, where high commuting costs became higher. The high transport fuel prices could have had two specific effects. One was on people already living there. Their household budgets, often already stretched by mortgage obligations, in some cases became unmanageable. The second effect was to depress the market for such homes. Jagadeesh Gokhale has argued, "energy cost increases meant households could no longer commit to lifestyles requiring high energy and commuting costs." Thus, there was no longer a robust market in which to sell the foreclosed properties of those who had defaulted. Gokhale concluded, "an oil price spike and a wealth shock in housing initiated the financial crisis [of 2008]." He also concluded that "unlike the earlier oil price shocks, the housing wealth shock has resulted in a near collapse of the financial sector."[56] There is indeed evidence from the US that high incidences of mortgage defaults and depressed property prices have both been more likely to occur in places with long commuting distances.[57]

High oil prices could also have contributed to problems in the US housing market through their effect on interest rates. Jeffrey Rubin, former chief economist for CIBC World Markets—who claims that "Oil prices...brought down the world economy"[58]—has argued that rising prices spurred inflation and high interest rates that in turn caused difficulties for homeowners renewing their mortgages.[59]

Both Hamilton and Rubin have argued that the main impact of the oil price rises on the US economy was via effects on the automobile industry rather than on the housing industry.[60] The impact was in phases, as illustrated in Figure 3.11. During 2007, sales were declining slightly, but sales of gasoline-hungry light trucks for personal use—including sport utility vehicles (SUVs), passenger vans and pick-ups—remained ahead of

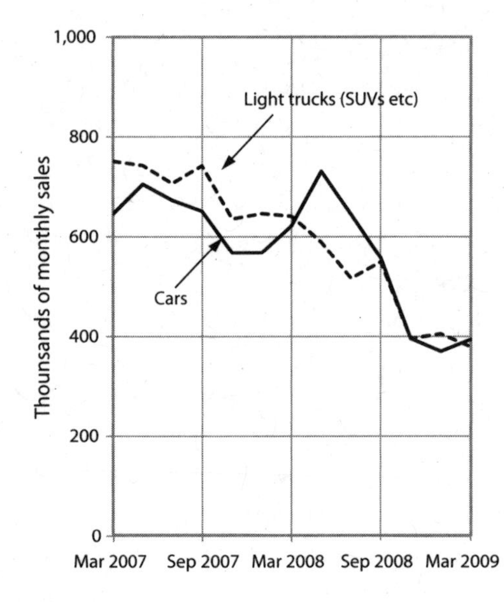

FIGURE 3.11 Sales of personal vehicles in the US, 2007–2009[61]

sales of regular cars. The 1.2-percent decline in total sales from the fourth quarter of 2006 to the fourth quarter of 2007 was a trend of concern to the industry but certainly within normal variation.

During the first part of 2008, as gasoline prices rose steeply (see Figure 3.9), sales of light trucks began to fall more steeply, but the decline was initially offset by increased sales of regular cars. For the first time since 2004 (see Figure 3.11), more cars were sold in the US than light trucks. According to Hamilton, "...the dominant story in the first half of 2008 was one in which American consumers were switching from SUVs to smaller cars and even more efficient imports."[62]

Then, in the second half of 2008, sales of both types of personal vehicle fell steeply, eventually bottoming out at the end of the year. Total sales in the first quarter of 2009 were an astonishing 38 percent below sales in the same quarter of 2008. Both types of personal vehicle were affected, suggesting a cause that would affect both types such as falling household incomes or loss of consumer confidence. Hamilton noted that the effect on the auto industry was similar to what had happened after previous oil shocks, as were the falls in overall consumer spending, particularly on durable goods. He noted too that after previous oil shocks there had been what might be considered to be irrationally large declines in consumer sentiment—i.e., consumer optimism about economic conditions—and suggested this fact was also at play in 2008.

Hamilton noted that the cause of the steep rise in oil prices, "booming demand and stagnant production...is fundamentally a long-run problem, which has been resolved rather spectacularly for the time being by a collapse in the world economy." He wrote that we could soon "find ourselves back in the kind of calculus that was the driving factor behind the problem in the first place. Policy-makers would be wise to focus on real options for addressing those long-run challenges, rather than blame what happened [in 2008] entirely on a market aberration."[63]

We find Hamilton and other economists who identify oil prices as a contributing or principal cause of the current economic malaise to be plausible. This analysis is pursued in Chapter 6, where we look at recent events in more detail and at how things might unfold in the shorter term. For the moment we draw three profoundly important lessons from the events of 2008.

The first is that, if production cannot rise to accommodate rising demand, oil prices can rise even more steeply than we had expected. This strengthens the argument we make in Chapter 5 that there should be anticipatory initiatives to reduce oil consumption in developed countries and moderate the expected growth in oil consumption in developing countries.

The second lesson is that oil is even more important to national economies and the global economy than we and many others had supposed. It's now clearer that high oil prices bring the very real risk of devastating economic depression. Again, the best solution is timely implementation of measures to reduce oil consumption, particularly for transport, already noted in Figure 3.3 as the major use of oil.

The third lesson is that we seem already to be at a point in the world production curve in Figure 3.7 where production constraints can produce major impacts on price and through this undermine economic well-being. We should not wait until production is clearly in decline before working to ensure that demand remains below available supply by a comfortable amount. Urgent action may be an imperative if oil use is to be reduced by other than a further collapse in the world economy.

We have not found a more compelling account of the recent economic turmoil than Hamilton's, but we should note, as we do again in Chapter 6, that conclusions such as his are at variance with the conventional wisdom about the causes of the current recession.[64] The prevalent view is that the recession was caused by the financial crisis, which in turn has its origins in "...the steep increase and subsequent sharp decline of housing prices nationwide in recent years, which, together with poor lending practices, led to large losses on mortgages and mortgage-related instruments at a wide range of financial institutions."[65] Although these factors may well have made things worse, the fact remains that what happened in the second half of 2008 was consistent with the results of earlier oil shocks, and can be mostly accounted for as a consequence of the steep run-up in oil prices earlier in 2008. The difference from earlier oil shocks lay in the cause of this shock, which had geological causes—global oil production limits were being reached—rather than the political causes of some earlier shocks.

Many economists remain skeptical about the role of supply constraints as a cause of high oil prices, as well as high oil prices being a cause of economic recession. There nevertheless seems to be some appreciation that "energy price shocks make themselves felt primarily through reduced demand for cars and new houses."[66]

As a postscript to the representation in Figure 3.10 of the strong link between *changes* in oil use and changes in GDP, it's worth noting that total oil use and total GDP are *not* necessarily so linked. Figure 3.12 shows that in the US total GDP and total oil consumption became unlinked following the oil shocks of the 1970s and early 1980s. Both rose from 1949 to 1972. GDP per capita continued to rise, with minor interruptions. Oil consumption per capita fell and since 1983 has been more or less constant at a lower level than it was in 1972.[67]

The significance of Figure 3.12 may be that it suggests the possibility of a further change in the relationship between oil use and GDP comparable to the change from that in the period 1949–1972 to that after 1983. Such a change would mean that US oil consumption per capita would begin to fall by about 2.5 percent per year while GDP per capita would continue to rise by about 2.5 percent per year. By 2025, after allowing for expected population increase, this change would result in a total decline in oil consumption of about 20 percent. We shall propose in Chapter 5 that a consumption cut of about twice this amount will be needed in richer countries such as the US to match the expected fall in oil production

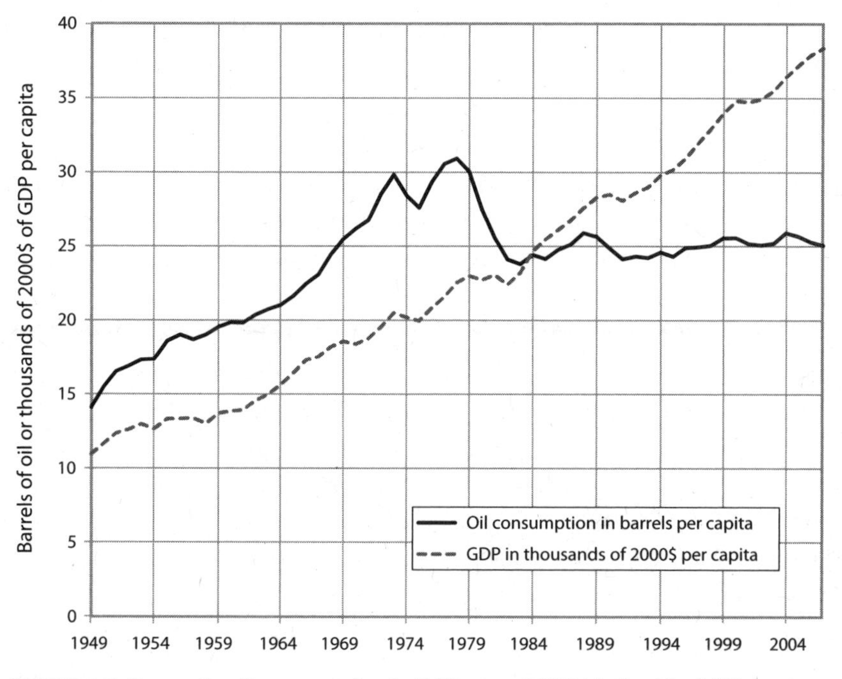

FIGURE 3.12 Per capita oil consumption (solid line) and GDP (dashed line), US, 1949–2007[68]

while allowing for some growth in developing countries. The challenge will be to achieve this without economic devastation, and we show how this might be done for transport. There is much to do in the years ahead and the events of 2008 may have served as a salutary warning, if recognized for what they are.[69]

Alternatives to Oil Products

Natural Gas

Several alternatives to the usual petroleum-based transport fuels also involve use of fossil fuels. One involves direct use of natural gas (mainly methane, the simplest hydrocarbon), which can be carried in compressed or liquid form in an appropriate fuel tank and, with little modification, used as the fuel in spark-ignition ICEs. With more modification, diesel engines can use natural gas, although some diesel fuel must also be used. About 90 percent of the diesel fuel can be replaced by natural gas.[70] Natural gas burns more cleanly than the gasoline or diesel fuel it replaces, and it can be a lower-cost fuel, especially if it is taxed less heavily. The disadvantages are that the fuel tanks are bulkier and heavier, and the range for each refill tends to be shorter. Carbon dioxide is still produced, although less than when gasoline or diesel fuel is burned.

Natural gas can be used to make liquid fuels using the Fischer-Tropsch process, as can coal—discussed below. Substitutes for both diesel fuel and gasoline can be produced, although with considerable cost and, especially in the case of coal, considerable environmental impact. This process provided a large share of the transport fuels used in Germany in the period 1935–1945, and in South Africa to this day.[71] In Qatar, which has the world's third largest reserves of natural gas and no easy way of getting it to market, at least two gas-to-liquids (GTL) plants are under development.[72] In the US there is much consideration of replacing imported oil with liquid fuels produced from coal,[73] a matter discussed below.

Natural gas burns more cleanly than the gasoline or diesel fuel it replaces, and it can be a lower-cost fuel, especially if it is taxed less heavily.

Natural gas is presently the main feedstock for the production of methanol and hydrogen, both of which can serve as fuels for ICEs—although major modifications are required—and as fuels for fuel cells, also discussed below.

One challenge for plans to increase the use of natural gas as a transport fuel, directly or indirectly, is that it too is a finite resource that will experience a world peak in production as will oil, albeit perhaps a few decades later than oil.[74] Natural gas is difficult to transport between continents, and so availability within a continent can be of more importance.

Until 2007, natural gas production appeared to have peaked in North America in 2001.[75] The US Energy Information Administration (US EIA) has reported increased production from 2007 onward,[76] the result of application of new techniques for extracting gas from shale. This has caused natural gas prices to fall, generated considerable excitement in energy industries, and renewed proposals for massive use of natural gas as a transport fuel.[77] Some industry analysts say the boom in US natural gas production is a mirage resulting from faulty representation of production data by US EIA.[78]

Until the recent boom in natural gas production, terminals for importing the *liquefied natural gas* (LNG) into North America were being developed apace. There is still considerable interest in LNG in Western Europe, which depends increasingly on imports of natural gas from Russia that have been subject to interruptions and unexpected price changes.[79] LNG is the only form in which natural gas can be shipped in quantity across oceans. Capital costs are high for liquefaction and regasification plants and for LNG tankers. However, a more important barrier to expansion of LNG could be that the tankers are considered to present an extreme hazard.[80]

LNG is being tried as a fuel for heavy-duty trucks and freight locomotives, with a view to overcoming a major challenge in using natural gas as a transport fuel: the weight and size of its fuel tanks. When the fuel is compressed natural gas (CNG), the tanks are about four times larger than those for an equivalent amount of gasoline or diesel fuel. When LNG is used, the tanks and insulation require about twice the space of a conventional liquid fuel.[81] Barriers to use of LNG as a transport fuel include the high cost of on-vehicle and off-vehicle equipment, the relative unavailability of the fuel and safety concerns.

Natural gas will make no more than a modest contribution to the replacement of oil as a transport fuel, directly or indirectly.

One form of natural gas may be truly abundant. It is the form known as *methane hydrates* or *gas hydrates,* a crystalline solid with the appearance of ice in which each methane molecule is surrounded by a cage of water molecules. Methane hydrates are stable in ocean-floor sediments at depths below 300 meters and occur in vast quantities in layers up to several hundred meters deep. According to the US Geological Survey, "The worldwide amounts of carbon bound in gas hydrates is conservatively estimated to total twice the amount of carbon to be found in all known fossil fuels on Earth."[82] That was written in 1992. A 2003 review made a similar statement about prevalence and concluded, "During the next decade, gas production will begin from permafrost hydrates associated with conventional gas reservoirs. However, efficient production of ocean hydrates is problematic and requires an engineering breakthrough to be economically feasible."[83] A 2004 assessment concluded that the total resource may be between a twentieth and a quarter of previous estimates.[84]

According to a 2007 presentation, "Gas hydrates are mostly located in difficult terrain underneath the ocean floor or in permafrost. Exploitation of these resources carries considerable environmental risk, both from unintended release of greenhouse gases and from the possibility of triggering massive undersea slides."[85] Early in 2007, the US Department of Energy's methane hydrate newsletter reported demonstration of a "successful gas hydrate methodology" in Alaskan permafrost.[86] A 2009 issue of the newsletter suggested that the technically recoverable resource at this location may be less than 15 percent of what was previously estimated.[87]

Another petroleum product now used as a transport fuel, more than natural gas, is *liquefied petroleum gas* (LPG), also known in Europe as Autogas, and not to be confused with LNG (see above). LPG is a by-product of natural gas production, already noted in reference to "natural gas liquids" in Figure 3.7. LPG is chiefly propane, the fuel for barbecues (LPG is also known as liquid propane gas) or butane, the fuel for cigarette lighters, or a mix of the two. It can be readily stored in vehicles and, with some conversion, used in ICEs designed for gasoline. It burns more cleanly than gasoline and is used, for example, in taxicabs in Hong Kong.[88] The availability of LPG depends on the amount of natural gas that is produced, and also on the extent to which natural gas liquids are required for production of gasoline.

The foregoing suggests that natural gas will make no more than a modest contribution to the replacement of oil as a transport fuel, directly or indirectly. Massive exploitation of methane hydrates could alter this conclusion, although this seems less likely than it appeared a decade ago.

Coal

Coal is a potential factor in transport futures in at least three ways. It can be used as a solid fuel for external combustion engines of the kind that provided the traction for steam locomotives and ships in the 19th and 20th centuries. It can be converted into a liquid fuel using the Fischer-Tropsch process, as was done extensively in Germany in the 1920s to 1940s, and has been done in South Africa since the 1950s.[89] Coal can fuel electricity generating stations that can charge batteries and power grid-connected vehicles. In 2004, 40 percent of the world's electricity was made in this way, and the share appears to be increasing.[90] Any use of coal, particularly conversion to liquid fuel,[91] produces copious amounts of carbon dioxide, which can be sequestered only at considerable financial and energy cost.[92]

Coal's availability is often assumed to be limitless or at least sufficient to allow expanded use for decades. For example, a report from the

Massachusetts Institute of Technology suggests that consumption of coal in energy terms could rise by 348 percent between 2000 and 2050 (i.e., 3 percent per year).[93] The IEA suggests that proven reserves of coal could allow for 164 years of consumption at current rates, compared with 64 years for natural gas and 42 years for oil.[94]

Other sources suggest that mineable global coal resources are much smaller. These include a report by Germany's Energy Watch Group that points to the unreliability of data on proven reserves of coal.[95] The most extreme example given was that of Germany itself, reported as having downgraded her proven hard coal reserves in 2004 by 99 percent from 23 billion tons to 0.183 billion tons. Botswana and the UK have also downgraded reserves by more than 90 percent during the last two decades. Only Australia—the major coal exporter—and India have reported growing reserves. China—by far the biggest producer and user (39 percent of world consumption in 2006)—has reported exactly the same reserves of coal each year since 1992, even though subsequent consumption and loss through uncontrollable fires amount to about a quarter of this total. The report suggests that China's coal production will peak in about 2015.[96]

The US is the second major consumer of coal and has by far the largest proven reserves (about 27 percent of the world total). It uses less than 0.5 percent of its reserves each year, but production of high-quality hard coal (anthracite and bituminous coal) has already peaked. Coal production has risen in volume—by about 1 percent a year in the last decade—but the growth had been in less energy-dense sub-bituminous and lignite coals. Production *in energy terms* reached a peak in 1998 and has since fallen by about 4 percent.[97] The report's authors suggest that US production volumes could be further increased, but only until about 2025, when they will inevitably decline.[98] Production in energy terms could begin to increase again but would reach a maximum—before the volumetric peak—that would be no more than about 20 percent above the current value. World volumetric production of coal would also peak in about 2025, with an earlier peak in energy terms.

Much methane occurs with coal. Recovery of coal-bed methane from worked-out mines and other coal resources provides the equivalent of about 10 percent of natural gas production in the US.[99] Proved US reserves of coal-bed methane total about 12 times production or about a tenth of proved natural gas reserves.[100] Production is much higher in the US than elsewhere, but the potential for exploitation of this energy source elsewhere may be comparable. The major challenge in recovering coal-bed methane arises from the need to remove the underground water whose pressure holds the methane in the coal seam. The removal can cause subsidence and other geologic effects. Bringing what is often polluted water to the surface can result in pollution of land and water. Exploitation of coal-bed methane is an active topic of research and development.[101]

Energy recovery from coal could be substantially enhanced by underground gasification. This involves injecting steam and air or oxygen into a coal seam, igniting the coal, and using the hydrogen, carbon monoxide, methane and other resulting gases as fuel or as feedstock for the production of liquid fuels and other chemicals. Industrial coal gasification has been practiced since the 19th century to provide piped coal gas (town gas) for cooking, space heating and lighting. Gasification of the coal *in situ* presents the possibility of many more hazards including explosions, uncontrollable fires, emissions of noxious gases, contamination of water resources, and land subsidence. The potential, however, seems huge in that coal resources (what is in the ground) are usually several times proved reserves (what can be feasibly mined) and much of the unmineable coal could be amenable to relatively low-cost gasification.[102] Underground coal gasification (UGC) is an active area of research and development.

If large-scale UGC proves economically feasible and safe, perhaps an even more dangerous undertaking will be ventured: gasification of otherwise unrecoverable oil by injecting oxygen and steam into old wells, setting fire to the oil, and capturing the gaseous product. Such extraordinary risks in energy production would illustrate Orr's paradox, noted in Box 3.1

Biofuels

Liquid and other fuels for ICEs can be produced from plant material. Notable among them are *ethanol,* made from fermentation of sugars derived from corn (maize) or sugarcane and subsequent distillation; *biodiesel,* made from vegetable oils through a process known as transesterification; and *biogas,* made from anaerobic digestion of waste plant material. Biofuels are carbon-based fuels, but their use does not necessarily result in additions to the atmospheric burden of carbon dioxide (CO_2), a green-

BOX 3.1 Orr's paradox[103]

David Orr highlighted a paradox in an argument often made in industrial societies against reducing energy use:

> On the one hand technologically advanced societies are portrayed as so inflexible that any leveling or reduction in energy consumption would cause disaster. On the other hand, these same societies are portrayed as flexible enough to withstand the acknowledged certainty of periodic catastrophes, including large oil spills, nuclear meltdowns, liquid natural gas explosions and global climate change.

Orr's paradox may have resolved somewhat in the decades since 1979. Tolerance for human-induced disasters may have waned, and acceptance of the need to reduce energy use may have increased. But if future oil shortfalls are not anticipated, the temptation to pursue risky forms of energy production could arise again.

house gas (see Chapter 4). Combustion of biofuels releases only the carbon taken up from the atmosphere by the plant matter during its growth. However, production and distribution of biofuels can require substantial amounts of fossil fuel, and petroleum-derived fertilizer and pesticide, thereby resulting in net additions to atmospheric CO_2, as shown below. Often, more important issues associated with biofuels are the amounts of land required and depletion of soil nutrients, and the resulting competition with food production.

Biofuels generally burn more cleanly than their petroleum equivalents, although evidence is growing that the use of ethanol in particular can contribute more to some kinds of local and regional pollution—including production of ground-level ozone (photo-chemical smog)—than use of an equivalent amount of gasoline.[104] Here the focus is on the extent to which biofuels can replace petroleum products as an energy source.

World biofuel use in transport in 2005 was equivalent to about 20 million tonnes of petroleum oil, or 1.0 percent of total transport fuel. Only in Brazil (13.3 percent), Cuba (6.2 percent) and Sweden (2.2 percent) was biofuels' share of transport fuel higher than 2 percent. Ethanol comprised 85 percent of this biofuel consumption, and biodiesel almost all the remainder. Biofuel use in transport has more than doubled since 2000 and, according to the IEA, could more than double again by about 2020. However, even in its "Alternative Policy" scenario, the IEA expects that biofuels will provide only 7 percent of the world's road transport fuel in 2030,[105] although there are estimates that ethanol alone could provide a larger share.[106]

Ethanol was one of the fuels used in the first ICE patented in the US, in 1826, and in Henry Ford's first car, built in 1896. His first Model T, introduced in 1908, could run on ethanol or gasoline.

Ethanol has been used on a large scale as a transport fuel in Brazil since the 1970s. The cost of production from sugarcane is about $0.26 per liter, equivalent to about $0.38/liter for gasoline, when the lower energy yield of ethanol is taken into account.[107] Fermentation and distillation of the sugarcane is fueled by burning bagasse, the plant's fibrous material. No other energy input is required. The remaining third of the plant (after removal of the sugar and fibrous material) remains in the field as soil nutrient. Sugarcane and sugar are produced more cheaply in Brazil than anywhere else, but the cost of their production nevertheless represents 60 percent of the cost of producing ethanol. Even in Brazil, ethanol is subsidized in that it is taxed at a lower rate than gasoline, lower by the equivalent of $0.20 per liter. The world average production cost of sugar is four or more times Brazil's cost, and so crude oil prices would have to rise considerably for most ethanol made outside Brazil to become competitive as a transport fuel.

In the US, rapidly increasing amounts of ethanol are produced from corn.

In the US, rapidly increasing amounts of ethanol are produced from corn. In 2005, the US total of some 18 billion liters was about the same as Brazilian production. Together they comprised about 95 percent of world production. In both Brazil and the US, almost all of the ethanol is used as a transport fuel. In the US, about 12 percent of the corn crop is used to make ethanol from about 3.5 million hectares of land, or about 2.5 percent of the land used for crops.[108] Even the more modest proposals for expansion of the use of ethanol as a fuel speak to sevenfold or greater increases in this use.[109] Some speak to fifteenfold or greater increases, which could require up to 40 percent of US cropland.[110]

Adding annual US ethanol production of more than 200 billion liters would require construction of more than 1,000 industrial plants similar to the Goldfield plant in Iowa, which uses 100,000 tonnes of coal to produce 190 million liters of ethanol annually from about 500,000 tonnes of corn.[111] The energy input from the coal is 70–75 percent of the energy output in the form of ethanol. A full reckoning—including, for example, inputs for farming the crop and moving the coal and the corn to the plant—would bring the process closer to although not necessarily into a negative energy balance.[112]

In the US, 99 percent of non-food ethanol is used as E10, a mixture of 10 percent ethanol and 90 percent gasoline by volume, usable as gasoline by vehicles manufactured since 1986. The ethanol is added chiefly to oxygenate the gasoline during combustion, providing what in many respects—chiefly reduced emissions of carbon monoxide—is a cleaner-burning fuel.[113] It also displaces gasoline and additives made from petroleum products.

The rapidly growing production and use of ethanol in the US is a matter of considerable controversy.

About six million of the roughly 230 million light-duty vehicles in use in the US can use gasoline or E10 or E85, a mixture of 85 percent ethanol and 15 percent gasoline. Sensors in the fuel line detect which fuel is used and adjust the engine accordingly. Pure (100 percent) ethanol is not used in the US because gasoline is needed for starting on cold days. Indeed, the E85 blend can contain as little as 70 percent ethanol in northern states in the winter. A provision in the legislation that mandates fuel consumption limits favors production of these "flex-fuel" vehicles, particularly as SUVs and vans. Flex-fuel vehicles are deemed in the legislation to have lower fuel consumption than they actually have, thereby—if they are produced—improving a manufacturer's performance in relation to its required Corporate Average Fuel Economy (CAFE).[114] Less than 5 percent of flex-fuel vehicles actually use E85, which has limited although growing availability. Prices per liter of E10 and gasoline, where both are available, are about the same on average, and E85 is about 20 percent lower. To achieve the same value to the user, E85 has to cost about 30 percent less than gasoline because of its lower energy content; E10 has to cost about 4 percent less.[115]

The 2005 cost of ethanol production from corn in the US was only a little more than production from sugarcane in Brazil: $0.30 vs. $0.26 per liter.[116] It attracts a US government subsidy of about $0.13/L and additional subsidies from several state governments. Ethanol becomes profitable as a substitute for gasoline when the crude oil price is $35–$50 a barrel, according to how much it is subsidized. Oil has been mostly above $50/barrel since late 2004 and ethanol's subsidies will mostly continue until 2011, promising large profits for US producers.

To overcome the energy and land challenges inherent in current ethanol production, proposals have been made to use *cellulosic ethanol*, produced by enzymatic processing of the presently unused fibrous components of plant material. If this process were to become commercially available, it could require less energy and less agricultural land to produce a given amount of ethanol from corn and other plant sources.[117] However, such ethanol production could result in severe depletion of soil nutrients and unsupportable requirements for fertilizer. There would also be the high energy and financial cost of preprocessing the plant material, and the challenge of maintaining sterility in fermenter vessels across the relatively long dwell times required for enzymatic action. One analysis suggests production of cellulosic ethanol from wood waste may require *more* energy on a full life-cycle basis than conventional production from corn.[118]

The rapidly growing production and use of ethanol in the US is a matter of considerable controversy on several counts. Advocates see ethanol as a home-grown alternative to rising oil imports. Opponents characterize encouragement of ethanol production as a support program for agribusiness that will waste energy, accelerate climate change, deplete soil and displace food production.[119] The fact that the US presidential campaign season begins with the Iowa caucuses meant that most candidates, including the eventual winner, pledged to replace some amount of foreign oil in America's gas tanks with ethanol brewed from Midwestern corn.

Biodiesel, used mainly in Europe, is less controversial than ethanol, in part because producing the fuel generally requires less energy per equivalent unit than producing ethanol.[120] Biodiesel is usually blended with petroleum diesel, five parts per 100—known as B5—but can be used in proportions up to 100 percent (B100). Biodiesel contains a little less energy than petroleum diesel, but engines run more efficiently on it; the net result is that a liter of biodiesel takes a vehicle as far as a liter of petroleum diesel.[121]

Biogas has been used as a transport fuel in Sweden since 1996, although in 2005 it comprised less than 5 percent of Swedish biofuel use (the remainder was mostly ethanol) and a negligible amount of world biofuel consumption for transport. Biogas as produced from anaerobic digestion of vegetable matter is about 50 percent methane and 50 percent carbon dioxide. The latter is removed and the methane mixed in roughly equal proportions with natural gas. In compressed form, the mixture is used in spark ignition engines as natural gas is used (see above).[122] Biogas production in Sweden is expanding and was expected to make a growing contribution to the possibility that Sweden would break its dependence on petroleum oil by 2020, a matter discussed in Box 3.2. Several other countries, chiefly in Europe, are introducing or exploring the introduction of biogas as a transport fuel.

Biofuels promise renewable transport fuels. They are a means of harnessing the sun's energy in the form of liquid or gaseous fuels useable in vehicles of the kind in common use today. Their production can require less energy in the form of fossil fuels than is released on their use. It can also require more, in which case the process is an energy sink rather than an energy source. Production of ethanol could nevertheless be useful as a means of producing a liquid fuel with use of energy from, say, coal.

We believe ethanol production will be limited by the availability of land and thus by competition with the more important requirement to grow food. Such competition would be reduced if the numerous issues noted above concerning large-scale use of plant cellulose are resolved.

Hydrogen

For much of the automotive industry and many governments, hydrogen is the transport fuel of the future, powering fuel cells that drive electric motors.[124] At least it was considered so until recently. Both hydrogen fuel cells and their fuel present challenges. Hydrogen fuel cells are expensive and unreliable.[125] Hydrogen is expensive to distribute and store.[126] Today it is mostly made from natural gas, the future availability and price of which, as noted above, may be almost as problematic as those of oil.

However, the main challenge to prospects of a "hydrogen economy" is its inherent inefficiency, especially when the hydrogen is to be produced using energy from renewable resources. These resources—chiefly sun, wind and falling water—produce electricity that would power electrolytic production of hydrogen that would be used to produce electricity in fuel cells. The transition from electricity to hydrogen and then back to electricity involves energy losses of between 57 percent and 80 percent.

BOX 3.2 Sweden's short-lived goal of ending oil dependency by 2020[123]

In September 2005, Swedish Prime Minister Göran Persson made a remarkable announcement: "Sweden will seek to end it dependency on fossil fuels by 2020." Early in 2006, the Sustainable Development Minister, Mona Sahlin, said, "A Sweden free of fossil fuels would give us enormous advantages, not least by reducing the impact from fluctuations in oil prices." The suggestion that the country was aspiring to be oil-free was reinforced by the title of the June 2006 report of the government's Commission on Oil Dependence: "Making Sweden an Oil-Free Society."

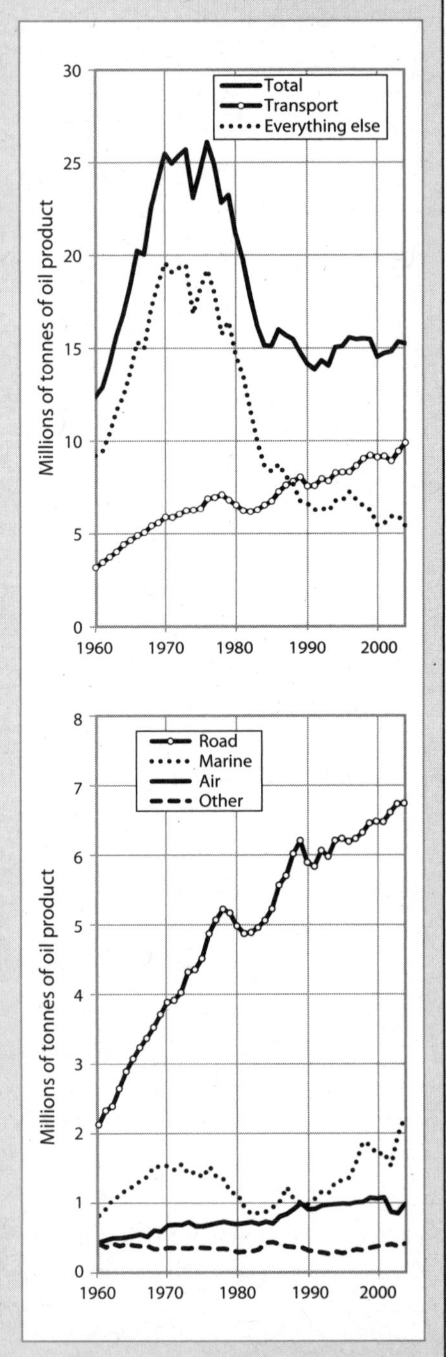

In 2004, fossil fuels comprised 39 percent of Sweden's energy use. The remainder comprised nuclear energy (34 percent), hydroelectric power (11 percent), and biomass or waste burned for electricity generation and district heating (16 percent). Of the fossil fuels, the largest share (82 percent) was oil. Of all oil consumption, 45 percent was used for road transport. Other oil uses were in industry, as a fuel and feedstock (27 percent of all oil consumption), and for air and marine transport (19 percent). Apart from a little coal, all fossil fuels used in Sweden are imported. The remaining energy sources are indigenous, except for imports of uranium and for purchases of electricity that are roughly offset by sales.

Oil use in Sweden has declined by 42 percent since 1976, the year of peak consumption, when it was used chiefly for home heating and industrial purposes. These uses of oil have fallen dramatically, but consumption for transport has increased by 44 percent and is still increasing at the rate of about 1 percent per year, shown in the upper chart to the right. Consumption for other purposes continues to decline, although more slowly than during the 1980s. The lower chart shows that the increase in oil consumption for transport has almost entirely comprised an increase in the amount used for road transport.

Sweden's oil-dependency challenge was thus mostly a matter of reducing oil use for transport, particularly use of gasoline and diesel fuel for road transport. The proposal of the Commission on Oil Dependence—which was chaired by the prime minister—was not to move off oil completely by 2020 but to reduce fuel use for transport by 40–50 percent and ensure that motorists would always have the option of using a renewable fuel. Present trends would have transport fuel use grow by 17 percent, mostly because use per person is growing but also because modest growth in population is expected. Thus, the target requires a reduction to more than 50 percent below expected use.

Exacerbating Sweden's challenge is its position as among the heaviest oil users for transport per capita in Europe. Only Ireland, Austria and Belgium have higher rates. Cars in Sweden tend to be larger, heavier and more fuel-consuming than the European average, car ownership is higher (even higher than in Canada), and distances driven per car are longer.

Measures proposed included:

- Improve road vehicle fuel efficiency—reduce average liters per 100 kilometers by 25–50 percent—by more use of diesel fuel, of hybrids vehicles, and of smaller, lighter vehicles, achieved by taxing vehicles and fuels according to their net carbon dioxide emissions.
- Increase biofuel production to 12–14 terrawatt-hours (TWh) per year by 2020, equivalent to about 15 percent of current use of oil products for transport.
- Educate car purchasers about likely rising fuel prices and label new cars according to their fuel use.
- Improve traffic planning and route optimization, and teach driving for reduced fuel use (ecodriving).
- Use procurement for public purposes to achieve early adoption of vehicles with no or low fossil-fuel use.
- Improve freight logistics—that is, reduce fuel use per tonne-kilometer of payload—and intermodality—that is, the ease with which goods can be moved between modes, so as to take advantage of the most fuel-efficient mode available.
- Encourage use of public transport for the local and longer-distance movement of people, and rail and water for the movement of freight; promote alternatives to aviation.
- Replace mobility by accessibility chiefly through greater use of information technology to obviate work-related travel.

Successful application of the first two measures alone would result in the target being met, if higher fuel prices were to restrain an increase in vehicle movement that could result from increased efficiency. Strong application of all measures would allow for further reductions in fossil fuel use or at least compensate for shortfalls in use of the first two measures.

A new government took office in October 2006. According to a spokesperson for Fredrik Reinfeldt, the present prime minister, the previous government's policy direction toward breaking Sweden's dependence on fossil fuels by 2020 is "no longer valid."

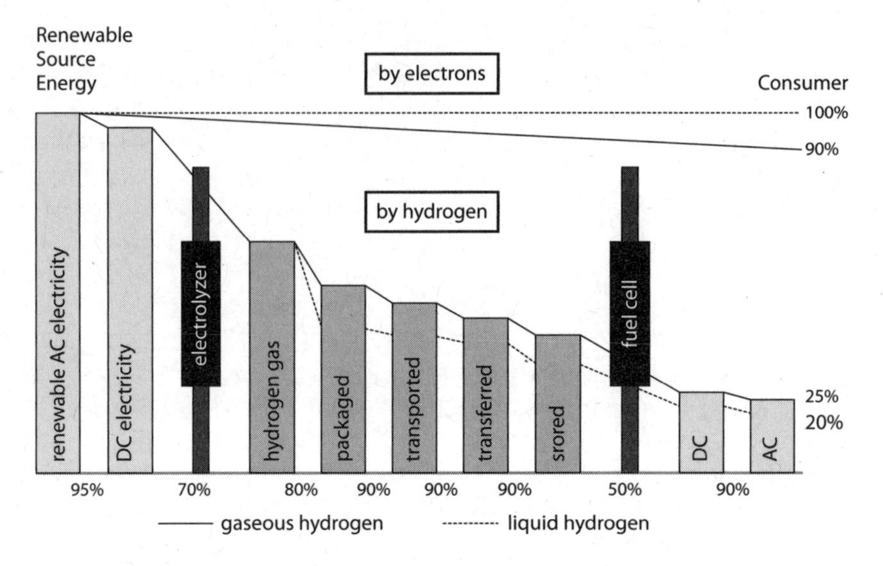

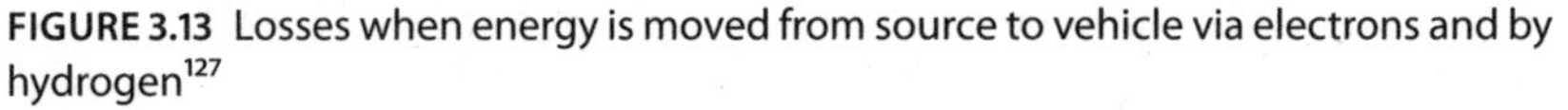

FIGURE 3.13 Losses when energy is moved from source to vehicle via electrons and by hydrogen[127]

One representation of these losses is in Figure 3.13, which tracks the movement of energy from, say, a wind turbine to the kind of alternating-current electric motor used in many vehicles. The numbers below the axis show the efficiency of each transformation. The cumulative efficiency is shown to the right of the bars. There are losses of 75 percent or 80 percent when hydrogen is made from electricity by electrolysis and then converted into electricity again in a fuel cell. The higher estimate of loss applies when the hydrogen is liquefied for transmission.

> The main challenge of a "hydrogen economy" is its inherent inefficiency, especially when the hydrogen is to be produced using energy from renewable resources.

Figure 3.13 also contrasts these losses with those from delivery of energy directly from the turbine to the motor in the form of electric current, i.e., as electrons. An example is Calgary's light rail system, portrayed in Figure 3.14, which is powered directly by wind turbines near the Rocky Mountains. In this case, there is only a distribution loss, which is generally about 10 percent. About four times as much energy is available at the motor compared with conversion to hydrogen and back. Thus, a vehicle could go four times as far on the same initial amount of generated electricity. In an energy-constrained era, such differences could be of profound significance.

FIGURE 3.14 Wind-powered light-rail train in Calgary, Alberta[128]

An estimate more favorable to hydrogen fuel cell systems puts their typical loss at 57 percent rather than 75–80 percent.[129] In this case, the direct-delivery mode allows only about twice as much vehicle movement as the hydrogen fuel cell mode, but this difference could still be significant in an energy constrained world. Direct delivery of electricity to vehicles is discussed below.

> Electric motors (EMs) will substantially replace ICEs during the next two or three decades as a means of propelling land vehicles.

We believe this inherent inefficiency is the main reason the hydrogen-fuel-cell vision is unlikely to be realized. Hydrogen, as an energy carrier, always has to compete with its own energy source, in this case electricity from a wind turbine, which will usually have a cost advantage. Fuel cells are often promoted as being more efficient users of energy than ICEs at the vehicle. This is generally true, but if energy is expensive, and initially available in the form of electricity, there will be a strong cost advantage in using the electricity more directly. Another feature of fuel cell vehicles is their complexity, which can lead to higher energy costs for materials production, assembly, vehicle distribution and scrappage.[130]

The main reason the hydrogen-fuel-cell vision had been popular is that it provides for vehicles that are similar to those of today in two important features: quick refueling and the ability to carry enough fuel for a range of several hundred kilometers. These features appear likely to carry too heavy an energy price to be maintained in an era of energy constraints. Perhaps another appeal of hydrogen fuel cells has been the long time frame for their deployment and implementation, meaning that new political and corporate leaders will be in place when the success or failure of this alternative energy path becomes clear. It could be time to redirect the massive public and private investment of the hydrogen-fuel-cell vision to more efficient ways of providing electric mobility.[131]

Electricity

The basic thesis of this book is that electric motors (EMs) will substantially replace ICEs during the next two or three decades as a means of propelling land vehicles. This transport revolution will occur chiefly because of the decline in world oil production anticipated earlier in this chapter to begin soon after 2012, because electricity from renewable and other sources will be the most feasible alternative fuel, and because of the appeal of electric traction.

Using an EM to provide traction rather than an ICE can be appealing for several reasons. Electric vehicles are quiet, energy-efficient, require little maintenance, have good acceleration at low speeds, and emit essentially no pollution at the vehicle. The challenge for electric vehicles has always been and still is getting electricity to the motor or motors. There are basically three ways of doing this.

GCVs are responsible for the most person-kilometers of travel by electric vehicle today.

In the first type of electric vehicle, the electricity is generated elsewhere, stored on board, and delivered to the motor from the storage device, usually a battery. In the second type, which includes fuel cell vehicles, the electricity is generated on board the vehicle. In the third type, the electricity is generated elsewhere and delivered directly by wire or rail to the motor as the vehicle moves, as in the case of the Calgary light-rail train depicted in Figure 3.14. Mixed versions can occur. For example, a trolley bus, which is a vehicle of the third type, could have a battery for off-wire operation, which would make it also a vehicle of the first type.

The next section of this chapter addresses each of these three types of electric vehicle in turn, beginning with battery-electric vehicles, then moving to vehicles with on-board generation. Of this second type, fuel cell vehicles have already been touched on, and we focus below on vehicles in which the energy for electricity generation comes from an on-board ICE. Increasingly the most common of these are what are known as hybrid cars, which have both an EM and an ICE. Then we consider the third of the above three types of electric vehicle, which we call "grid-connected vehicles" (GCVs). GCVs are responsible for the most person-kilometers of travel by electric vehicle today and we expect this lead to continue, even as many more electric vehicles come into use, because of GCVs' especially high energy efficiency.

In this chapter's final section we consider how sufficient electricity might be generated to support transport systems in which electric motors provide most of the propulsion.

Electric Vehicles

Battery-Electric Vehicles

Vehicles relying on on-board storage of electricity usually have *electrochemical storage* in the form of lead-acid, nickel metal hydride (NiMH), lithium or other batteries. They can have *mechanical storage* in the form of a flywheel and *electrostatic storage* in the form of a supercapacitor (also known as an ultracapacitor). We focus here on batteries, but other storage systems should not be overlooked.[132]

Battery electric vehicles (BEVs) have been available for as long as ICE vehicles. At the beginning of the 20th century, the few available automobiles were as likely to be steam- or battery-powered as reliant on ICEs.[133]

ICEs won out chiefly because of BEVs' limited range, low power and lengthy recharging. These disadvantages were all the result of batteries' low energy density, that is, the amount of energy that can be delivered per kilogram or liter. Energy densities for lead-acid, NiMH and lithium batteries are 0.14, 0.25 and 0.45 megajoules per kilogram of battery (MJ/kg).[134] By contrast, the energy density of gasoline is about 45 MJ/kg or 180 times that of an NiMH battery, the type now mostly used in BEVs. Because the efficiency of an ICE system is about 14 percent and the comparable efficiency of a BEV system is about 61 percent,[135] the effective ratio of the energy densities is about 40:1.

BEVs may have a resurgence as oil prices rise.

Put another way, for NiMH batteries to have the energy capacity of a 50-liter tank full of gasoline (13.2 US gallons), a total of about 1.5 tonnes (3,300 pounds) of batteries would be needed. If lithium batteries could

TABLE 3.1 Features of modern internal combustion engine (ICE), battery (BEV) and fuel cell automobiles[137]

	ICE Honda Civic 2.2 i-CTDi (diesel)	BEV Mitsubishi Lancer Evolution MIEV	Fuel cell Honda ZC2
Length (m)	4.25	4.49	4.17
Width (m)	1.76	1.77	1.76
Height (m)	1.46	1.45	1.65
Unladen weight (kg)	1,400	1,590	1,670
Seats	5	5	4
Drive (2 or 4 wheels)	2	4	2
Max torque (Nm)	340	518	272
Max power output (kW)	103	50	86
Max speed (km/h)	205	180	150
Range (km)	980	250	430
Rate of use of energy at the vehicle (MJ/100km)	197	69	124

be used, the required battery weight would be just over 0.8 tonne. As well as weight, the required amount of electrical storage brings in the factor of cost, which was another reason why ICEs won out a century ago. Recharging times are much shorter than they were 100 years ago, but can still be unacceptably long.[136]

Nevertheless, BEVs may have a resurgence as oil prices rise. A possible sign of this is the development by Mitsubishi of the Lancer Evolution, a prototype BEV whose specifications are in Table 3.1, together with comparable ICE and fuel cell vehicles. This BEV uses a lithium battery to give a range of 250 km. It powers four motors, one in each wheel. Its low energy use is of particular note, equivalent in gasoline terms to 1.8 L/100km. The corresponding fuel use by the other two vehicles is 5.1 L/100km (ICE) and 3.2 L/100km (hydrogen fuel cell).

Of equal interest is the Tesla Roadster, which has a single, much larger electric motor and is now in production. Fuel consumption equivalent to 1.7 L/100km gasoline is claimed, as well as acceleration from 0 to 100 km/h in about four seconds.[138]

The BEV's low energy use compared with the comparable hydrogen fuel cell vehicle—the BEV uses 44 percent less energy to move the vehicle the same distance—arises because moving electricity through a battery to a motor is more efficient than making electricity in a hydrogen fuel cell.

TABLE 3.2 Actual and proposed vehicle efficiencies[139]

	Country/vehicle	L/100km
Average ratings for 2002 light passenger vehicles (cars, vans, SUVs)	US	9.8
	Canada	9.2
	Australia	8.1
	EU15	6.3
	Japan	5.1
Data for vehicles in Table 3.3	ICE	5.1
	BEV	1.8
	Fuel cell	3.2
Proposed ultra-efficient ICE vehicles	Loremo GT	2.7
	Loremo GS	1.5

If the hydrogen for the fuel cell is made by electrolysis, the BEV's advantage is greater. The BEV system uses about 60 percent less of the original output from the electricity generator. This is because more energy is lost converting electricity to hydrogen (about 40 percent) than storing it in a battery (10–30 percent, depending on the battery and the charging method).

The BEV's low energy use compared with the comparable ICE vehicle (the BEV uses 65 percent less energy) has two main contributing factors: (i) the inherent inefficiency of ICEs compared with electric motors, particularly at low speeds; and (ii) the availability of regenerative braking in the BEV, whereby the electric motor can act as a generator and capture kinetic energy for later use, slowing the vehicle in the process.

How these three vehicles—ICE, BEV and fuel cell vehicle—compare with *average* new personal vehicles sold in several places—all ICE vehicles—is shown in Table 3.2. The average vehicle sold in Europe is a little larger than the reference ICE vehicle in Table 3.2 (a European Honda Civic), and the average vehicles sold in North America and Australia are much larger. In 2002, as Table 3.2 indicates, new light-duty passenger vehicles sold in Europe used 36 percent less fuel per kilometer than US vehicles; new Japanese vehicles used 48 percent less. Also shown in Table 3.2 are two versions of a prototype ICE vehicle—the Loremo—that may be in production by 2011.

The information presented in Table 3.2 suggests there is a lot of scope for reducing energy use, by reducing the size of vehicles and by switching to electric motors. Even without using electric motors, further reductions could be achieved, as suggested in the last two rows of Table 3.4. These diesel-powered, very low-weight and low-resistance vehicles follow a path forged by Volkswagen's Lupo 3L, which achieved 3.0 L/100km but was pulled from the market in 2005 because of "inadequate customer demand." Volkswagen also abandoned its prototype 1L vehicle, so slim that its two occupants sat in tandem as in a small airplane. During 2006, Volkswagen introduced its BlueMotion car, a five-seat diesel-fueled car that is said to average 3.9 L/100km.

Many of the changes that lead to higher fuel efficiency in ICE vehicles can also be applied to BEVs. Thus, although significant improvements in ICE vehicles can be expected, BEVs' inherent advantage—due to its electric drive—will likely be sustained.

Hybrid ICE-Electric Vehicles, Including Plug-in Hybrids

These cars have an ICE and an electric motor (EM). Usually both can propel the car. The ICE also charges the battery that powers the EM. The battery can be charged as well by regenerative braking. The electric motor is used more at low speeds, when it is more effective than the ICE. The ICE is used more at higher speeds. Both can be used to provide acceleration. An on-board computer sorts it all out. The result is a vehicle that uses considerably less fuel, especially under urban driving conditions.[140]

The hybrid is more powerful and capable of greater acceleration even though it is heavier; yet it uses 30 percent less gasoline.

Table 3.3 compares similar hybrid and regular ICE vehicles. The hybrid is more powerful and capable of greater acceleration even though it is heavier; yet it uses 30 percent less gasoline, chiefly because it is more frugal when driven in the stop-start conditions of cities. These conditions nominally embrace 55 percent of all driving in the US, but in reality are a little over 60 percent.[141] Note that the road test consumption shown in Table 3.5 is about 15 percent higher than the consumption during the standard tests conducted by the US Environmental Protection Agency (EPA).[142] Note too that the hybrid vehicle has slightly worse braking performance, which may result from its higher weight or from the use of regenerative braking.

Sales of hybrid vehicles have been rising rapidly in the US—by about 75 percent a year from 2000 to 2005 and then by 25 percent—notwithstanding their considerably higher price (see Table 3.3), but they comprise

TABLE 3.3 Comparison of ICE and ICE-electric hybrid vehicles[143]

	Camry (ICE) (SE automatic)	Camry Hybrid	Difference
ICE and electric motor, power			
2.4 L ICE (kW)	118	110	-7%
Electric motor (kW)		30	
Total/combined (kW)	118	140	19%
Torque			
ICE at ~4000 rpm (N)	218	187	-14%
Electric motor at 0–1500 rpm (N)		270	
US Environmental Protection Agency (EPA) fuel use ratings			
City (L/100km)	9.8	5.9	-40%
Highway (L/100km)	7.1	6.2	-13%
Combined	8.7	6.0	-31%
***Consumer Reports* road test fuel use**			
City (L/100km)	14.7	8.4	-43%
Highway (L/100km)	6.5	5.7	-12%
Overall	9.8	6.9	-29%
Other			
Acceleration, 0–96.5 km/h (sec)	9.6	8.4	-13%
Braking (dry), 96.5–0 km/h (m)	42.4	44.2	4%
Curb weight (kg)	1,500	1,669	11%
Manufacturer's US list price	$22,140	$26,200	18%

only about 1.5 percent of sales of new personal vehicles (cars, vans, SUVs). In 2006, there were about 255,000 sales of new hybrid vehicles in the US and another 60,000 or so in Japan, with no more than about 50,000 elsewhere in the world.[144]

What might the 30 percent fuel saving showing in Table 3.3 mean in practice? It amounts to a cost saving of about $315 a year when a car is driven 15,000 km and the gasoline price is $0.75 per liter (about $2.80 per US gallon). The 18 percent difference in purchase price is thus equivalent to 12.9 years of savings at the pump, perhaps beyond the lifetime of the vehicle. If the price of gasoline were at or above $2.00 per liter, as it is today in some European countries, the payback period for recovering the higher purchase price from fuel cost savings would be less than five years, usually well within a vehicle's lifetime.

Hybrid vehicles seem to be a new feature of the automotive world, but they were among the first automobiles. In the 1890s, Ferdinand Porsche—later the developer of the Volkswagen "beetle"[145]—designed a hybrid racing car with a one-cylinder gasoline engine that powered an electricity generator. The generator drove four wheel-mounted EMs that provided the vehicle's traction. This vehicle won races in Europe taking advantage, as do all later hybrids, of EMs' superior performance and gasoline's high energy density. Porsche's car was a *series* hybrid, in which only the EMs moved the wheels. Current road hybrids are mostly *parallel* hybrids in which the ICE or the EM(s), or both, can move the wheels, an arrangement that provides for more power output from the hybrid system. An exception is the Chevrolet Volt, a series hybrid due to come into production during 2010 or perhaps 2011.[146]

The main application of series hybrid systems over the years has been in diesel-electric locomotives. Diesel engines cannot drive the wheels of such massive vehicles directly because no available gearbox could cope reliably with the enormous forces at play. The most frequently used solution is for the engine to power a generator that powers one or more EMs that drive the wheels. The EMs have inherently high torque across a wide ranges of speeds and require much less or no gearing. The usual diesel-electric locomotives have little or no power storage.

Another hybrid application is shipping. The *Queen Mary 2*, the world's largest passenger liner when it was launched in 2004, has four diesel engines and two gas turbines, all producing electricity.[147] Propulsion depends on electric motors in four submersed "pods," two fixed and two movable. This arrangement allows for much more precise steering and produces much less noise and vibration than convention drives in which propellers are connected directly to diesel engines. It also helps reduce emissions by allowing more operation of the ICEs at close to their optimal outputs. The *Queen Mary 2*'s drive is a serial hybrid without storage, similar to that of the usual diesel-electric locomotive. A parallel hybrid tugboat began operation in a southern California port in 2009, operating on EMs alone when not moving large ships and on EMs and a diesel drive when working.[148]

Plug-in hybrid vehicles have larger batteries that can be charged from the grid when stationary and plugged in, as well as from their ICEs when in motion. The larger battery capacity allows a substantial amount of battery-only driving and battery-assisted ICE use. In North America, the best known example is the Chevrolet Volt, noted above. In China, at least one plug-in hybrid appears to be in production, the F3DM produced by

BYD Auto.[149] Several other manufacturers have plans to produce plug-in hybrid vehicles.

Plug-in hybrids are also known—confusingly for this book—as "grid-connected hybrids." They are known too as PHEVs, or as HEV-30, -50, and so on, with the number denoting the distance, usually in miles, that the vehicle can travel on its battery alone. Here, we use the term "grid-connected vehicle" (GCV) for a vehicle that can be powered from the grid both while stationary *and while moving*. In our use of terms, plug-in hybrids are not GCVs.

Plug-in hybrids have had high-level support in the US. They were promoted in an April 2006 speech on energy policy by President George W. Bush.[150] A press release issued in connection with a March 2009 speech by President Barack Obama included the following.

> Today, President Barack Obama announced the availability of $2.4 billion in funding to put American ingenuity and America's manufacturers to work producing next generation Plug-in Hybrid Electric Vehicles and the advanced battery components that will make these vehicles run. The initiative will create tens of thousands of U.S. jobs and help us end our addiction to foreign oil. Americans who decide to purchase these Plug-in Hybrid vehicles can claim a tax credit of up to $7,500....
>
> The Department of Energy (DoE) is offering up to $1.5 billion in grants to U.S. based manufacturers to produce these highly efficient batteries and their components. The DoE is offering up to $500 million in grants to U.S. based manufacturers to produce other components needed for electric vehicles, such as electric motors and other components. The DoE is offering up to $400 million to demonstrate and evaluate Plug-In Hybrids and other electric infrastructure concepts—like truck stop charging station[s], electric rail, and training for technicians to build and repair electric vehicles....
>
> DoE will also support demonstration, evaluation, and education projects to help develop the market for advanced electric drive vehicles. These vehicles will get up to 100 miles per gallon [2.4 L/100km], achieve a driving range of up to 40 miles [64 km] without recharging and run much like today's hybrids beyond that

> 40-mile range. Under this program, the DoE will also demonstrate other electric vehicle technologies such as truck stop electrification to reduce idling, electric rail, and necessary infrastructure. The solicitation covers demonstration projects that test a variety of vehicles, including small off-road vehicles, passenger vehicles, and over-the-road trucks, in geographically and climatically diverse locations.[151]

Plug-in hybrids have been characterized by Joseph Romm as "the car of the future." He outlined such a car as allowing 30–60 km on battery only, fueled while moving by E85 (see above) and from the grid while stationary. It would travel "500 miles on 1 gallon of gasoline [about 0.5 L/100km] and 5 gallons of cellulosic ethanol."[152] Romm also promoted plug-in hybrids—which he called "E-hybrids"—as load-levelers for the electricity grid. Charged at night, but also connected during the day while their users were at work, PHEVs' batteries could provide power to the grid during peak hours. Other advantages of such "vehicle-to-grid" (V2G) systems are discussed below.

We endorse the interest in plug-in hybrid cars as a useful expansion of the role of electric traction in road transport. We do not see them as the car of the future but rather as a bridge to GCVs, which could be the true "cars of the future."[153] Some of the issues posed by PHEVs are their complexity, cost (which could fall with mass production), safety (an issue in any vehicle with advanced batteries), ability to start when cold (batteries often do not work well at low temperatures) and battery disposal.[154]

Grid-Connected Vehicles (GCVs)

However good a BEV may be, an electric vehicle connected to the grid while moving would be better in two respects. First, the GCV would not have to carry a large weight of batteries, which can amount to several hundred kilograms, even more than a tonne. The batteries take up space and their weight increases the vehicle's energy consumption during acceleration and hill-climbing. Second, for the GCV there would be only distribution losses in moving the electricity from its source (e.g., a wind turbine) to the motor. In a BEV, as well as distribution losses there are losses when charging and discharging the battery. These losses total about 37 percent, that is, almost four times more than for a comparable GCV.[155]

Thus, on the same amount of generated electricity, the BEV would travel no more than 81 km for every 100 km traveled by the GCV, a difference that could be significant in an energy-constrained world. The actual distance could be considerably less, because of the additional weight of the batteries.[156] (A GCV could also have a battery allowing it to make short off-wire trips, thereby reducing its total energy efficiency a little.)

GCVs have been in use for at least as long as vehicles using ICEs. Electric trams and trains were operating in many cities at the end of the 19th century. Today, about 150 cities around the world have or are developing electric train (metro) systems running at the surface, elevated or, most often, underground. More than three times as many cities—about 550 in total—have tram systems, also known as streetcar or light-rail systems, including 72 cities in Russia and 70 in Germany.[157] There are also 353 active trolleybus systems.[158]

Where they are available, GCV systems generally provide the backbone of public transport systems. In Canada, for example, five of the six largest cities have one or more metro, tram or trolleybus systems, and these GCVs carry most of the public transport passengers served in these cities.

Electrification of intercity rail routes began early in the 20th century, although most of it occurred after 1950. Now most routes in Japan and Europe are electrified. Russia has the most extensive system of electrified rail, approximately half of the total of 85,000 km. They are mostly main lines including the whole length of the 9,258 km Trans-Siberian Railway, for which electrification was begun in 1928 and completed in 2002.[159] China's rail system is being rapidly electrified and it now boasts the second most extensive electrified system: 49 lines totaling about 24,000 km. As in Russia and elsewhere, these are mostly main routes and thus carry a disproportionately large share of passengers and freight.[160] The revolution caused by the advent of high-speed electrified passenger rail was discussed in Chapter 1.

As well as freight trains, other types of GCV have been or are being used to move goods. These include diesel trucks with trolley assist, such as those used in the Quebec Cartier iron ore mine from 1970 until the mine was worked out in 1977. These trucks were in effect hybrid vehicles with electric motors powered from overhead wires that provided additional traction when heavy loads were carried up steep slopes. A diesel generator provided the electricity. The reported result was an 87 percent decrease in fuel consumption and a 23 percent increase in productivity.[161]

The iron ore mine example illustrates a profoundly important point. When there are heavy loads, hill climbing or frequent starts and stops, using a fuel to generate electricity that powers a vehicle's electric motor from a grid can be more efficient than using the fuel to power a vehicle's ICE.

The main benefit of GCVs, and BEVs, could be their ability to function with a wide range of sources of electricity.

Several direct comparisons of energy consumption by GCVs and comparable vehicles with diesel-engine drives are available. Energy use *at the vehicle* is invariably lower. For example, trolley buses in the US use an average of 0.85 megajoules per person-kilometer; the average for diesel buses plying routes in urban areas is 2.40 MJ/pkm.[162] (The trolley buses tend to run in more congested traffic—and thus start and stop more, increasing energy use—but this is roughly offset by their higher average occupancy.) If the electricity for the trolley buses were produced by a diesel generator operating at 35 percent efficiency, with a 10 percent distribution loss, they would still use less energy overall than diesel buses. When electricity is produced renewably, what counts is energy use at the vehicle.

However, the main benefit of GCVs, and BEVs, could be their ability to function with a wide range of sources of electricity. Such electric vehicles use electricity produced from hydroelectric sources, wind turbines and photovoltaic panels, and from steam-turbine generating stations fuelled by coal, natural gas, oil, enriched uranium, wood waste or solar energy, all with more or less equal ease. Thus, whatever the exact paths of the transitions toward renewable generation of electricity, transport systems based on these vehicles can readily adapt. They will not have to be changed each time a primary energy source changes. Equally, consideration of transport systems will not stand in the way of evolution of energy production systems. And energy production systems will not have to change in nature as transport systems evolve.

Another substantial benefit is that several kinds of GCV system comprise familiar, well-proven technology. Some of this technology, as well as having a long pedigree, is also advancing rapidly. Today's new trams bear as little similarity to trams of the 1950s as do today's cars to cars of that period. One difference is that trams of the 1950s can still be found in regular operation, as they are in Hong Kong and San Francisco—although rebuilt to modern safety and other requirements.

Evolution and Assessment of GCVs

The advantages of grid-connection are such that in an energy-constrained world its scope could well be expanded beyond the present public-transport-based systems (trolley buses, trams, light rail, intercity rail). The major disadvantage of GCVs is the requirement for infrastructure that can provide electricity to vehicles moving along the route, and the corresponding inflexibility of a system that, except for limited battery-powered operation, allows motorized travel only along such routes.

The inflexibility of GCV systems is especially apparent in comparison with today's cars and trucks, which can move wherever there are roads and occasional refueling stations. An alternative to the car, offering more flexibility than conventional public transport, could be a widespread system of personal GCVs, usually known as a personal rapid transport (PRT) system. Such systems comprise fully automated, one- to six-person vehicles on reserved guideways providing direct origin-to-destination service on demand.

PRT has been debated for decades[163] and may now be poised for implementation. A system is about to open for moving passengers between parking areas and Terminal 5 at London's Heathrow Airport, and systems may be under development in the Dubai International Financial Centre and in Abu Dhabi's Masdar project. At least one town in the UK is exploring installation of a comprehensive system that could replace regular public transport.[164]

An assessment of PRT, conducted for the European Commission's Fifth Framework Programme, Energy, Environment, and Sustainable Development, concluded,

> The ideal target of cities is a self financed public transport system. PRT has a relatively low capital and operating cost, e.g., lower operating cost per passenger-km and lower capital cost per track-km than light railway. PRT can cover its operating costs, and has the potential to even cover capital costs depending on the type

of network, the discount rate and a reasonable fare (corresponding to the increased efficiency and quality). In the longer-term, large-scale implementation and mass production of PRT will lead to reductions in cost.

The total investment costs for guideway, vehicles and stations of a PRT system have been compared with different public transport systems. The investment cost in million Euros per track-km of three PRT systems has an average of 6 million euro per track-km. This value is lower than for all other systems considered (Automated Guided Transit, light rail transit, bus ways and trolley bus routes).[165]

Two possible pathways toward implementation of a GCV-based land-transport system are these: One is via the plug-in hybrid car described above. Extensive operation of such vehicles could lead drivers to want more use of their electric motors. To facilitate this, governments or entrepreneurs could provide means of powering them along major routes, accessible by appropriately equipped vehicles while in motion. When such en-route powering is sufficiently extensive, EVs with only batteries and retractable connectors could prevail over plug-in hybrids. As the grid-connection system expands, the need for off-grid movement would decline. Roads could be supplemented and even replaced by lower-cost guideway infrastructure. At the same time, vehicles would evolve to move only on the guideways. They would be as light as possible and, where appropriate, be assembled into trains. They would comprise PRT.

Another pathway could involve the evolution of public transport toward supplementation of or even replacement by PRT. This could be driven by PRT's low energy cost and, perhaps even more, by its potentially low infrastructure cost. If fuel prices for cars increase steeply, civic administrations will be pressed to provide alternative means of local travel. PRT could prove to be an attractive option. An analysis for Corby, a community of some 55,000 residents about 150 km north of London UK, compared costs of PRT and light rail. For similar initial investment, operating costs and fare structure, the PRT system would carry almost twice as many passengers annually, resulting in coverage from revenues of both operating and capital costs. Revenues from the light rail system would cover operating costs only, resulting in a net loss per rider of about 25 percent of the fare paid.[166]

PRT has not been well regarded by public transport experts. Perhaps the most critical among them has argued that "the basic concept of PRT is inherently unsound," that "the PRT mode is impracticable under all conditions," and that

> ...the main objective of the PRT concept—to match the positive features of the private auto such as privacy and direct station-to-station travel—cannot be achieved without also experiencing the major drawbacks of the auto/highway system in urban areas, such as high costs, large space requirements, low capacity, and poor reliability. Consequently, PRT is an unrealistic solution rather than a "future transit mode."[167]

Notwithstanding such criticism, some urban experts have become enamored with PRT. Peter Hall, among the best-known of the UK's urban planners, has said, "...if [the Heathrow PRT system] is as successful as I think it will be, this could be a big breakthrough in developing new kinds of totally personalised rapid transit, which could transform our cities in ways that we can't yet see."[168]

Peter Calthorpe, among the best-known of American architects and urban planners, has been reported as saying,

> One of my pet peeves is that we've been dealing with 19th-century transit technology. We can do better than LRT [light rail transit]. We can have ultra light, elevated transit systems (personal rapid transit) with lightweight vehicles. Because the vehicles are lighter, the system will use less energy. I used to be a PRT sceptic, but now the technology is there. It won't be easy to develop PRT technology and get all the kinks out, but it is doable. If you think about what you'd want from the ideal transit technology, it's PRT: a) stations right where you are, within walking distance, b) no waiting.[169]

PRT is one potential manifestation of GCV systems. Trolley trucks are another, not just in mines as in the example noted above but on regular roads along which wires have been strung from which suitably equipped vehicles may purchase electricity.[170] A less likely application could be electric barges on rivers and canals, for which on-board solar and wind generation of electricity is supplemented by opportunities for grid connection while moving and stationary.

Disadvantages of GCV Systems

A major disadvantage of GCVs is their infrastructure requirements. At a minimum, they require wires above existing roads, and the means to power them. According to the type of vehicle, they could also require new rails or other guideways.

A similar challenge confronted automobiles 100 years ago. They were mostly confined to summer travel on roads within urban areas. In 1910, the only paved highway in Canada, for example, was a 16 km stretch from Montreal to Ste.-Rose. Present levels of route flexibility took many years to develop. Indeed, an automobile was not driven across Canada until 1946, and the Trans-Canada Highway was not completed until the 1960s.[171] Today's automobiles and trucks may be more likely to drive on paved roads than those of a century ago, but the road system is extensive, reaching most parts of southern Canada.

A major disadvantage of GCVs is their infrastructure requirements.

Widespread adoption of GCVs for the next transport revolution could well involve continued use of the present road system, with the addition of powered overhead wires that can be shared by all. However, vehicles run more efficiently on rails or tracks than on roads, and energy constraints may favor trains and other vehicles confined to special-purpose rights-of-way.

Another disadvantage is that GCV systems are more expensive to install. The simplest present type of grid-connected installation is that of trolley buses powered from wires strung over a regular road. This is nevertheless more expensive than providing the means to operate diesel buses along the same road. A trolley bus presently costs 1.5–3.2 times as much as a diesel bus.[172] Capital costs—assuming availability of the road in both cases—are about $1 million more per route kilometer for the trolley bus. On the other hand, trolley buses last longer, are easier to maintain, and have lower energy costs. On the last point, a comparison for Landskrona,

Sweden, showed that fuel costs per vehicle-kilometer for trolley buses are 56 percent lower than for similar diesel buses, even though the cost of electricity per megajoule is 12 percent higher than the cost of diesel fuel ($0.085/kWh vs. $0.76/L).[173]

The higher purchase cost of trolley buses compared with diesel buses is a matter of scale. Trolley buses are simpler vehicles and are intrinsically less costly to produce. However, worldwide, there are presently hundreds of times more diesel buses than trolley buses, resulting in much lower production costs for diesel buses. If hundreds of times more trolley buses were sold, their prices would likely fall below those of diesel buses.

Under present circumstances, trolley buses become financially advantageous when daily vehicle distance exceeds about 150 km.[174] Building on this analysis, the conclusion can be drawn that if diesel fuel and electricity prices were to double and quadruple, trolley buses would be financially advantageous for daily distances above about 90 km and 40 km, respectively, well under typical daily bus movement.[175]

Such calculations may not mean much to a public transport operator in a poorer country who has little access to capital and seeks to provide the maximum possible amount of service at the lowest possible cost. Nevertheless trolley buses—now found mostly in Europe and cities of the former USSR—are being deployed in developing countries. In Kathmandu, Nepal, a new trolley bus service is proposed, with support by the Kyoto Protocol's Clean Development Mechanism, which allows industrialized countries to offset requirements to reduce GHG emissions by financing reductions in developing countries.[176] Such support could be a critical factor in enabling the development of grid-connected systems.

Yet another criticism is that GCV systems require continuously available, centrally provided power. Toronto's streetcars and subway trains stopped during the major blackout that affected eastern North America on August, 14 2003, but cars and trucks kept on rolling, at least for a time. Then the cars and trucks were stopped in traffic jams caused by non-functioning traffic signals and by line-ups at non-functioning gas stations, both out of order because there was no electric power.

It is nevertheless true that cars and trucks have some additional resilience compared with grid-connected systems because they carry their own fuel. However, both depend ultimately on heavily centralized systems of energy distribution.

Generating Enough Electricity for Transport

How Much Electricity Would Be Required

How much more electricity would have to be generated if *all* cars and other personal vehicles were to become BEVs? Estimates for Belgium and the UK are in the order of 15 percent of respective total electricity consumption.[177] The share for North America might be about the same given that both car use and electricity use per capita are roughly double that of European countries. The energy required for generation, renewable or otherwise, could similarly increase by some 15 percent.

A more critical issue concerns when the electricity is used. During peak periods, a system's generating capacity often operates near its limit, whereas at nights and on weekends there can be substantial spare capacity, as much as 100 percent of what is used. If the electricity is stored on board, most charging can be at night or otherwise outside of peak periods, and time-of-use pricing could help ensure that this happens.

Thus, whereas an all battery-electric personal-vehicle system could increase the requirement for electrical energy by about 15 percent, the requirement for additional generating *capacity* could be much lower, perhaps zero.

Extensive use of V2G would increase the amount of electricity delivered to vehicles, and thus the overall energy use, but could reduce the requirement for additional generating capacity.

One advantage of having a large battery in each of so many vehicles has already been noted: it is the opportunity for V2G arrangements whereby a vehicle's battery is charged when there is a surplus of generating capacity and discharged to the grid—through suitable converters—when generating capacity is needed. As well as leveling loads in current systems, V2G could be used to bring on systems, notably wind and solar, whose intermittent generation presents challenges for electrical utilities.[178] During windy or sunny periods, power would be stored in connected batteries, to be returned at other times, providing what one analyst has described as an "unexpected synergy" between wind power and electric vehicles.[179]

Extensive use of V2G would increase the amount of electricity delivered to vehicles, and thus the overall energy use, but could reduce the requirement for additional generating capacity. Effective time-of-use pricing could help both generators and users of electricity strike appropriate balances among several factors. Generators could balance capacity and energy consumption. Users could balance vehicle battery size, kilometers driven and the net cost of electricity consumption less sales to the generating utility. However, from the perspective of total system efficiency, storing electricity to shave peak loads might be better achieved with stationary arrays of batteries, such as are already being tested.

Several kinds of large-scale stationary batteries are in use, for shaving peak loads on the grid, storing surplus wind-generated electricity for use when there is insufficient wind, and providing back-up protection against unreliability of transmission. What may be the most powerful such battery in the world is used in Fairbanks, Alaska, chiefly for the last purpose. It is an array of 13,760 nickel-cadmium cells, weighing a total of 1,360 tonnes, and capable of providing 27 megawatts (MW) of power for up to 15 minutes (or less power for longer).[180] Two sodium-sulfur batteries in Japan store more energy: each a total of 58 megawatt-hours (MWh)—vs. about 7 for the Fairbanks battery—but with a maximum power output of 8 MW that can be sustained for several hours.[181] In Halton Hills, Ontario, a 100 kWh sodium-nickel-chloride battery helps the local electricity supplier improve power quality and reduce peak demand on the Ontario grid.[182] What may be the most promising system is on King Island, off the northwest tip of Tasmania, where vanadium flow batteries have allowed the average proportion of wind-derived electricity in the island's grid to rise from about 12 to more than 40 percent.[183] What will be Europe's largest energy storage system for wind power, in Donegal, Ireland, will use a vanadium flow battery.[184]

GCVs, if much more widely used, would introduce different considerations. These vehicles are more likely to consume electricity from the grid during periods of peak consumption. Overall consumption would be lower unless reductions were achieved in other sectors, more generating capacity—or more storage capacity—would be needed to meet the "on-demand" requirements of grid-connected transport.

There is an alternative to providing more generating capacity for a transport system that relies on heavy use of electric motors. It is to reduce consumption for non-transport purposes. The scope for such reductions is evident from Table 3.4, which reveals large differences in consumption among countries and large increases within countries across time. If, for example, Canada were to attain the UK's current per capita use, or even revert to its own per capita use in 1973, there would be much scope within

TABLE 3.4 Annual per-capita electricity consumption, selected countries, 1973, 1990 and 2004[185]

	kWh/person			Increase	
	1973	1990	2004	1973–1990	1990–2004
Australia	3,855	7,525	9,837	95%	31%
Belgium	3,518	5,819	7,740	65%	33%
Canada	9,788	15,098	15,750	54%	4%
Denmark	3,207	5,520	6,105	72%	11%
Finland	5,777	11,826	15,911	105%	35%
France	2,790	5,192	6,691	86%	29%
Germany	3,964	5,736	6,224	45%	9%
Iceland	9,733	15,339	26,631	58%	74%
Japan	3,822	6,090	7,570	59%	24%
Norway	15,353	22,835	23,936	49%	5%
Sweden	8,508	14,066	14,499	65%	3%
UK	4,147	4,796	5,685	16%	19%
USA	7,870	10,530	12,388	34%	18%

existing generating capacity for accommodating additional loads from transport activities.

How Enough Electricity Could Be Produced

The ultimate goal should be to generate all required electricity from renewable sources—sun, wind, tide, etc.—whether used for transport or other purposes. This goal is desirable to ensure that provision of a form of energy so essential to what we know as civilization is not dependent on a diminishing resource. Achieving the goal is also desirable because producing electricity from renewable sources usually has less cumulative environmental impact than generation from fossil fuels, uranium and other non-renewable resources.

The main shorter-term goal could be to prevent an increase in the use of non-renewable generation as increasing amounts of electricity are used to provide for replacement of ICEs by electric motors as the main means of propelling vehicles. A particular fear could be that increases in demand for electricity will be met by construction of generating stations that use coal as a fuel. This appears to be happening now in China, the US and other places. Coal consumption tripled in China between 1980 and 2004, and is expected to more than double again by 2030. Coal consumption in the US increased by 56 percent between 1980 and 2004 and is expected to increase by another 27 percent by 2030. China consumed less coal than the US in 1980 and is projected to consume more than three times as much by 2030.[186]

A report on coal's future availability, discussed above, questions whether increases in consumption can be sustained beyond about 2025, but there seems little doubt that coal consumption could grow until then.[187] Much ongoing research is directed toward reducing the local and global impacts of burning coal,[188] but such reductions promise to be costly and may well not be implemented even if available.

> The ultimate goal should be to generate all required electricity from renewable sources — sun, wind, tide, etc. — whether used for transport or other purposes.

Generation of electricity from nuclear power is another concern, particularly the long-term storage of radioactive wastes. The IEA has noted that 443 nuclear reactors in 31 countries provided 15 percent of the world's electricity in 2005. This share is expected to fall to 13 percent by 2015 and to 10 percent by 2030, but actual production of nuclear power would rise slightly by 2015 and then fall.[189] A German report notes that only Canada has high-grade uranium deposits (ore grade more than 1 percent), although lower-grade ores are known to exist in some 35 other countries.[190]

Uranium is unusual among fuels in that less than two thirds of consumption is met from current production; the remainder is met from stockpiles accumulated before 1980. These stockpiles are being rapidly depleted and production must rise substantially if present levels of consumption are to be sustained. However, uranium prices do not yet reflect scarcity. They have fallen by more than half from a June 2007 peak. The German report concludes that after about 2020 severe uranium supply shortages will limit the expansion of nuclear energy. In the meantime, until about 2015, rapid expansion will be limited by the long lead times of new reactors and the decommissioning of aging reactors.[191]

Thus there is some potential for nuclear expansion, particularly in the period 2015–2020 and for longer if new uranium ore is found or if the use of available radioactive material is extended. A further possibility is the use of another fuel, notably thorium, of which India or Brazil appears to have the largest reserves.[192]

Nuclear energy is among the most controversial of energy topics, in part because its use is advocated by some people believed to be strong supporters of actions to protect the environment, and opposed by others. Notable among advocates is James Lovelock, best known as formulator of the Gaia hypothesis, which holds that the Earth's surface and its biological systems function as if they were a single organism.[193] In a 2004 article Lovelock wrote, "...I am a Green and I entreat my friends in the movement to drop their wrongheaded objection to nuclear energy... We have no time to experiment with visionary energy sources; civilisation is in imminent danger and has to use nuclear—the one safe, available, energy source—now or suffer the pain soon to be inflicted by our outraged planet."[194]

Notwithstanding Lovelock's view, several analysts have concluded that renewable energy sources could be developed with considerable dispatch, given the right economic and policy contexts. For example, one analysis suggested, "The renewable energy potential within Europe's borders would actually be almost capable of satisfying current electrical energy demand."[195] This analysis then considered the transferability of the analysis to other world regions, including China, and concluded the following:

- An entirely renewable and thus sustainable electricity supply is possible even if only current technologies are used.
- The costs of electricity don't have to lie far above today's costs even if very conservative assumptions are made.
- The costs are dependent on the future system configuration, and could be reduced by ongoing technical progress, or be negatively influenced by wrong energy policies.
- The general results of the scenarios for Europe and neighbors can be transferred to other world regions even if in some cases some more detailed information—especially on the local wind conditions—would be welcome to reduce uncertainties. The latter, for example, applies to China where the rough, mountainous terrain causes problems with resource assessment.
- The problem of converting our electricity system to one that is environmentally and socially benign is therefore much less a financial or technical issue, being instead almost entirely dependent on political attitudes and governmental priorities. There is more than enough evidence to justify a confident call for a comprehensive transition to a sustainable electricity supply, bearing in mind that a broad variety of

solutions is possible. Responsible political decisions are now imperative for allocating the necessary technical, scientific and economic resources to achieve this goal.[196]

What is particularly appealing about the above analysis—from which the quotation was taken—is the way in which it demonstrates the complementarity of energy resources distributed over a wide area. For example, in Europe wind energy is available mainly in the winter. Along the Atlantic coast of Morocco, which is as close as 14 km to Europe, wind energy is available mainly in the summer.

Another appealing feature of the above is the focus on solar thermal energy and the huge resource it can provide. The desert regions of northern Africa could, with available technology (parabolic mirror arrays powering steam turbines using desalinated sea water) produce about 500 times the electricity used in the European Union.[197] Similarly, desert and other areas in the southwest US and in western China could respectively supply much more than the present electricity usages of these two countries.[198] High-temperature solar thermal energy—also known as concentrating solar power—is especially appealing because it can be combined with relatively inexpensive thermal storage to provide continuous production of electric power.[199]

Electricity is the ideal transport fuel for an uncertain future.

Yet another means of renewable generation of electricity is from the kinetic energy of ocean tides and currents. A comprehensive review of such opportunities for the UK suggested that *marine energy* could make "a reasonable target contribution to UK electricity supplies" of 80 TWh annually.[200] Of this, 20 TWh would come from tidal barrages such as the one at La Rance in France,[201] 10 TWh would come from using the energy in marine currents in much the way the energy in wind is used,[202] and 50 TWh would come from making use of the energy in waves. The UK's estimated gross generation from all sources in 2005 was 400 TWh.[203]

A fundamental prerequisite for the major transport revolution we anticipate—moving from ICEs to electric motors—will be provision of sufficient electrical energy. How much would have to be achieved by 2025 is discussed in Chapter 5, where detailed analyses for the US and China are presented. Whether the energy transformation required for the transport revolution will be achieved in a timely manner will depend mostly on the extent to which the need for it is anticipated. At present, the need to change our electricity supply, perhaps radically, is not widely appreciated. We hope this book will enhance understanding of the need for action.

Finally, the most important reason for a switch to electric transport is worth restating. In a world characterized by changing sources of energy, electric transport systems can stay the same whether the electricity is produced from coal, nuclear energy, natural gas, wind, solar thermal or marine currents, or some or all of these sources. Transport systems will not have to change to be compatible with changes in generation. Equally, changes in generation will not be held back by features of a transport system. The move toward renewable sources can be incremental or radical, according to what becomes appropriate. Electricity is the ideal transport fuel for an uncertain future.

4

Transport's Adverse Impacts

Introduction

This chapter discusses the adverse impacts of present motorized transport. These are mostly but by no means entirely the result of the use of internal combustion engines (ICEs) to propel today's vehicles. We could have included a chapter on transport's benefits, but they hardly need stating. We noted in Chapter 2 how effective transport gave advantage to particular peoples in history, and how motorized transport has facilitated and even stimulated just about everything now regarded as progress. What should be added is the suggestion that beyond a certain point the costs of increased mobility may outweigh the benefits:

> Near the end of the 20th century, the belief in the desirability of perpetual growth in mobility and transport has started to fade. In many countries, highway accessibility is so ubiquitous that transport cost has almost disappeared as a location factor for industry. In metropolitan areas, the myth that rising travel demand will ever be satisfied by more motorways has been shattered by reappearing congestion. People have realized that the car has not only brought freedom of movement but also air pollution, traffic noise and accidents. It has become obvious that in the face of finite fossil fuel resources and the need to reduce greenhouse gas emissions the use of petroleum cannot grow forever. There is now broad agreement that present trends in transport are not sustainable, and many

conclude that fundamental changes in the technology, design, operation, and financing of transport systems are needed.[1]

Figure 4.1 provides an illustration of what may happen as the level of motorization increases. Benefits from growing mobility—in terms of greater access to people, goods and services—grow more steeply at first. Congestion and the costs of managing it grow with increasing motorization, perhaps less steeply at first. Environmental and social costs grow in proportion to the level of motorization. Beyond a certain point—"A" in Figure 4.1—*net* benefits begin to decline. At a higher level of motorization—"B" in Figure 4.1—the costs of increasing mobility outweigh the benefits. A good question to ask is whether we have too much mobility when point B is reached. Or does the condition of what has been called "hypermobility"[2] begin at point A, when net benefits begin to decline?

> Beyond a certain point the costs of increased mobility may outweigh the benefits.

Another approach has been to note that access to goods, services and people has not kept pace with mobility. Figure 4.2 suggests that people in

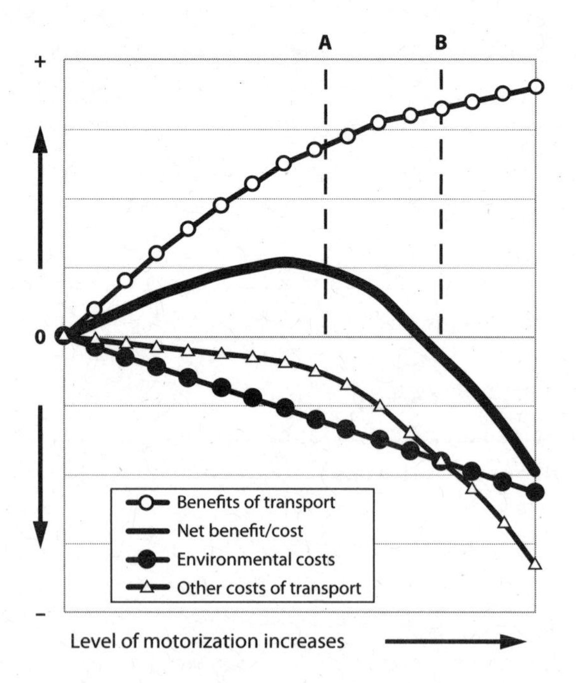

FIGURE 4.1 Schematic illustration of mobility benefits and costs[3]

what used to be West Germany had hardly more access to destinations in 1990 than they did in 1960, when there were much lower levels of mobility. This was chiefly because car ownership and use levels were much lower. The main change was that in 1990 access was much more often achieved by car, whereas in 1960 it was achieved more by public transport, walking and cycling. Destinations were probably on average farther away in 1990 than in 1960.

This chapter begins by considering actual and potential global environmental impacts of transport, including climate change, ozone depletion, the proliferation of persistent organic pollutants, and ocean acidification. We next consider transport's local and regional environmental impacts. The focus is on atmospheric pollution and air quality, but impacts on land and water are also considered. Finally, there is some consideration of adverse social and economic impacts.

Global Environmental Impacts

Climate Change

The average temperature of the planet's surface has been rising and the principal cause is believed to be emissions of greenhouse gases (GHGs) resulting from human activity.[5] Other causes of the rising temperature are posited, including reductions in the amount of low-level cloudiness resulting from solar deflection of cosmic radiation (see Box 4.1). However,

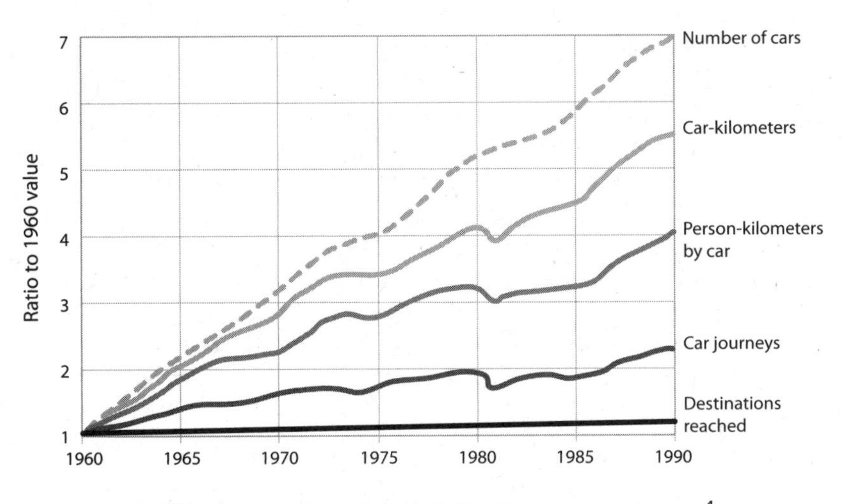

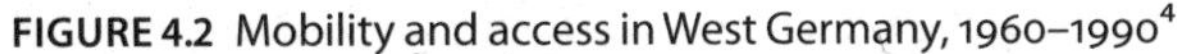
FIGURE 4.2 Mobility and access in West Germany, 1960–1990[4]

BOX 4.1 Cosmoclimatology[6]

The prevailing account of ongoing climate change, particularly the rising average temperature of the Earth's surface, is that it is caused chiefly by anthropogenic (i.e., human-produced) emissions of radiatively active gases, usually known as greenhouse gases (GHGs). A coherent alternative account is that the warming can be attributed to a decline in the amount of cosmic radiation penetrating the Earth's atmosphere.

Cosmic rays are the result of stellar activity, notably the explosive "deaths" of stars. They comprise charged subatomic particles, mostly protons. In air containing water vapor, these particles release electrons that initiate cloud formation, particularly over oceans at heights below about 3 km. Such low-level clouds cool the Earth, offsetting some of the warming produced by the Earth's blanket of water vapor and other GHGs. Increases in the sun's magnetic activity reduce the amount of cosmic radiation reaching the Earth, and vice versa. According to the alternative account of global warming, increased solar activity over the last century has reduced cosmic ray penetration, resulting in less low-level cloudiness and consequently less cooling.

Henrik Svensmark, the Danish physicist who proposed this account, claims that it can explain all the 0.6°C of global warming between 1900 and 2000. He allows that the increase in anthropogenic GHG emissions may also have had an impact, albeit minor. Svensmark argues that his theory accounts for many other features of climate change, including the "snowball" and "hothouse" Earths of the distant past, the medieval warm period and subsequent low temperatures, and the current cooling of Antarctica (which is opposite from what might be expected if changes in atmospheric GHGs were the main cause of global warming).

Svensmark and co-author Nigel Calder note that, "To correct apparent overestimates of the effects of carbon dioxide is not to recommend a careless bonfire of the fossil fuels that produce the gas...there are compelling reasons to economize in the use of fossil fuels that have nothing to do with climate: to minimize unhealthy smog, to conserve the planet's limited stocks of fuel, and to keep energy prices down for the benefit of poorer nations." Svensmark and Calder use the term *cosmoclimatology* to embrace investigation of the effects of cosmic rays on the Earth's climate.

most scientists concerned with climate believe that anthropogenic emissions of GHGs are the main reason why many of the warmest years worldwide in the last century have occurred during the last two decades.[7] Here is this majority view, as set out in a document of the Intergovernmental Panel on Climate Change (IPCC):

> *Most* of the observed increase in globally averaged temperatures since the mid-20th century is *very likely* due to the observed increase in anthropogenic greenhouse gas concentrations. This is an advance since [IPCC's 2001] conclusion that "*most* of the observed warming over the last 50 years is *likely* to have been due to the increase in greenhouse gas concentrations." Discernable human influences now extend to other aspects of climate, including ocean warming, continental-average temperatures, temperature extremes and wind patterns.[8] [emphasis added]

In parts of the world, some warming could be considered desirable.[9] Nevertheless, warming by even a few degrees could have numerous adverse effects. They include an increase in extreme weather events, changes in land and marine growing seasons and species composition, sea level rise and consequent flooding, changes in water availability, infrastructure damage and health effects from heat stress and changes in disease patterns.[10]

To reduce potential effects of human activity on climate, 168 countries have ratified the Kyoto Protocol, which requires 40 of them—as listed in Annex 1 of the Protocol—to reduce their emissions of six GHGs so that during 2008–2012 these emissions average about five percent below the overall 1990 level.[11] Many countries are not on target to reach their Kyoto commitments, often because of increases in emissions from transport. Information about the performance of 33 of the 40 Annex 1 countries is in Table 4.1. In 29 of these countries—all but Australia, Belarus, Finland and Turkey—the rate of change in GHG emissions from transport has been higher, in some case substantially higher, than the overall rate of change in GHG emissions.

Table 4.2 shows changes in GHG emissions from all sources for the 15 countries of the European Union (EU) as it was before 2004 and for three of the other countries in Table 4.1. Also shown in Table 4.2 are the

TABLE 4.1 Changes in greenhouse gas emissions by country, 1990–2004, from transport and from all sources (except land use and forests)[12]

Country	Change in GHG emissions, 1990–2004, from: All sources	Transport	Country	Change in GHG emissions, 1990–2004, from: All sources	Transport
Australia	25.1%	23.4%	Ireland	23.1%	143.8%
Belarus	–41.6%	–66.2%	Italy	12.1%	27.6%
Finland	14.5%	9.9%	Japan	6.5%	19.8%
Turkey	72.6%	56.8%	Latvia	–58.5%	12.1%
Austria	15.7%	86.8%	Netherlands	2.4%	33.8%
Belgium	1.4%	34.0%	New Zealand	21.3%	61.6%
Bulgaria	–49.0%	–32.1%	Norway	10.3%	27.5%
Canada	26.6%	29.9%	Poland	–31.2%	16.1%
Croatia	–5.4%	35.7%	Portugal	41.0%	99.4%
Czech Republic	–25.0%	113.6%	Romania	–41.0%	95.2%
Denmark	–1.1%	26.8%	Slovenia	–0.8%	57.3%
Estonia	–51.0%	–20.3%	Spain	49.0%	77.3%
France	–0.8%	20.8%	Sweden	–3.5%	9.0%
Germany	–17.2%	5.1%	Switzerland	0.4%	6.9%
Greece	26.6%	42.6%	United Kingdom	–14.3%	12.5%
Hungary	–31.8%	25.8%	United States	15.8%	28.1%
Iceland	–5.0%	16.7%	Medians	0.4%	27.5%

respective Kyoto[13] targets and per capita GHG emissions in 2004. These countries and group of countries all increased their GHG emissions from transport by 20–30 percent even though they have large differences in overall and in per capita emissions. If the GHG emissions from the EU15's transport had declined at the rate of its other emissions (i.e., by 6 percent between 1990 and 2004) EU15 would be on a path to reach its Kyoto target for 2008–2012. Canada and Japan would not be on a path to meet their targets even if their transport emissions had declined at the rate of their other emissions. This is also true of the US, which has not ratified the Kyoto Protocol and is thus not obliged to meet its target.

GHG emissions from transport result almost entirely from the burning of oil products in ICEs. For example, in the US in 2005, 94 percent of GHGs from transport comprised CO_2 from the burning of carbon-based fuels. Another 3 percent comprised N_2O, produced during the operation of the three-way catalytic converters used in vehicles to reduce noxious emissions.[15] Most of the remaining emissions of GHGs from transport comprised HFC-134a, a hydrofluorocarbon used as the refrigerant in ve-

TABLE 4.2 Greenhouse gas emissions, Canada, EU15, Japan and the US[14]

	Changes in GHG emissions 1990–2004			Kyoto commitment (reduction by 2008–2012)	Per capita emissions 2004 in tonnes of CO_2 equivalent.		
	All sources including LULUCF	All sources except LULUCF	Transport only		All sources including LULUCF	All sources except LULUCF	Transport only
Canada	62%	27%	30%	–6%	26.2	23.7	6.0
EU15	–3%	–1%	26%	–8%	10.3	11.0	2.3
Japan	5%	7%	20%	–6%	9.9	10.6	2.0
US	21%	16%	28%	–7%	21.4	24.1	6.4

Note: LULUCF means "land use, land use changes and forests." Their emissions—or absorption of emissions—are required to be reported to the Kyoto Protocol Secretariat, but they do not count toward reduction targets.

hicle air conditioners, discussed below.[16] The burning of a liter of gasoline produces about 2.36 kg of CO_2; burning a liter of diesel fuel or jet kerosene produces about 2.73 kg.[17] The higher CO_2 emissions from diesel fuel—about 15 percent higher—are usually more than offset by the greater efficiency of diesel engines, which can use about 35 percent fewer liters of fuel per 100 km.[18]

It follows that transport's GHG emissions are closely correlated with transport's fuel use, with slight adjustments according to the mix of fuels used. A minor reason why EU15 reports low per-capita GHG emissions from transport compared with North America (see Table 4.2) is that a larger share of EU15 road vehicles have diesel engines.[19]

> GHG emissions from transport result almost entirely from the burning of oil products in ICEs.

Aviation is an exception to the close relationship between transport fuel use and GHG emissions. As well as the production of CO_2 from fuel burning, the exhausts of aircraft flying at a height of about 10 km (where commercial jets cruise) have a variety of additional effects estimated to add about 90 percent to the CO_2 effect. This estimate does not include cirrus cloud enhancement, whose extent and effect are less clear but which could add almost twice the CO_2 effect.[20] Thus, compared with travel in a car that burns the same amount of fuel per 100 person-kilometers (pkm), a plane journey could result in two to four times the radiative activity.[21]

Aviation has sometimes been characterized as the fastest growing transport mode, and thus the worst offender in terms of GHG emissions.

Such a statement about growth may have been true for the movement of people, but does not appear to have been true overall or in terms of GHG emissions. For example, worldwide GHG emissions from road transport increased by 27 percent between 1990 and 2000, while GHG emissions from aviation increased by "only" 24 percent. GHG emissions from water modes increased by 21 percent (but see below) and those from rail *fell* by 12 percent.[22] However, aviation has been the worst offender in terms of GHG emissions per pkm, largely because of the "altitude effect" described in the previous paragraph. It is also the worst offender in terms of GHG emissions per tonne-kilometer (tkm) of freight.

Less well researched than the "altitude effect" for aviation is an opposite effect whereby ships' emissions may counteract global warming. Burning the high-sulfur bunker fuels used in ocean-going ships produces CO_2 but also gives rise to low-level clouds that reflect the sun's radiation. According to one analysis, a warming effect from the CO_2 emissions could be more than offset by a cooling effect from reflection of the sun's radiation out to space.[23] Another analysis suggests that the clouds and other effects could partially but not completely offset warming from ships' CO_2 emissions.[24]

There are several ways in which oil depletion—the decline in availability of oil after the peak in production—and climate change could interact. Climate change could add to the costs of extracting fossil fuels. The impact of hurricanes provides a possible example, to the extent that they are becoming more intense as a result of climate change.[25] Hurricanes Katrina and Rita in 2005 caused damage to oil and natural gas installations in the Gulf of Mexico estimated at $17 billion.[26] The short-term impact was to induce shortages that raised the prices of these fossil fuels, perhaps reducing consumption of them and reducing the impact of later production peaks. The longer-term impact could also be higher prices, as the repair costs are passed on to consumers or facilities are abandoned as becoming too risky. Climate change could also reduce costs, for example by making exploitation of polar resources more feasible.

A critical link between oil depletion and climate change is that mitigation of the impacts of climate change could require additional use of oil. For example, construction of levees to protect coastlines from rising sea levels would require much use of heavy-duty equipment that runs on diesel fuel, and much use of concrete, the production of which is energy-intensive, although not necessarily requiring oil. Such uses could compete with requirements for additional fuel to construct means of alternative energy generation—for example, hydroelectric dams and tidal power stations—in preparation for replacement of oil by electricity.

Another example could be the development of a light-rail system powered by wind energy, such as that of Calgary. Oil would be required during construction of the wind turbines, building or expanding the distribution grid, laying track and providing vehicles. The energy costs of infrastructure development can be considerable, and are often neglected.[27]

Reducing Greenhouse Gas Emissions from Transport

Concern to reduce GHGs has become a major reason—perhaps the major reason—for reducing oil consumption. This is because climate change has moved to the top of environmental agendas and the oil products burned as transport fuels are seen as significant sources of GHGs. Here we review some of the programs in place or proposed that have the goal of reducing GHG emissions from transport or oil use by transport, or both. Our own proposals as to how oil consumption for transport should be reduced are set out in Chapter 5.

Programs to reduce oil use, if effective, usually have the effect of reducing GHG emissions, and vice versa. Increasingly, the two are addressed together.

> The longest-standing regulations concerning vehicle fuel use are the Corporate Average Fuel Economy (CAFE) standards of the US.

Thus, the current EC proposal to limit GHG emissions[28] was issued with a proposal for a new EU energy policy.[29] With respect to transport, the early focuses of the proposed energy policy are to achieve increases in the use of biofuels, in the energy efficiency of vehicles and in the use of public transport. Regarding the first focus, the proposal noted,

> ...biofuels are today and in the near future more expensive than other forms of renewable energy, [but] over the next 15 years they are the only way to significantly reduce oil dependence in the transport sector.... The Commission therefore proposes to set a binding minimum target for biofuels of 10 percent of vehicle fuel by 2020 and to ensure that the biofuels used are sustainable in nature, inside and outside the EU.[30]

Shortly after proposing an EU energy policy, the EC proposed a strategy to reduce CO_2 emissions from light-duty vehicles.[31] Its overall goal was to reduce emissions from new cars to an average of 120 grams per vehicle-kilometer (g/vkm) by 2012. The strategy would have two components: a reduction to 130 g/vkm through improvements in engine technology, and, if necessary, a further reduction by 10 g/vkm through improvements in air conditioning, tire management, fuels and other features. New light-duty commercial vehicles (vans) would have to achieve 175 g/vkm by 2012.

These are to be legislated requirements, replacing an existing agreement with the main association of car manufacturers "...in view of growing concerns regarding the progress made by the industry under this voluntary approach."[32] Under that agreement, new vehicles' average emissions of CO_2 were to decline from 185 g/vkm in 1992 to 140 g/vkm in 2008. CO_2 emissions are declining according to this trajectory, but not because of technical improvements in energy efficiency. The decline has occurred chiefly because of growth in the share of diesel vehicles in the new car fleet, from 24 percent in 1995 to 48 percent in 2003.[33]

Increases in energy efficiency were achieved during this period, enabling vehicles to do more for less fuel, but they were mostly directed toward increasing the power and weight of vehicles rather than reducing their fuel use.[34] Only the shift to diesel engines and fuel produced a significant reduction in fuel use. This may not have reduced the EU's GHG emissions as much as was expected because the shift to diesel has created a surplus of gasoline that is exported to North America.[35] Emissions from the refineries producing the surplus gasoline count toward the EU's total.[36]

The longest-standing regulations concerning vehicle fuel use are the Corporate Average Fuel Economy (CAFE) standards of the US. These came into effect in 1978 for cars and in 1979 for other four-wheel vehicles with a total weight—vehicle weight plus payload capacity—of less than 3,856 kilograms (stated as 8,500 pounds). The other four-wheel vehicles, known as "light trucks," include passenger vans, sport utility vehicles (SUVs) and pick-ups used as personal vehicles. For each model year, two CAFE "miles-per-gallon" standards are set, one for cars and one for the other vehicles. Each manufacturer's products for a model year must average better than the respective standard.

The evolution of the standards since the 1970s is shown by the faint lines in the top left panel of Figure 4.3, expressed as liters per 100 kilometers (L/100km). The 2006 standards were equivalent to 8.6 and 10.9 L/100km (actually set as 27.5 and 21.6 miles per US gallon). Manufacturers generally meet the CAFE standards, and are fined if they do not do so. Some foreign manufacturers disregard the CAFE standards and pay the fine, roughly $175 for each car in a manufacturer's fleet for

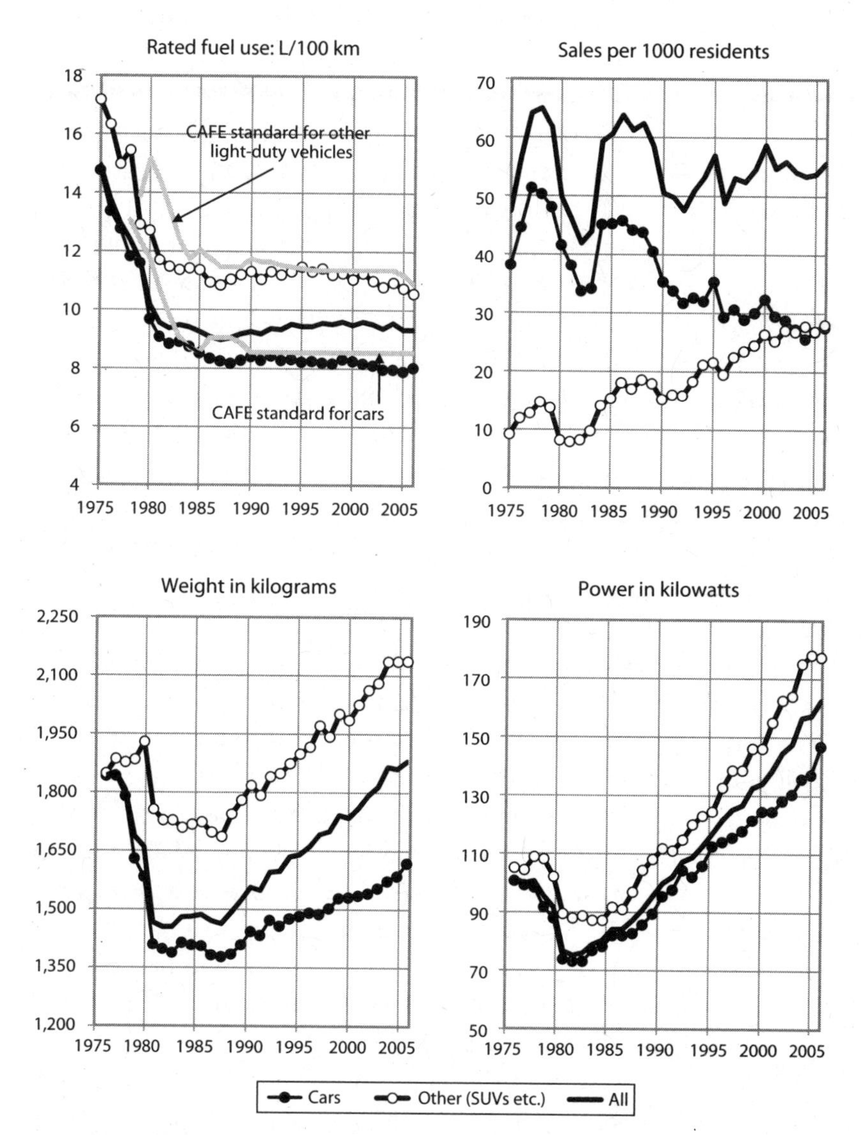

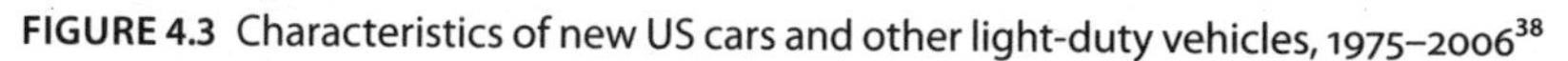

FIGURE 4.3 Characteristics of new US cars and other light-duty vehicles, 1975–2006[38]

each L/100km by which the fleet exceeds the standard. Porsche paid fines amounting to more than $3.5 million for its 2003 vehicles.[37]

The top left panel of Figure 4.3 also shows how new cars have conformed to the CAFE standards. New vehicles' average fuel use per 100 km was falling rapidly when the standards were introduced. The decline could have been a direct response by purchasers to fuel price increases in the early 1970s. It could also have been manufacturers' anticipation of the standards, which were legislated in 1975 in response to high oil prices, to come into effect in 1978 (cars) or 1979 (other light-duty vehicles). Implementation of the CAFE standards in the late 1970s may have pushed average new-vehicle fuel use down further. The CAFE standards have been essentially unchanged since the mid-1980s.

Figure 4.3 shows that new-vehicle fuel consumption per 100 km has increased a little in the US since the mid-1980s. This has happened chiefly because of the declining share of regular cars in the new vehicle mix in favor of more fuel-hungry SUVs, passenger vans and pick-ups, known collectively as "light trucks." The major change since 1975 in the kinds of vehicles sold in the US as personal vehicles is illustrated in the top right panel of Figure 4.3. In 1975, cars comprised 81 percent of these vehicles; in 2006, they comprised 50 percent. The remainder consisted of light trucks, labeled as "Other (SUVs etc.)" in Figure 4.3. The shares of the two types of vehicle in the new vehicle fleet appear to have stabilized since 2002 at about 50 percent for each category.

The top right panel of Figure 4.3 also shows the considerable constancy in sales per US resident across the years, albeit with substantial fluctuation from year to year. The growth in vehicles per person in the US—illustrated in Figure 2.4 of Chapter 2—has occurred because vehicles, particularly passenger cars, are lasting longer, as discussed in Box 4.2.

Meanwhile, there have been substantial improvements in fuel-use technology, so that more work is done for each unit of energy delivered to the vehicle. Because fuel use per 100 km has changed little—it has not been required to change—the technological improvements have been directed instead toward increasing vehicles' weight and power. The bottom left panel of Figure 4.3 shows how vehicle weight increased by 17 percent (cars) and 26 percent (other light-duty vehicles) between 1986 and 2006. The bottom right panel shows how vehicle power increased by 78 and

BOX 4.2 Vehicles are lasting longer[39]

The upper panel on the right shows how personal vehicles in the US, particularly cars, have been lasting longer. This appears to have happened because they have been made to be more durable, rather than for other reasons such as changes in driving habits, maintenance arrangements or climate.

A necessary consequence of the growth in longevity is that new vehicles form an increasingly smaller share of vehicles on the road, shown in the lower panel.

Another consequence is that new technology—which comes chiefly with new vehicles—takes longer to penetrate the total vehicle fleet. It's harder to achieve dramatic changes in vehicle characteristics than it was three decades ago.

A remedy could be incentives for retiring older vehicles such as the SCRAP-IT program in place in the Lower Mainland of British Columbia. Vehicles produced before 1993 that fail the mandatory air quality test may be traded for one of a variety of incentives including rebates on the purchase of new vehicles and public transport passes. A transient example of such a program was the US Car Allowance Rebate System (Cash for Clunkers program), in effect for 32 days in July and August 2009. The initial authorization of $1 billion was exhausted during the first seven days, and the program was terminated when an authorization of a further $2 billion was exhausted. Cash allowances averaging near $4,000 per vehicle were provided toward replacing almost 700,000 "clunkers" with new vehicles.

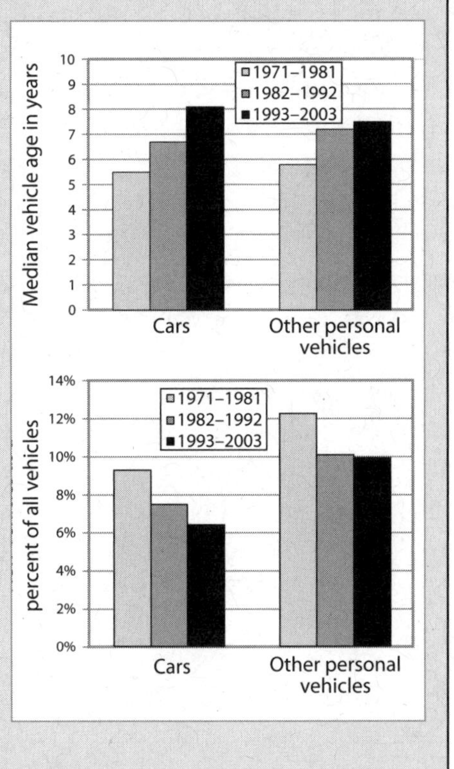

94 percent. Vehicle power is reflected above all in acceleration, but also in hill-climbing and towing performance.

SUVs in particular, but also passenger vans and pick-ups, are often maligned as the "gas-guzzlers" responsible for the high level of use of transport fuel by individuals in North America. Their growth has certainly contributed, but not as much as the increase in weight and power of *all* personal vehicles. Indeed, if power had not increased over the past 20 years, personal vehicles would use 14 percent less fuel. If vehicle weight had not increased, they would use 9 percent less fuel. If the share of SUVs etc., had not increased, personal vehicles would use only 5 percent less fuel. To put this another way: if power and weight stayed as in 1986, and had there been no further shift to SUVs, technological advances in engines and fuels would have resulted in a reduction of fuel use of 20–25 percent.[40]

A manifestation of the trend to use technology improvements to support increased power rather than reduce fuel consumption has been the use of hybrid technology to enhance acceleration. In such vehicles, the electric motor provides added power during acceleration rather than substitute for a smaller ICE. An example of this is the Honda Accord, which accelerates more rapidly than its non-hybrid equivalent, but achieves only slightly lower fuel consumption.[41] In North America, hybrid systems are being used increasingly in SUVs rather than regular cars.

An extraordinary feature of Figure 4.3 is what it reveals about how rapidly transport fuel consumption can change when the circumstances are right. The top left panel shows that in the US between 1975 and 1985, new personal vehicles' rated fuel use declined from an overall average of 18 L/100km to 11 L/100km. This was a decline of 39 percent in 10 years, or 4.8 percent a year. The rapidity of this decline is as extraordinary as the wartime decline in car use described in Chapter 1. Transport efficiency and transport activity can both change quickly when circumstances demand such adjustment.

If oil use for transport were to fall again at this rate—4.8 percent a year—but for 15 years (e.g., from 2010 to 2025), the total decline would be more than 50 percent and our challenges concerning both oil depletion and climate change would be considerably resolved. As in the 1970s and 1980s, some of the decline could be achieved through improved technology and some would be achieved through traveling and moving goods differently. Today, however, substitution of ICE-propelled transport by electric traction would be a major feature. Our proposals for such revolutions are detailed in Chapter 5.

CAFE standards for regular automobiles have been unchanged since 1990; those for light trucks have risen slightly, from 20.0 to 23.5 mpg for the 2010 model year (11.8 to 10.0 L/100km). Legislation passed in 2007 requires that the combined standard for both types of vehicle be 35.0 mpg (6.7 L/100km) by 2020.[42] In May 2009, President Obama announced plans for a further tightening that would require slightly better than attainment of the 2020 combined target by 2016.[43]

Had this book been written 20 years ago, the global impact given the most attention would have been ozone depletion.

The May 2009 announcement was remarkable not only for the stringency it proposed but also because it was made with the apparent support of auto manufacturers and unions.[44] (At the time of the announcement, the federal government mostly owned the largest US manufacturer; the largest union mostly owned the third largest manufacturer.) It also had the apparent support of the State of California, whose special position in these matters is discussed below.

The announcement was also remarkable because for the first time it proposed regulation of vehicles' greenhouse gas emissions as well as their fuel economy. Soon after, the agencies responsible for issuing these regulations announced they would coordinate their respective rulemaking activities.[45] Each manufacturers' personal vehicles—i.e., regular automobiles and light trucks, including SUVs, etc.—sold in the US for the 2016 model year would have to have average rated fuel consumption of 35.5 mpg (6.6 L/100km) or better, and average rated carbon dioxide emissions of 250 grams/mile (155 g/km) or better. Less stringent standards would apply to the 2012–2015 model years.[46]

The US has also been a leader in matters concerning vehicles' local environmental impacts, as we discuss below. The pace has been set by the State of California, which legislated new-vehicle emission standards before the federal government and has been allowed to maintain more stringent standards and greater autonomy in setting them than the rest of the US. In 2002, the state legislature enacted standards for emissions of GHGs for the first time. They were scheduled to come into effect for the 2009 model year and require that by 2016 most new personal vehicles emit no more than 127 grams of GHGs per kilometer, equivalent to consumption of about 5.4 liters of gasoline per 100 km.[47] However, until June 2009 the US Environmental Protection Administration used its authority to prevent California—and 13 other states following California's lead—from enforcing regulations concerning GHG emissions. California and the other states can now regulate GHGs. California appears to have agreed that automakers showing compliance with the federal fuel economy and greenhouse gas emission standards announced in May 2009 shall be deemed to be in compliance with state requirements.[48]

Several Asian jurisdictions have rules that limit the fuel consumption or GHG emissions of new vehicles, including China, Japan, South Korea and Taiwan. Japan's is of special note. Her requirements concerning GHG emissions are already the most stringent in the world,[49] and may soon be superseded by new fuel consumption requirements. Current regulations are complex but in essence are equivalent to requiring that 2009 model-year personal vehicles use on average no more than 4.9 liters of gasoline per 100 km. The regulations under discussion would require that this average be no more than 4.1 L/100km for the 2015 model year.[50]

Other Global Effects of Transport

Had this book been written 20 years ago, the global impact given the most attention would have been ozone depletion, caused chiefly by the release of chlorofluorocarbons (CFCs) and similar compounds into the atmosphere.[51] These chemicals rise to the stratosphere where they dissociate into elements that combine with and deplete ozone. Depletion of this stratospheric ozone allows passage of more medium-wavelength ultraviolet light, which except in small amounts is harmful to life.

The 1987 Montreal Protocol and subsequent amendments restrict production and deployment of *ozone-depleting substances* (*ODSs*),[52] which were used chiefly as refrigerants and solvents. Emissions of ODSs have declined. Ozone depletion appears to have stabilized and may be beginning to reverse.[53] The relatively successful Montreal Protocol served as a model for the less successful Kyoto Protocol. The success of the Montreal Protocol depended on the availability and rapid introduction of non-ozone-depleting alternatives. The Kyoto Protocol's relative failure reflects the lack of readily available alternatives to fossil fuels, particularly oil.

The main transport use of ODSs was in vehicle air conditioning systems. Use of CFCs as a refrigerant in these systems was common until 1993, with consequent release of CFCs to the atmosphere during filling, use and disposal. Now, non-ODS refrigerants are used, chiefly a hydrofluorocarbon known as HFC-134a. This compound is a potent GHG, and efforts are under way to provide alternatives.[54] Meanwhile, there has been a large increase in the share of new vehicles fitted with air conditioners. Before the 1990s, air conditioning was usual only in North America. Now it is common in Europe and elsewhere. In Germany, for example, only 19 percent of new cars had air conditioners in 1994. In 2002, 87 percent of new cars had this feature.[55]

Because of strict rules about the management of refrigerants, the likely major global impact of vehicle air conditioning arises from the increase it causes in vehicle energy use and consequent GHG emissions. When used fully during normal driving, air conditioning adds about 20 percent to the fuel use of US vehicles. Overall it raises the fuel consumption of vehicles in the US by about 6 percent, and the fuel consumption of vehicles in Europe and Japan by about 3 percent.[56]

Persistent organic pollutants (POPs) are "chemical substances that persist in the environment, bioaccumulate through the food web, and pose a risk of causing adverse effects to human health and the environment."[57] POPs are of global significance because they can be transported over long distances. The best-known POPs are the pesticide known commonly as DDT and the classes of compounds known as PCBs (polychlorinated biphenyls), dioxins and furans. Many POPs have been found in the blood of polar bears, thousands of kilometers from where the chemicals were produced or used.[58]

Dioxins, which can be a by-product of the combustion of petroleum fuels, appear to be the main emissions of POPs from transport. They have been little studied, and available estimates of transport's contribution to all emissions of dioxins vary widely. An Australian report concluded that transport accounted for 0.03–16.2 percent of dioxins emissions in that country.[59] The report noted that emissions inventories elsewhere have estimated contributions from transport of between 0.2 and 12 percent of all dioxins emissions.

These potential global environmental impacts of transport are almost all related to the use of ICEs and the burning of fossil fuels in them. A different kind of global impact is the accidental or deliberate discharge of solid and liquid materials by ships at sea. This can be human waste from cruise ships, general waste including packaging, and spills—notably of oil. Where the discharged materials do not degrade well in water, there can be long-term adverse effects that can extend over long distances because of the nature of ocean currents.[60]

Yet another global impact of transport is *ocean acidification*, sometimes known as "the other carbon dioxide problem."[61] Here are some of the conclusions from a comprehensive review of the topic:

- The surface ocean currently absorbs approximately one third of the excess CO_2 injected into the atmosphere from human fossil fuel use and deforestation, which leads to a reduction in pH [i.e., acidification] and wholesale shifts in seawater carbonate chemistry.
- Acidification will directly impact a wide range of marine organisms that build shells from calcium carbonate. Evidence of the impacts is largely from laboratory and other experimental studies; the response of individual organisms, populations, and communities to more realistic gradual changes is largely unknown.
- The potential for marine organisms to adapt to increasing CO_2 and the broader implications for ocean ecosystems are not well known; an emerging body of evidence suggests that the impact of rising CO_2 on marine biota will be more varied than previously thought, with both ecological winners and losers.
- Acidification impacts processes so fundamental to the overall structure and function of marine ecosystems that any significant changes could have far-reaching consequences for the oceans of the future and the millions of people that depend on its food and other resources for their livelihoods.[62]

Conclusion Concerning Global Environmental Impacts

Current concern about climate change has pushed transport's global environmental impacts to the forefront of agendas for policy change. In every country, transport is a major—often *the* major—source of the GHGs that are believed to be raising the temperature of our planet's surface. Moreover, transport's share of all GHG emissions has been growing in most countries, and in many cases growth from transport has prevented the meeting of international commitments to reduce GHG emissions.

For most transport, potential climate change impacts are strongly correlated with energy use. The salient exceptions are aviation, for which the impacts could be more than would be expected from energy use, and electric vehicles, for which the impacts can be unrelated to energy use. Much policy making recognizes strong links between climate change and energy use. We would urge that concerns about energy availability be given priority. This should be done because oil depletion could be a more urgent issue, and also because effective prevention and mitigation of climate change could require appropriate allocations of what could well be increasingly scarce oil. We return to this matter in Chapters 5 and 6. We also urge that other actual or potential global impacts of transport not be overlooked. The impacts we have touched on—ozone depletion, pollution by persistent organic compounds and ocean acidification—have been overshadowed by consideration of climate change. They deserve more attention, as may the more local effects of transport, to which we now turn.

> Much policy making recognizes the strong links between climate change and energy use.

Local and Regional Environmental Impacts

Concern about air quality continues to be prominent in many parts of the world, even as concern about climate change has grown.[63] The two are potentially related. Air quality could be affected by temperature change, precipitation levels, cloud cover and other manifestations of climate change. The strongest relationship between air quality and climate change could be in their common cause: both are attributed to the burning of fossil fuels, including oil for transport. The link in the case of air quality is well established. Poor air quality arises chiefly from the burning of coal for industry and electricity generation and from the use of oil products in ICEs. As we discuss further below, the greatest concern is possibly in Asia, as reflected in the following:

> The rapid economic development in Asian countries has been associated with growing urbanization, motorization, industrialization and an increased use of energy. Together, these processes have resulted in increased pressure on urban environmental systems, including urban air quality. Urban air pollution is a serious threat

> to the health and well-being of people in the region. The World Health Organization estimates that urban air pollution causes over half a million premature deaths yearly in Asia and that it affects the lives of millions of people negatively. A recent ADB and CAI-Asia study estimates the economic costs of urban air pollution to range from 2 to 4 percent of GDP.[64]

In Asia, industrial and other sources of poor air quality have predominated, although they are being replaced by transport. Elsewhere, transport—specifically emissions from internal combustion engines—is usually the main cause of air pollution, particularly in urban areas.

A possible paradox in considering transport as a primary cause of air pollution lies in the claims by the automotive industry and others that today's vehicles produce 99 percent less emissions than three decades ago.[65] One sympathetic observer even suggested that today's vehicles can be so good they can clean the air rather than pollute it.[66] This statement is an evident exaggeration.[67] Moreover, we shall show below that the statement about a 99 percent reduction is not justified.

This section provides information about transport-related air pollution, chiefly in cities, and about manufacturers' efforts to reduce it. Other kinds of transport-related pollution are also touched on, including pollution of land and water, and also noise pollution. By "local effects" we mean effects that are pretty much confined to the vicinity of transport activity or the infrastructure that supports it. Noise is generally considered to have only local effects, as are emissions of carbon monoxide, at least on human health, because it disperses rapidly from its source and is harmful only above certain concentrations. Other pollutants are considered to have regional effects. For example, through wind action sulfur oxides can acidify water bodies hundreds of kilometers from the emission source.

Transport—specifically emissions from internal combustion engines—is usually the main cause of air pollution, particularly in urban areas.

There is no simple dividing line between the global effects discussed in the previous section and the local and regional effects discussed here. The global effects are not consistent over the entire planet. For example, the extent of climate change may differ between the Northern and Southern Hemispheres because some emissions that could have an effect on climate do not travel well.[68] Similarly, the local and regional effects are not entirely local or even regional. For example, ozone precursors originating in US cities and blown across the Atlantic have been said to raise concentrations of ground-level ozone (smog) in the UK by 20–30 percent.[69]

Transport Emissions in the US

Air pollution had been a prominent feature of many cities long before transport became mechanized, chiefly on account of the use of fossil fuels for heating and industry. As motorization expanded, first in the US, and most of all in California, so did concern about this new source of pollution. The prevailing winds and topography of southern California result in accumulations of emissions. In response to concerns in the 1940s and 1950s, the Motor Vehicle Pollution Control Board was established in 1960 and issued the world's first motor vehicle emissions standards in 1966.[70] Passage of the federal Clean Air Act in 1970 led to national vehicle emissions standards. California, as the pioneer and because of the extreme nature of its problem at that time, was alone allowed to set more stringent standards. It still has what may the most stringent standards in the world.

As a consequence of this history—and a truly laudable dedication to collecting and publishing information—the US has data on vehicle emissions over the longest period. Key features of these data are in Table 4.3, which shows the average performance of vehicles on the road between 1970 and 2002. Data are shown separately for light- and heavy-duty vehicles and for five pollutants. *Carbon monoxide (CO)* results from incomplete combustion of carbon fuels. CO interferes with the uptake of oxygen in the blood, and in high concentrations is lethal. *Nitrogen oxides* (NOx) are formed whenever combustion occurs in the presence of air.[71] NOx are harmful to organisms—they contribute to "acid rain"—and are an ingredient in the formation of ozone, discussed below. *Volatile Organic Compounds* (VOCs) mostly arise from evaporation of fuel and are another ingredient in ozone production. Many VOCs are harmful, notably benzene, which is carcinogenic. *Particulate matter* (PM10) is a product of combustion, especially but not only from diesel engines. PM10 refers to particulate matter with a diameter of less than 10 micrometers (millionths of a meter, microns or μm), small enough to penetrate into the lungs. PM10 has been implicated in respiratory and heart disease.[72] *Sulfur dioxide* (SO_2) results from the oxidation of sulfur in fuel during

TABLE 4.3 Actual emissions from on road vehicles in the US, 1970 and 2002[73]

Type of vehicle and pollutant	Emissions in grams/kilometer from all vehicles on the road			Change in total emissions
	1970	2002	Change	
Light-duty vehicles				
Carbon monoxide	76.4	12.5	−84%	−59%
Nitrogen oxides	5.45	0.76	−86%	−65%
Volatile organic compounds	7.98	0.88	−89%	−72%
Particulate matter (PM10)	0.17	0.02	−90%	−75%
Sulfur dioxide	0.08	0.03	−59%	+3%
Heavy-duty vehicles				
Carbon monoxide	197.8	9.4	−95%	−84%
Nitrogen oxides	23.16	9.98	−57%	+49%
Volatile organic compounds	19.38	1.07	−94%	−81%
Particulate matter (PM10)	1.43	0.32	−78%	−23%
Sulfur dioxide	1.09	0.31	−72%	−3%

combustion. It causes respiratory disease, damages the fabric of buildings and contributes to acid rain. Sulfur levels in fuel have been greatly reduced by refiners because sulfur compounds impede the operation of pollution control devices.

Table 4.3 shows that emissions per kilometer from light-duty road vehicles have fallen substantially, mostly by 85–90 percent (although in no case by the 99 percent claimed by the industry). These vehicles include cars and motorcycles and also "light trucks": SUVs, vans and pick-ups. The total distance moved by light-duty vehicles increased by 153 percent between 1970 and 2002, from 1.68 to 4.30 trillion kilometers. Except for SO_2, this was not enough to offset the reduction in emissions per kilometer and thus reduce total emissions, shown in the right-hand column of Table 4.3. For example, emissions of NOx per kilometer from light-duty vehicles fell by 86 percent between 1970 and 2002 from 5.45 to 0.76 grams. Even with the 153 percent increase in distance driven, total annual emissions of NOx from these vehicles fell by 65 percent (from 9.2 to 3.3 million tonnes).

Table 4.3 also shows corresponding data for heavy-duty vehicles, chiefly trucks with six or more wheels using gasoline or diesel fuel. The pattern of changes in emissions is different from that for light-duty vehicles, chiefly because of the growth in the use of diesel fuel by heavy-duty vehicles, but perhaps also because heavy-duty vehicles have been less closely regulated. Distances moved increased even more for heavy-duty vehicles, by 245 percent, from 100 to 345 billion kilometers. In the case of emissions of NOx, this large increase in distance moved more than offset the reduction in NOx emitted per kilometer, resulting in an overall 49 percent *increase* in NOx emissions from heavy-duty vehicles. For the other four pollutants, the growth in distance moved did not offset the reduction per kilometer.

The fall in total NOx emissions from light-duty road vehicles was enough to offset the increase from heavy-duty road vehicles. This is illustrated in Table 4.4, which shows emissions from road vehicles and other sources for the five pollutants, and specifically that total NOx emission from road vehicles declined by 42 percent between 1970 and 2002. The main feature of Table 4.4 is the comparison of changes in emissions from road vehicles with those from other transport and non-transport sources. For each pollutant except SO_2, road vehicles showed the greatest improvement (and for SO_2—and PM10—road vehicles produced only a small part of total emissions of this pollutant). Of note are the increases in emissions, sometimes large, from other transport, including aviation, rail, shipping and the most important of these sources: off-road land transport including much agricultural activity. The hefty increases in emissions from other transport—which are mostly unregulated or more lightly regulated—have not offset the declines in emissions from road transport.

Recent regulations in the US for emissions from *new* light-duty vehicles are shown in Table 4.5, based on what is set out by the US Environmental Protection Agency (US EPA) and the California Air Resources Board. These are simplified versions; the actual standards and their application are so complex as to almost defy comprehension. For California, the LEV (low emission vehicle) is the basic standard. SULEV (super low emission vehicle) is the strictest standard apart from ZEV (zero emission vehicle), which applies to electric vehicles. Several hybrid ICE-electric vehicles meet the SULEV standard, as do some ICE vehicles.[75]

The actual emissions from light-duty vehicles in 2002 in Table 4.3 differ greatly from the standards for these vehicles in Table 4.5. There are several reasons. The most important is that the estimates in Table 4.3 represent *all* vehicles on the road up until 2002, but the standards in Table 4.5 are for *new* vehicles only. In an era of progressively tightened standards,

TABLE 4.4 Actual US emissions and sources, 1970 and 2002[74]

	(millions of tonnes)		
Pollutant and source	**1970**	**2002**	**Change**
Carbon monoxide (CO)			
Road vehicles	148.1	56.4	–62%
Other transport	10.3	22.2	115%
All other sources	26.7	23.1	–14%
All sources	185.1	101.6	–45%
Nitrogen oxides (NOx)			
Road vehicles	11.5	6.7	–42%
Other transport	2.4	3.7	55%
All other sources	10.5	8.8	–17%
All sources	24.4	19.1	–22%
Volatile organic compounds (VOCs)			
Road vehicles	15.3	4.1	–73%
Other transport	1.5	2.4	66%
All other sources	14.6	8.4	–42%
All sources	31.4	15.0	–52%
Particulate matter (PM10)			
Road vehicles	0.4	0.2	–58%
Other transport	0.1	0.3	90%
All other sources	11.2	19.6	75%
All sources	11.8	20.1	70%
Sulfur dioxide (SO_2)			
Road vehicles	0.2	0.2	1%
Other transport	0.3	0.4	51%
All other sources	27.8	13.3	–52%
All sources	28.3	13.9	–51%

as has been the case for four decades, new vehicles have the lowest emissions. Another factor is that emissions can increase as a vehicle ages. Thus, even if standards were not being tightened, new vehicles would have lower emissions.[77] A third factor—possibly the most important—is that Table 4.3 represents "real-world" emissions and Table 4.5 represents standards to be achieved under test conditions.

According to an authoritative source, "real-world emissions from cars exceed tailpipe standards by a large margin."[78] Tests on 1993 model cars in the US showed that CO and VOCs emissions during real-world driving were *four to five times higher* than the prescribed upper limits set by US EPA's standards. NOx emissions were twice those of standards.[79] A

TABLE 4.5 US and California emissions standards for new light-duty vehicles in 2007[76]

Pollutant	US	California LEV	California SULEV
Carbon monoxide	2.11	2.11	0.62
Nitrogen oxides	0.09	0.03	0.01
Volatile organic compounds	0.06	0.05	0.01
Particulate matter (PM10)	0.012	0.006	0.006

Note: Emission limits are shown in grams/kilometer (published as grams/mile)

later assessment suggested that the ratio of actual to required emissions could be increasing and could increase more at least until 2010. The NOx standard is set to be 97 percent lower in 2010 than in 1970, but the actual, real-world emissions of the cars to which the standards applied would fall by "only" 85 percent.[80]

Conventional wisdom—and common-sense logic—suggests that actual emissions converge toward applicable standards, and not increasingly diverge from them. For example, an authoritative review of vehicle technologies concluded,

> The discrepancy between emissions standards and actual vehicle emissions should diminish with time due to improved onboard diagnostic and catalytic control systems, improved durability of catalysts, longer standards lifetime (120 000 miles in the US), test cycles more representative of real world driving, and cleaner fuels (less likely to foul emission control systems).[81]

In the absence of good data, it's difficult to reconcile this position with the observation that the discrepancy may be increasing. Such data may become increasingly available with the growing deployment of portable emissions measurement systems of the kind the US EPA began using in 2003.[82]

> The simple truth about the US at least is that emissions from light-duty vehicles are down dramatically, notwithstanding the growth in the number of vehicles on the road.

Even though statements that emissions from cars are 99 percent lower than 30 years ago may be exaggerations, the simple truth about the US at least is that emissions from light-duty vehicles are down dramatically, notwithstanding the growth in the number of vehicles on the road. The main reason for the improvement has likely been the introduction and continued tightening of new-vehicle emissions regulations.

Regulations for heavy-duty vehicles in the US—most with diesel engines—have not been so strict. Although emissions per kilometer have declined, this has not always been enough to offset the growth in activity, as shown in Table 4.3. Part of the problem has been the high sulfur content of diesel fuel, which has prevented use of advanced emission control devices. The sulfur issue changed dramatically in 2006 when the allowable amount of sulfur in diesel fuel was lowered from 500 to 15 milligrams per kilogram (mg/kg). This was accompanied by a stringent tightening of emissions standards for heavy-duty vehicles—including the few with gasoline engines—set out in Table 4.6.

It's too early to determine whether a dramatic reduction in emissions, and consequent improvement in air quality, will result from deployment of heavy-duty vehicles that meet the new standards.

Air Quality in the US

Emissions from road vehicles, other transport sources, and factories and other sources all serve to raise the concentrations of pollutants in the air.

TABLE 4.6 Current and previous US emissions standards for new heavy duty vehicles[83]

Engine type and pollutant	Previous standard	2007 or 2008	Change
Gasoline (2008)			
Carbon monoxide	49.8	19.3	-61%
Nitrogen oxides	5.36	0.27	-95%
Volatile organic compounds	2.55	0.19	-93%
Particulate matter (PM10)		0.013	
Diesel (2007)			
Carbon monoxide	20.8	20.8	0%
Nitrogen oxides	5.36	0.27	-95%
Volatile organic compounds	1.74	0.19	-89%
Particulate matter (PM10)	0.134	0.013	-90%

Note: Emission limits are shown in grams/kilowatt-hour (published as grams per brake horsepower-hour)

Pollutants are substances that are unwanted because they can be hazardous or otherwise noxious. Air is of good quality to the extent it is free from pollutants.

How air quality has been changing in the US since 1980 is shown in Table 4.7,[85] which illustrates changes in five of the principal air pollutants, also known in the US as *criteria pollutants.*[86] These do not correspond precisely to the emissions from vehicles and other sources shown in Table 4.3 and Table 4.4. Here are the differences:

- The emissions tables (Table 4.3 and Table 4.4) show nitrogen *oxides* (NOx), while the air quality table (Table 4.7) shows nitrogen *dioxide,* to which just about all nitrogen oxides emitted from sources such as vehicles have been converted by the time they reach monitoring stations.
- Volatile organic compounds (VOCs) are shown in the emissions tables but not in the air quality table. In the US, VOCs are not among the criteria pollutants, but are included among pollutants known as *air toxics,* for which concentrations of individual compounds (e.g. benzene) are reported, but not total VOCs.[87]
- The most important difference concerns ozone, which is not shown in the emissions tables but is shown in the air quality table. This is because ozone is not an emission but is formed by the action of sunlight

TABLE 4.7 Atmospheric concentration of five principal pollutants, US, 1980 and 2005[84]

Pollutant	Number of sites	Measure	US national standard in 2005	Average concentrations			How much below standard in 2005
				1980	2005	Change	
Carbon monoxide	152	Annual 2nd maximum 8-hour average	10,000	9800	2500	–74%	–75%
Ozone	286	Annual 4th maximum 8-hour average	160	200	160	–20%	0%
Nitrogen dioxide	88	Annual arithmetic average	100	51	32	–37%	–68%
Particulate matter (PM10)	435	Seasonally-weighted annual average	50	32	24	–25%	–52%
Sulfur dioxide	163	Annual arithmetic average	80	31	12	–63%	–86%

Note: Atmospheric concentrations are in micrograms per cubic meter on a mixture of NOx and VOCs. Ozone is the principal ingredient of what is often known as *smog*, specifically *photochemical smog* or *summer smog*.[88] It is a highly reactive form of oxygen that damages living matter, and is implicated in crop damage and in respiratory illness in humans.

Shown too in Table 4.7 is the number of monitoring sites from which data for each pollutant were collected consistently across the period 1980–2005. In general these are located in population centers and thus represent human exposure to the pollutants. Also shown is the measure that underlies each standard. There are usually several standards for each pollutant, varying mostly according to the averaging period. Standards are often expressed in parts per million and have been converted to micrograms per cubic meter ($\mu g/m^3$) to provide for consistency and comparability among standards.

Table 4.7 shows that air quality improved considerably across the period 1980–2005. This should not be surprising in view of the declines in emissions noted in Table 4.3 and Table 4.4.

Ozone showed the smallest decline in the concentration of an air pollutant, and in 2005 ozone was the only pollutant the concentration of which was not within its air quality standard. The lower decline in emissions of VOCs than NOx reported in Table 4.4 could have limited the decline in ozone. Overall the concentration of VOCs may be the stronger limiting factor in the formation of photochemical smog, although in particular locations at particular times NOx concentrations may be the limiting factor.[89]

Although the US was a pioneer in the development and use of air quality standards, its current standards tend to be less stringent than those elsewhere.

Although the US was a pioneer in the development and use of air quality standards, its current standards tend to be less stringent than those elsewhere. This is illustrated in Table 4.8, which shows air quality standards (known as limits in the EU) for several jurisdictions and also the guidelines produced by the World Health Organization (WHO). To the extent that that the standards, limits and guidelines can be compared among these jurisdictions, the US would appear to have the most relaxed standard in the case of four of the five pollutants, that is, for all but CO. The standards in Table 4.8 were chosen for their comparability. The US

TABLE 4.8 Current air quality standards, limits and guidelines[90]

Pollutant	Standard	US	EU	Japan	China	WHO
Carbon monoxide	8-hour average	10,000	10,000	22,900	10,000	10,000
Ozone	1-hour average	240	80	120	160	
Nitrogen dioxide	Annual average	100	40		40	40
Particulate matter (PM10)	24-hour average	150	50	100	150	50
Sulfur dioxide	24-hour average	365	125	104	150	20

Note: Atmospheric concentrations are in micrograms per cubic meter

standards shown, except for nitrogen dioxide, are for assessments across different time periods from those represented in Table 4.7. Standards may say little about air quality. A critical factor is the extent to which standards are met, which may depend in turn on the extent to which they are enforced. Data on conformity are few and data on intensities of enforcement are almost non-existent.

Transport Emissions and Air Quality in Europe and Elsewhere

Table 4.9 sets out emissions from road transport and other sources for the 25 countries that were members of the EU before 2007. They are shown in the manner of Table 4.4 for the US except that the data in Table 4.9 are for 1990 to 2003 rather than 1970 to 2002 (the years in each case for which data are available).

Comparison of Table 4.4 and Table 4.9 shows that emissions in the US and EU25 have had similar trends. In both places, emissions from road vehicles have declined the most (except PM10 in EU25), while emissions from other transport have declined the least. One difference concerns transport emissions of SO_2, for which the decline has been large in EU25 and not in the US. Low-sulfur transport fuels were mandated earlier in EU25. Consequently, sulfur emissions from road transport fell earlier, particularly between 1996 and 1997. Another difference is illustrated in Table 4.10, which sets out emissions per capita from road transport for the US and for EU25. Emissions per capita are much lower in Europe, notably for SO_2 for the reason just given. The main reason for the difference is that Europeans use only about a third as much transport fuel per capita, as shown in the bottom row of Table 4.10. Emissions of CO and VOCs are similarly lower. Emissions of NOx and PM10 are higher in

TABLE 4.9 EU25 emissions and sources, 1990 and 2003[91]

	(millions of tonnes)		
Pollutant and source	**1990**	**2003**	**Change**
Carbon monoxide (CO)			
Road vehicles	37.5	16.1	–57%
Other transport	2.2	2.1	–3%
All other sources	22.1	13.6	–38%
All sources	61.9	31.9	–48%
Nitrogen oxides (NOx)			
Road vehicles	7.2	4.5	–38%
Other transport	1.7	1.5	–12%
All other sources	7.2	4.9	–31%
All sources	16.1	10.9	–32%
Volatile organic compounds (VOCs)			
Road vehicles	7.1	3.1	–57%
Other transport	0.7	0.6	–19%
All other sources	8.7	6.0	–31%
All sources	16.5	9.6	–42%
Particulate matter (PM10)			
Road vehicles	0.4	0.3	–11%
Other transport	0.2	0.2	8%
All other sources	1.8	1.5	–18%
All sources	2.4	2.0	–15%
Sulfur dioxide (SO_2)			
Road vehicles	0.8	0.1	–85%
Other transport	0.5	0.3	–46%
All other sources	238.2	78.9	–67%
All sources	239.5	79.3	–67%

TABLE 4.10 Road transport emissions and energy use per capita, US and EU25, 2002[92]

	Kilograms per capita		
	US	**EU25**	**EU25/US**
Carbon monoxide	128.9	42.4	33%
Nitrogen oxides	24.7	11.7	48%
Volatile organic compounds (VOCs)	24.3	8.1	33%
Particulate matter (PM10)	1.3	0.9	68%
Sulfur oxides	2.6	0.3	12%
Road transport energy use (GJ/person)	79.2	26.0	33%

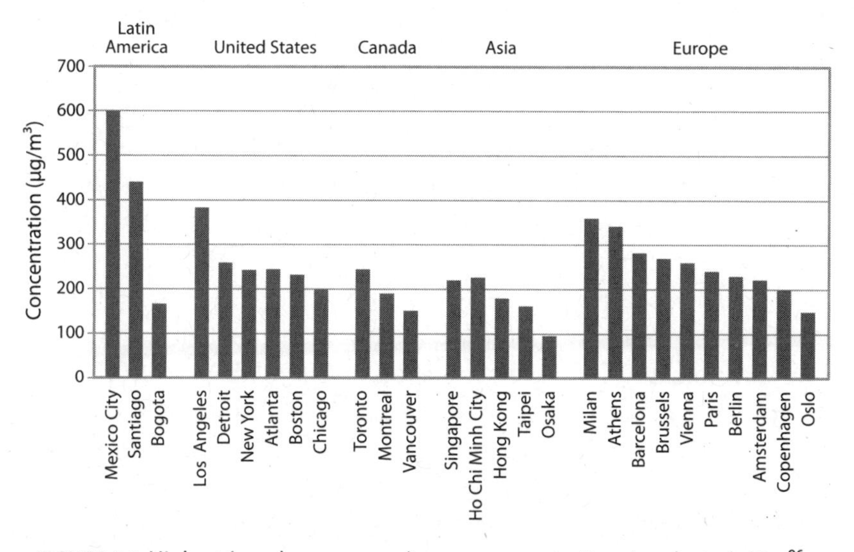

FIGURE 4.4 Highest (one-hour average) ozone concentrations in selected cities[96]

EU25 than might be expected from a straightforward comparison on fuel use because relatively more diesel fuel is used.

Urban air quality in Europe is not correspondingly low in comparison with the US. This is illustrated in Figure 4.4, which shows the concentrations of ozone in the air of what were described as "the most polluted cities."[93] In terms of the metric used (the recent, highest one-hour average concentration of ozone), European cities are as high or higher than US cities. This likely reflects their higher development densities. In Europe, a given amount of pollution may be emitted over a smaller area with the result that local atmospheric concentrations become higher. Moreover, air quality in Europe does not appear to have been improving to the extent shown in Table 4.7 for the US, although data for many fewer years are available for Europe and thus identification of a trend is more difficult.[94] A particular challenge in Europe is similar to one in the US: emissions of ozone precursors (NOx and VOCs) have fallen but ozone levels have not.[95]

A possible factor in Europe's relatively high air pollution could be a larger contribution in Europe from industry and other factors. Comparison of Table 4.9 with Table 4.4 suggests that this may not be so. For the three represented emissions that are more associated with transport activity—CO, NOx and VOCs—road transport's shares of total emissions were similar. If industry were a bigger factor in Europe, the shares of these emissions might well have been much lower.

The source for Figure 4.4—the WHO—also provides comparative data on atmospheric concentrations of PM10 for more cities than are represented in Figure 4.4, reflecting the greater availability of data on PM10.[97] Again, cities in Europe and the US have similar values. For PM10, other cities, particularly in Asia, have much higher values. Transport is usually less of a contributor to PM10 levels than to ozone levels. High levels of PM10 in Asia and other poorer countries result chiefly from the use of coal and biomass for domestic and industrial purposes.

However, in poorer countries, including China, transport is the fastest growing source of air pollution, especially in urban areas. In the Pearl River Delta region, which may have the world's most intensive concentration of manufacturing plants, an analysis of contributors to air pollution in 2001 found that transport was already the main source of high concentrations of nitrogen oxides, carbon monoxide and ozone.[98]

Similar conclusions can be drawn about other poorer countries and cities. For example, an assessment of air quality in Delhi, India, in 1995 showed that road transport was responsible for more than 80 percent of each of atmospheric concentrations of CO, NOx and VOCs, and also 39 percent of SO_2 and 16 percent of particulate matter. In some places, road transport's share of particulate matter in the air is higher. In Dhaka, Bangladesh, in 2001–2002, for example, road transport was responsible for more than 40 percent of particulate matter, more in the suburbs than in the inner city, and more of the finest particles (PM2.5) than of larger particles.[99]

A survey of air quality in Asian cities concluded the following:

> For many Asian cities, the main source of air pollution is vehicle emissions. In some cities, the sources may be diverse and include industrial emissions and area sources. The priority, however, in most cities is vehicle emissions. As Asian cities have developed and attention has been given to control emission sources, considerable success has been achieved with the control of SO_2 and coarse particulates (particle sizes larger than 10 microns), so the characteristics of emissions in some Asian cities are changing. Fuel use in vehicles is a major source of particles, carbon monoxide and NOx.

> With frequent traffic congestion, and large fleets of poorly maintained vehicles consuming fuel of poor quality, air quality in some cities in Asia is largely determined by vehicle emissions.[100]

Non-Road Modes and Air Quality

Transport-related emissions and air quality have been discussed so far almost entirely in relation to land transport. This is appropriate because, for air quality in particular, almost all of the concern is to do with road traffic. However, rail, water and air transport can contribute to poor air quality, particularly at stations and ports. A review of transport-related air pollution noted that few data are available.[101] The same source noted that at railway stations, sea ports and airports, (diesel) locomotives, ships and aircraft can contribute measurable amounts of air pollution, but the main source of each pollutant has been found to be the road traffic attracted to these places and otherwise in the vicinity of them.

Impacts of Transport-Related Poor Air Quality

The above discussion suggests that, in the US at least—and with the exception of ozone—there has been considerable improvement in air quality that can be reasonably related to reductions in emissions from transport. A recent WHO review concluded,

> ...transport-related air pollution affects a number of health outcomes, including mortality, nonallergic respiratory morbidity, allergic illness and symptoms (such as asthma), cardiovascular morbidity, cancer, pregnancy, birth outcomes and male fertility. Transport-related air pollution increases the risk of death, particularly from cardiopulmonary causes, and of non-allergic respiratory symptoms and disease.[102]

Thus, a reduction in adverse transport-related effects might have been expected. However, when a recent review posed the question, "How have health risks declined with declining levels of air pollution?" the answer was less than straightforward:

> Ambient air pollution levels have been generally declining over the last 20 years. Although this reduction in exposure to air pollution

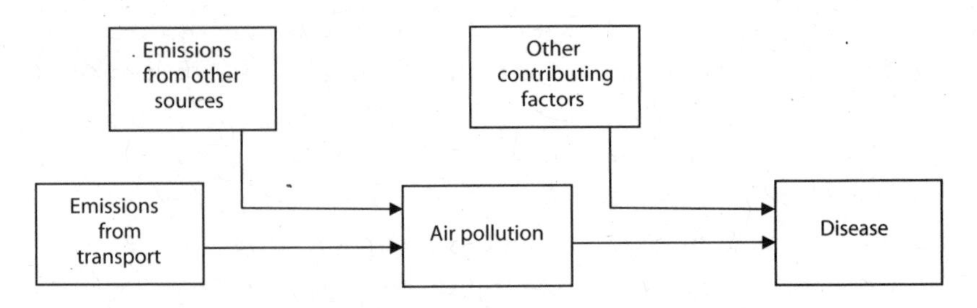

FIGURE 4.5 Schematic links between transport emissions and disease[104]

> may be expected to lead to reduced health risk, it is difficult to directly demonstrate concomitant reductions in mortality and morbidity in the general population because of the modest reductions in ambient pollution levels and the relatively small attributable risk associated with air pollution.[103]

The links between emissions from transport and air pollution are not straightforward, chiefly because there are other emissions that contribute to air pollution. Similarly, the links between air pollution and disease are complex, because there can be other contributing factors. These two kinds of uncertainty are illustrated in Figure 4.5. Thus, the lack of a decline in disease and death attributable to transport need not be surprising.

Some of the most compelling evidence that road transport emissions affect health outcomes comes from assessment of the health status of people who live near major roads. For example, a Swiss study of several thousand adults found that proximity to a major road is associated with "asthmatic and bronchitic symptoms, in particular with attacks of breathlessness, wheezing with breathing problems, wheezing without a cold, and regular phlegm."[105] The closer people lived to major roads, the more evident the symptoms. This study was unusual in that the same subjects were examined 11 years later, in 2002. Similar although mostly weaker effects were found. The changes were possibly the result of a decline in emissions per vehicle-kilometer, partly offset by increases in the amounts of traffic.

Studies of the concentrations of vehicle emissions with distance from the vehicle exhaust pipe show that elevated levels of pollution are detectable up to 1,500 meters from a major road (downwind from an expressway),[106] but that most of the change with distance occurs during the first

40 meters.[107] An Australian study looked at pollution concentrations resulting from passing trucks in the "breathing zones" of adult pedestrians on roadside paths. When the vehicles were moving at less than 45 km/h, pollution levels were on average *six* times higher when exhaust pipes were located on the curb side of the vehicle than when they were located on the other side, and 12 times higher when on the curb side than when located vertically.[108] At such close proximity to vehicles, even a meter or two appears to make a considerable difference in pedestrian exposure to pollution.

Curious as to manufacturers' preferences in exhaust pipe location, one of the authors conducted an informal survey of all vehicles parked one Sunday morning on the streets of a mixed residential-commercial area close to downtown Toronto. Of the 280 vehicles examined, one was a heavy duty truck with a vertical exhaust pipe, eight were medium-duty trucks all with curbside exhaust pipes, and 271 were light-duty vehicles—regular cars, light trucks, vans or SUVs—191 of which had their exhaust pipe on the curb side.[109] This sample suggests that more than two thirds of the vehicles on the road in Toronto may have their exhaust pipes located on the side that produces the greater exposure of pedestrians to their pollution.

A shift toward curbside location of exhaust pipes, if this has happened, could well be a factor in the worldwide increase in the prevalence of asthma and other respiratory conditions.[110] At the moment, manufacturers do not appear sensitive to the potential consequences of their practices. A magazine advertisement showed a hybrid ICE-electric car passing close to a woman pushing a child in a buggy. It was making the point that the woman and child's interests were served by the vehicle's low emissions. However, the sleeping child's nose was a few meters from the vehicle's curbside exhaust pipe and the child could well have been inhaling a harmful dose of pollutants.

Transport is the major but a mostly declining source of several atmospheric pollutants in richer countries and may be the fastest growing source of air pollution in poorer countries.

Concluding Remarks Concerning Transport's Impacts on Air Quality

Transport is the major but a mostly declining source of several atmospheric pollutants in richer countries and may be the fastest growing source of air pollution in poorer countries. Transport's pollution is chiefly the result of the burning of fossil fuels in ICEs. Replacement of ICEs with electric motors (EMs) could substantially improve air quality for two reasons. One is that there is no pollution *at the vehicle* when traction is provided by EMs. Thus, traffic need not be a source of pollution in urban areas even if coal is used to generate electricity for the EMs. The other reason is that EMs are more compatible with use of renewable, non-carbon sources of energy for transport—including solar, wind, tide and geothermal energy—the use of which would produce less pollution overall.

The main transport revolution proposed in this book—from use of ICEs to use of EMs for most land transport—would be accompanied by a reduction in transport-related pollution and resulting improvements in air quality and human health.

Other Impacts, Including Those of Noise

Transport noise, particularly road transport noise, is usually the major source of acoustic nuisance in urban areas. It affects people's well-being at lower levels. Higher levels of noise are detrimental to health. They contribute to sleep loss and disturbed sleep, and to high blood pressure and cardiovascular diseases. According to the WHO, day-time noise levels should be kept below 55 decibels (dB)—a measure of noise intensity—and night-time levels below 45 dB.[111]

Figure 4.6 shows exposure to noise from three transport modes in the EU in 1994, when the EU population was 370 million. It shows that about 15 percent of the population was exposed to what the WHO regards as dangerous levels of noise from transport (more than 65 dB). Road transport was the source of about 80 percent of this dangerous noise.

Noise exposure appears to be lower in North America—available data are not recent or good—but it is likely that road traffic is usually the major contributor to urban noise and is a significant cause of disturbance and ill-health.[113] The lower exposure may well be a positive effect of North America's less dense development patterns. Nevertheless, it has been claimed that in Canada "More people are affected by noise exposure than any other environmental stressor" and "...because its associated health effects are not as life-threatening as those for air, water and hazardous waste, noise has been on the bottom of most environmental priority lists."[114] Moreover, as large a share of the population seems to be annoyed by transport noise in Canada as in Europe.[115]

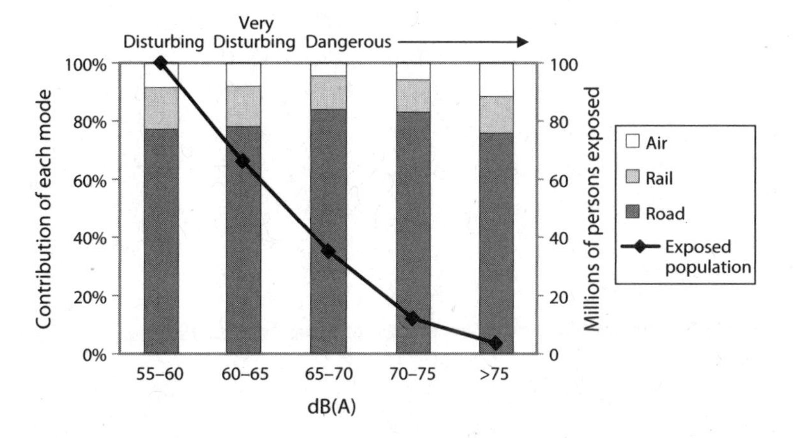

FIGURE 4.6 Population of EU15 exposed to excessive transport noise and contribution of modes[112]

Noise standards for new cars, commercial road vehicles and aircraft have been progressively tightened over the last 25 years.[116] For road traffic, this has resulted in reductions in engine noise, but tire noise in particular and traffic noise generally have remained largely unchanged or have even increased. Motor vehicles are permitted to produce much higher levels of noise than are consistent with health and comfort. For example, North American trucks are allowed to generate up to 80 dB, motorcycles up to 70 dB or more, railcars up to 93 dB and aircraft up to 95 dB or more.[117]

At low speeds, electric vehicles are generally much quieter overall than vehicles with ICEs. At higher speeds, vehicle noise comes mainly from wind and wheels and is less affected by the type of traction. The reduced noise that would come with a wholesale switch to electric vehicles would mostly be a positive feature. However, there have been complaints already that hybrid electric vehicles can be a safety hazard, particularly for blind pedestrians, on account of their quietness when moving on battery power only.[118]

Transport is mostly a contaminant of air, but it can also pollute water and land, directly and indirectly. Such pollution includes oil spills from damaged ships to leaks of fuel from underground storage containers. Most of it arises from error or lack of consideration rather than from the normal functioning of a transport system. Some pollution is clearly related to regular transport activity, for example acidification of lakes due to acidic transport-related pollution—nitric acid or sulfuric acid—in the air, and the accumulations of sodium chloride and other salts used to de-ice roads. Some has already been touched on as part of the discussion of global environmental impacts, specifically the unintentional or deliberate discharge of solid and liquid materials by ships at sea.

Figure 4.7 shows the ways in which road transport can pollute nearby waterways. The main inputs mostly comprise exhaust emissions, fuel and lubricant losses, and spillages, all more likely with ICE vehicles than electric vehicles.

Run-off from highways and other impermeable surfaces is the main way in which watercourses are polluted by transport activity. Such pollution can be avoided by adequate drainage and treatment of collected water, which is required in the EU and other jurisdictions.[120] The main methods

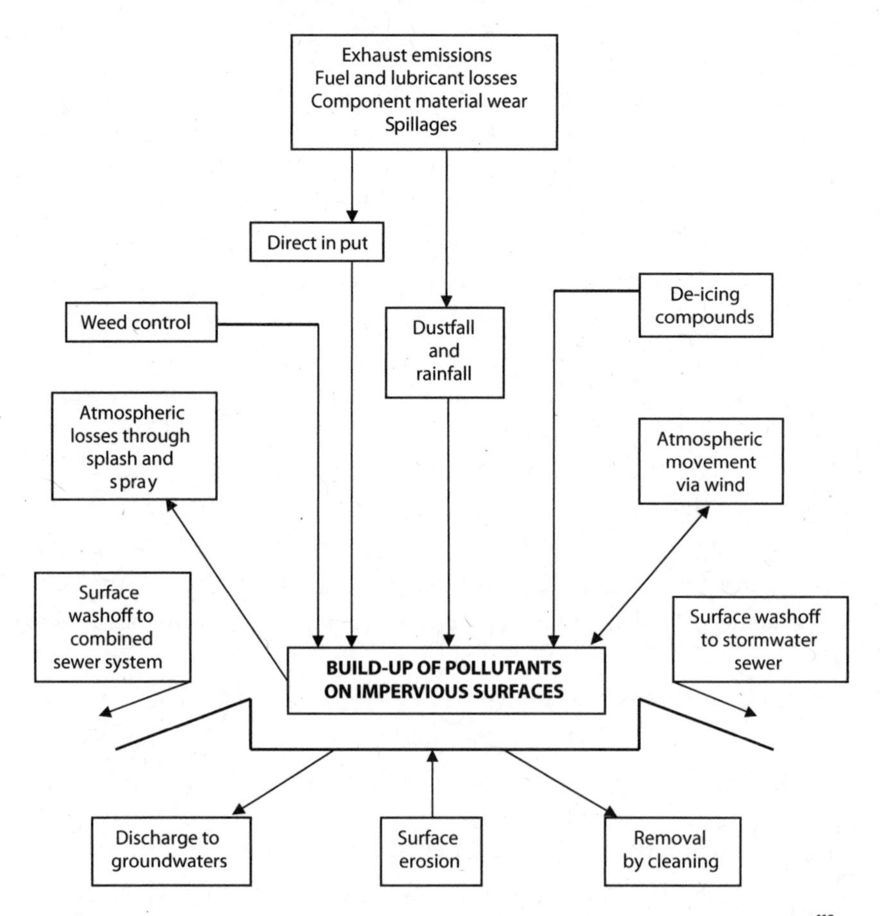

FIGURE 4.7 Pathways of transport-derived pollutants to the aquatic environment[119]

of treatment involve filtering or infiltrating the run-off through soil before it is discharged into waterways.[121] In doing so, land is polluted but usually with less consequence than pollution of groundwater or waterways. The biological impact of highway run-off seems usually slight, but this is not always the case.[122]

Sea ports and airports have particular environmental impacts. Sea ports produce disturbances of the aquatic zone that can result in major ecosystem disturbances.[123] Activity associated with airports generates local noise and some ground and water pollution impacts from de-icing and fuel leaks. Overall, airports appear to be relatively benign. Their land take is small in comparison with other modes—per million kilometers of air travel—although some airports are huge and have environmental impacts largely on account of their size.[124]

To this point, our consideration of transport's adverse impacts has mostly favored electric vehicles, chiefly on account of their lower emissions at the vehicle and overall. In the remainder of this section, we shall touch on impacts for which the means of traction does not pose a special benefit. However, the nature of the transport system could still be associated with differences in impact. For example, much less pollution of land and water by transport could be expected in an urban region such as Tokyo, where most motorized trips are made by rail, than a region such as Toronto where most motorized trips are made by road. This would likely apply even if diesel engines moved trains in Tokyo rather than the EMs that are actually used.

Transport is resource intensive even apart from the fuel that is used. The material flows into the production and use of vehicles and their infrastructure are considerable. They can be identified through a process known as Material Flow Analysis (MFA). The basic idea of MFA is that the most general indicator of a system's impact on the environment is the *quantity* of materials of all kinds—including fuels—moved in connection with the system during its whole life. The basic measure used in MFA is the material input per unit of service (MIPS).[125] Examples of MIPS relevant to transport would be kilograms of material moved per person- or tonne-kilometer.

MFA and its metric, MIPS, are associated with the notion of "ecological rucksack," which characterizes the ecological impact of a product or process in terms of what is left over after the product has been produced

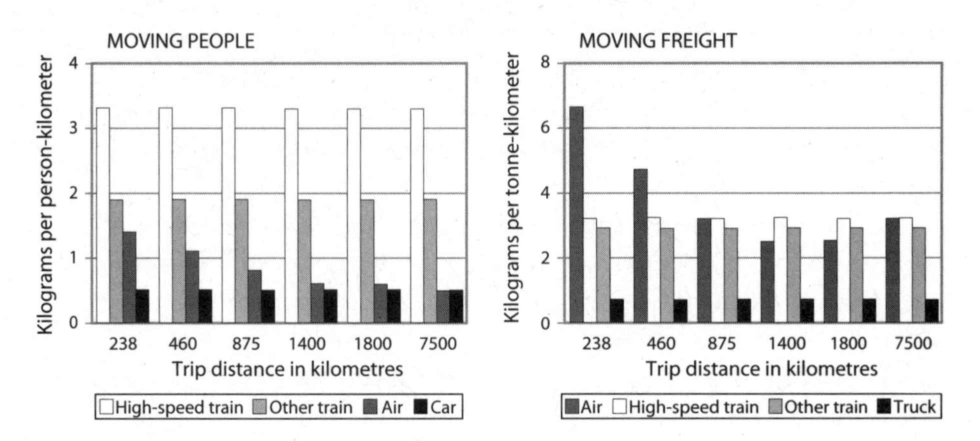

FIGURE 4.8 Life-time materials inputs for transport modes[127]

or the process is complete. The rucksack contains the leftovers, and is fuller or not according to the extent of the environmental impact. The ecological rucksack complements the "ecological footprint," which points to the area of land required to support a particular way of living.[126]

Material intensity can be an issue not only because of its indication of ecological impact but also because materials may not be renewable or may require large energy inputs to be renewable. Consumption of a scarce metal could mean that future generations will have less use of it. Their options could be thereby restricted, and a basic principle of sustainability—intergenerational equity—would be violated. MIPS as usually used does not take a material's scarcity or renewability into account, only the consumption of it.

Estimating MIPS for different transport modes can produce surprising results. The results of one such analysis are in Figure 4.8. It concerned moving people (left panel) and moving freight (right panel) mostly along Italy's Milan–Naples axis. In each case, the road mode (car or truck) was the *least* intensive in total materials use per unit of transport activity. The authors noted that "the material intensity for [movement] by highway is always lower than the other modalities because of the huge road traffic intensity that drastically reduces the material cost of infrastructure per transported unit."

Construction of aircraft is highly material intensive, but aviation infrastructure—chiefly airports—requires fewer materials than other modes. The variation of aviation's MIPS with distance in Figure 4.8 arises mostly from differences in fuel use rather than airport use. Aircraft fuel consumption is high for short trips because of the large amount of energy required to achieve take off and ascent. It then falls with distance but can begin to rise again for the longest trips because of the weight of fuel that must be carried.[128] A striking feature of Figure 4.8 is the lower material intensity of aviation than rail modes in every case for moving people and in some cases for moving freight.

Rail appears to perform less well than other modes in an MFA because of the material costs of maintaining a straight route and an even grade. This can be more true of high-speed rail than other rail. Of the approximately 800 km of tracks between Milan and Naples, 196 km of the high-speed route and 141 km of the other rail route are tunneled, with each kilometer of tunnel requiring over 12,000 tonnes of steel. As well, high-speed rail requires more energy per seat-kilometer—to overcome the higher wind resistance that comes with higher speeds—and generally has lower occupancy. Both rail modes have high material intensity at the vehicle. For example, a 10-coach high-speed train carrying 550 passengers weighs about 550 tonnes, that is, one tonne per passenger. Cars traveling between Milan and Naples weigh an average of about 1.3 tonnes and carry an average of 1.8 people, that is, about 0.7 tonne per passenger.

An MFA depends critically on its own inputs. For example, if aircraft occupancy were 80 percent, rather than the 50 percent assumed in the analysis represented in Figure 4.8, aviation would appear to be even less materially intensive. If cars had one occupant, rather than the average of almost two, their MIPs value would be considerably higher, although it could still be lower than for high-speed and other trains. Another factor is terrain. The Milan–Naples axis used in the above analysis is unusually hilly and thus favors road and air modes over rail.[129]

MFA and its metric, MIPS, are valuable tools for assessing overall environmental impacts, as is the notion of ecological footprint. Our main concern here, however, is energy consumption. The unexpected results portrayed in Figure 4.8 should not detract from this book's main objective, which is to anticipate looming energy challenges by dramatically changing transport systems. What counts above all is the non-renewable energy used by each system. Overall material use should also be reduced, to the extent that doing so does not add to consumption of non-renewable energy.

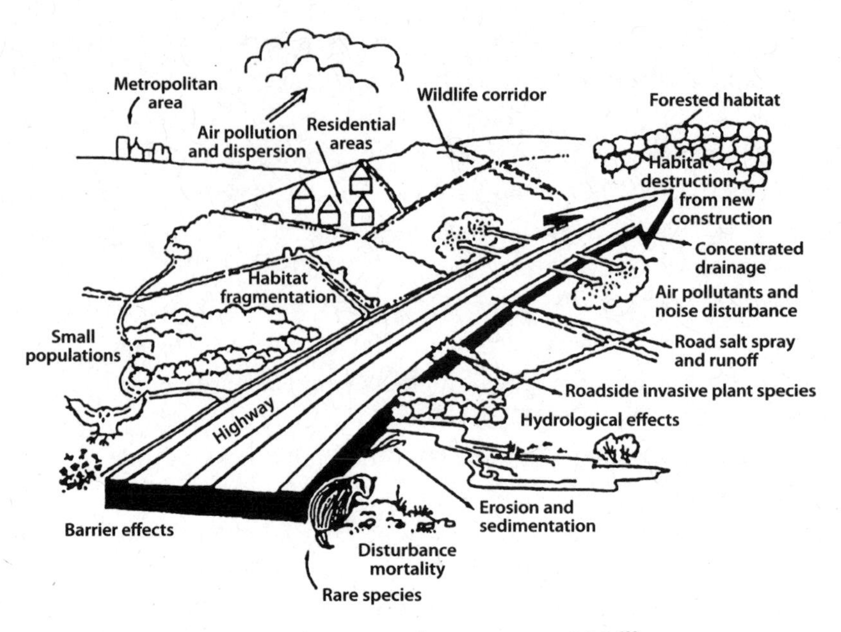

FIGURE 4.9 Impacts of transport on wildlife[132]

The last adverse impact of transport considered in this section arises from transport routes and their relationships to pathways used by other species for migration, foraging and other purposes. Roads and railways can block migration paths and fragment habitats.[130] Their design can be such as to reduce these impacts, for example, by constructing tunnels under highways for use by wildlife.[131] Figure 4.9 illustrates these impacts in the context of a wide range of transport's impacts on other species.

Adverse Societal and Economic Impacts

This section is about the adverse impacts of transport that do not readily qualify for inclusion in the previous two sections on global and local environmental impacts. It is shorter than those sections because there is much less to say about these other kinds of impact. The section's title—"Adverse Societal and Economic Impacts"—is imprecise, but may signal best what is covered. The common feature of the effects considered below is that they are not the result of direct or indirect contamination of the environment by transport activity. By "societal" we mean "relating to human society and its members." By "economic" we mean "relating to the financial standing of organizations and individuals."

TABLE 4.11 Deaths from transport crashes and collisions by mode, EU15, 2001/2002[135]

Mode	Deaths per billion person-kilometers	Deaths per 100 million person-travel-hours
Road (Total)	9.5	28
Motorcycle/moped	138.0	440
Foot	64.0	75
Cycle	54.0	25
Car	7.0	25
Bus and coach	0.7	2
Ferry	2.5	16
Air (civil aviation)	0.4	8
Rail	0.4	2

Transport-Related Fatalities and Injuries

Much of the concern about transport is related to its impacts on human health. Many of these impacts arise from environmental contamination, as we have noted above. The major transport-related impact on health, however, arises from what are usually known as "accidents."[133] Almost all transport-related fatalities and injuries from collisions and crashes happen on roads. For example, in the EU in 2001, road modes were responsible for 97 percent of such fatalities.[134] Actual rates for each mode by travel distance and time are set out in Table 4.11.

Fatalities and injuries from road collisions and crashes in higher-income countries have declined steeply. An example is in Figure 4.10, which provides Canadian data and shows per capita declines of almost 50 percent between 1985 and 2004. US and European data show similar steep declines.[136]

Figure 4.10 shows fatalities and serious injuries declining together, but this may obscure a different trend in outcome. Injuries may be increasing in severity, and increasingly include injuries that with less sophisticated treatment might have resulted in early death. One report suggested that rates of permanent impairment and disability from road traffic injuries could be increasing.[138]

Although road fatalities and injuries have been falling in richer countries, they appear to have been increasing worldwide, on account of considerable growth in lower-income countries. Table 4.12 shows the WHO's estimates of fatalities in 1990 and 2000, with projections for 2010 and

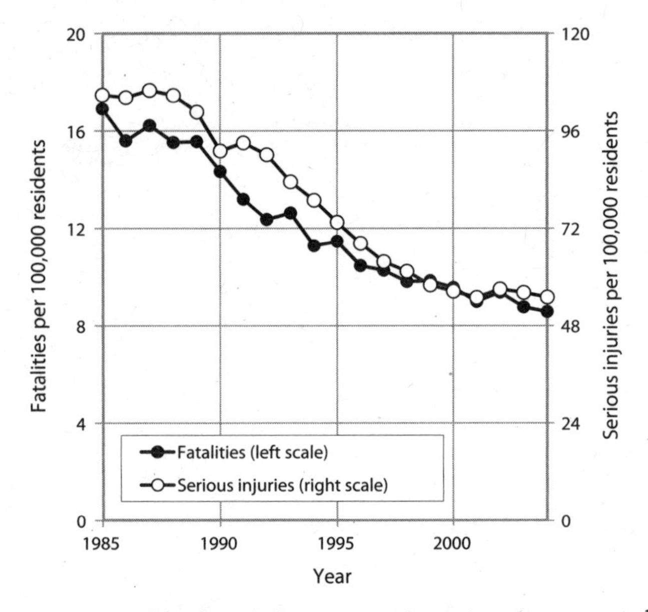

FIGURE 4.10 Road fatalities and serious accidents, Canada, 1985–2004[137]

2020. Nearly all road fatalities occur in lower-income countries (85 percent in 2000, expected to rise to 93 percent in 2020). During the 1990s, fatalities per capita in lower-income countries passed the rate in higher-income countries and the WHO expects them to be more than double the higher-income rate in 2020. A commentary on the WHO report noted, "Over the past 25 years, vehicle ownership in most developing countries grew more rapidly than fatalities per vehicle fell."[139]

There is much variation as to which road users are killed or injured in road traffic crashes. In higher-income countries, it is primarily people in cars. In the US, this category of road use accounts for 80 percent of fatalities; pedestrians and cyclists account for only 15 percent. Even in the Netherlands, where there is much more walking and cycling, and almost a third of fatalities are pedestrians and cyclists, most fatalities are of car occupants. In lower-income countries, by contrast, pedestrians and cyclists can be the most often killed. In Delhi, India, for example, they comprise over 50 percent of road fatalities. Often, riders of motorized two-wheeled vehicles account for the most fatalities. In Thailand, over 70 percent of fatalities are of this type.[141]

A theme of discussions of road crashes is the need for reliable data. Only a minority of countries report fatalities and injuries and then not

TABLE 4.12 Estimates and projections of road fatalities in lower- and higher-income countries[140]

	1990	2000	2010	2020	Change by decade To 2000	To 2010	To 2020
Fatalities (thousands)							
Lower-income	419	613	862	1124	46%	41%	30%
Higher-income	123	110	95	80	−11%	−14%	−16%
Total	542	723	957	1204	33%	32%	26%
Fatalities per 100,000							
Lower-income	9.6	12.1	14.9	17.3	26%	23%	16%
Higher-income	13.6	11.3	9.3	7.5	−17%	−17%	−19%
Total	10.3	12.0	14.1	15.9	16%	18%	13%

always in a usable manner.[142] In many cases, the data are not available or evidently incomplete. One study, for example, suggested that in Ghana only 8 percent of pedestrian injuries are reported to the police.[143] The estimates noted above, particularly for lower-income countries, were often derived through indirect means.

Even more questionable may be estimates of the costs of road traffic crashes. One admittedly crude estimate suggested that the world total of such costs in 1997 was $518 billion.[144] Lower-income countries incurred only about 12 percent of this total, even though, as indicated in Table 4.12, they may have had over 80 percent of fatalities, and perhaps a similar share of injuries. The discrepancy is likely the result of differences in the way lives are valued.

There is a particular concern about road crashes involving children and young people. A WHO report noted that road crashes are worldwide the leading cause of death among 15- to 19-year-olds, the second major cause of death among 10- to 14-year-olds and 20- to 24-year-olds, and the third major cause of death among 5- to 9-year-olds. For all ages, they are the 11th leading cause of death.[145]

Children and young people have lower overall road fatality rates than older people, as shown in Figure 4.11. Lower exposure to opportunities for road crashes—in terms of kilometers traveled or hours traveling (see Table 4.1)—may be a factor, but few data are available. The lower vulnerability of young people could be another factor.[146] Figure 4.11 also shows the large differences in fatality rates between males and females. These differences occur at all ages, but are more pronounced in adults.

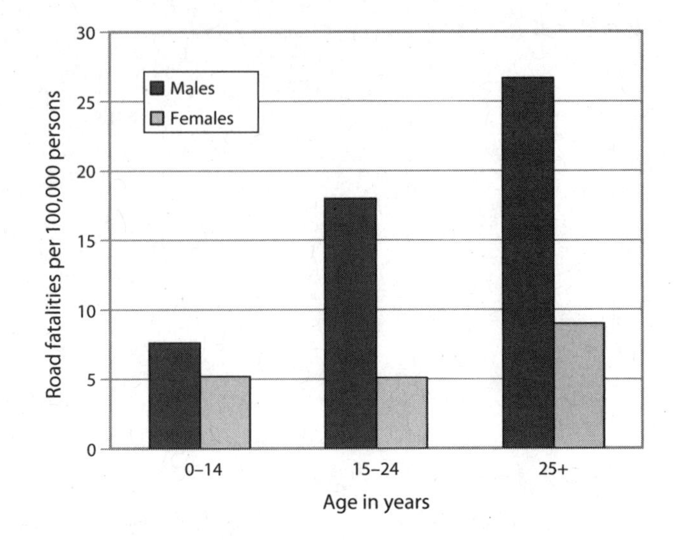

FIGURE 4.11 Road fatalities by gender and age, world, 2002[147]

Traffic-related fatalities and injuries, particularly deaths resulting from road crashes and collisions, have been presented in some detail because they are in many ways the most extreme manifestation of social and economic factors in transport. Also, compared with many other such factors, good data are available. An indication of the relative severity of road fatalities can be gained from comparison with rates for homicide, the most severe form of anti-social behavior. The best available data suggest that worldwide, at or near the beginning of this decade, there were about 19 road fatalities and 8.8 homicides per 100,000 persons.[148] To the extent that road fatalities are not accidents, but are thus intentional or at least have known causes, they may thus be ranked as among the most extraordinary of social phenomena.

Further details of road fatalities and homicide rates are in Table 4.13. As well as showing again—as in Table 4.12—that lower-income regions have higher road fatality rates, Table 4.13 shows the considerable variability among regions and even more among countries. In Western Europe, China and Japan, there are ten or more times as many road fatalities as homicides. In lower-income Americas and Russia, homicide rates exceed road fatality rates. The variation in road fatality rates—ranging from 2.6 to 24.2 per 100,000 among the countries featured in Table 4.13—is almost as extreme as the variation in homicide rates. This high variation sup-

TABLE 4.13 Comparisons of road fatalities and homicide rates[149]

	Mortality rate per 100,000 persons		
	Homicide	**Road crash**	**Ratio**
Selected higher income regions or parts of regions			
World	2.9	12.6	4.3
Americas	6.5	14.8	2.3
Europe	1.0	11.0	11.0
Selected lower-income regions or parts of regions			
World	10.1	20.2	2.0
Africa	22.2	28.3	1.3
Americas	27.5	16.2	0.6
Europe	14.8	17.4	1.2
South-East Asia	5.8	18.6	3.2
Selected countries			
France	0.7	12.1	17.3
Japan	0.6	7.4	12.3
China	1.8	19.0	10.6
Canada	1.4	9.3	6.6
Chile	3.0	10.7	3.6
US	6.9	15.0	2.2
Sweden	1.2	2.6	2.2
Mexico	15.9	11.8	0.7
Russia	21.6	9.7	0.4
Colombia	61.6	24.2	0.4

ports the argument that road fatalities are not accidents but arise from the particular circumstances of road traffic.

The point should be made again that road fatalities are merely the most readily definable part of a very much larger set of problems arising from road crashes and collisions. As was noted for Canada in Figure 4.10, for each fatality there can be many serious injuries (about six times more in the case of Canada). Many of these serious injuries result in long-lasting disability and illness that can be as challenging for families and society as fatalities.

If for no other reason, avoidance of road modes in the planning of transport facilities would appear to be a wise step toward reduction of the stresses associated with fatalities and injuries. Rate of road-related incidents can be kept relatively low, as in the case of Sweden in Table 4.13.

Sweden's road fatality rate is nevertheless several times higher than its rates for other transport modes.

Adverse Societal Impacts

Matters become much less clear when other social impacts of transport are considered. Work on environmentally sustainable transport by the Organization for Economic Cooperation and Development (OECD) attempted to address these challenging matters by asking panels of experts to comment on propositions as to the effects of continued growth in motorization. There was substantial agreement as to 5 of 18 offered propositions as to what further growth in car use would be associated with:

- increased land-use sprawl, that in turn will increase dependence on the car, and increase the disadvantage of those without cars;
- reduced street life (i.e., street-related social activity);
- a more anonymous society in which fewer people know their neighbors;
- greater disadvantages for children than other age groups, in part because their independent mobility will be constrained;
- deterioration of health due to lack of exercise.[150]

The impacts of sprawl—low-density development at the edge of urban areas—have become controversial. Sprawl is believed to be facilitated by car ownership and use and also to contribute to it, in a positive feedback loop that reinforces both low-density development and motorization. There is no doubt it is a phenomenon. Concern has mostly focused on North America and Australia, but a report on sprawl in Europe included the following, "Sprawl threatens the very culture of Europe, as it creates environmental, social and economic impacts for both the cities and countryside of Europe. Moreover, it seriously undermines efforts to meet the global challenge of climate change."[151]

In North America, the controversy is exemplified by two books. One, described as the "first book to make the connection between suburban sprawl and the violent breakdown of American society," argued that "suburbia empties our souls through the isolation and alienation it creates, picks our pockets through increased taxation and automobile dependency, and is bankrupting the nation."[152]

The other book on sprawl noted that "just fifty years ago sociologists were describing how the central cities caused alienation and how residents of suburban areas were such compulsive joiners and volunteers." It concluded with these words: "In its immense complexity and constant change, the city—whether dense and concentrated at the core, looser and more sprawling in suburbia, or in the vast tracts of exurban penumbra that extend dozens, even hundreds of miles into what appears to be rural land—is the grandest and most marvelous work of mankind."[153]

One of the few quantitative assessments of sprawl and social relationships suggested that the strength of the social bonds of adults in a neighborhood varies inversely with the extent of the residents' car dependence, but not with the extent of sprawl per se, that is, not according to how thinly the neighborhood was populated.[154] The study thus highlighted car ownership and use as the key factor. Sprawl was a factor only to the extent that it was associated with car ownership and use.

This finding is consistent with Swiss work on the residents of streets that are dominated by traffic and those that are not. The primary focus was children's play, but the study also found that parents on well-trafficked streets had fewer social contacts with other parents on the street and were less able to meet child-care needs. Five-year-old children who lived on such streets rarely played outside, while those who lived on "adequate" streets played outside for more than two hours a day. The authors concluded, "Unsuitable living surroundings, that is, mainly those living surroundings dominated by street traffic, prevent the development of lively neighborhoods capable of mutual help."[155]

Noise, also discussed above, could be one factor in the adverse impacts of traffic on children. A study of over 2,000 children aged 9–10 years in 89 schools in the Netherlands, Spain and the UK found that reading ability was impaired according to the extent of aircraft noise, whether the noise was at school or at home.[156] Road traffic noise did not affect reading. In explaining the difference, the authors noted that aircraft noise can be more intense and less predictable than road traffic noise. In the same project, there was an assessment of the effects of road traffic noise on sleep quality, which may be a significant factor in mood and performance and perhaps in social interactions. Sleep quality declined with increasing traffic noise in 160 Swedish children and their parents, although the children were less affected and were less likely to report being affected even when their sleep quality was poor.[157]

The above few paragraphs give only the briefest of hints as to a considerable literature on some of transport's adverse societal impacts. The literature is confusing and often contradictory, and does not provide a basis for strong conclusions. Notably absent is consideration of whether and at what point transport's adverse societal impacts might outweigh its evident benefits, including the benefits that arise from experiencing different places and culture and expanding the scope of personal contacts.

Adverse Economic Impacts

There appears to be more certainty that transport has net economic benefits, although a frequent assumption supporting this certainty—that transport facilitates economic development—has been questioned. The simple fact—noted in Chapter 2—is that increases in transport activity, freight transport in particular, are associated with increases in economic activity. The easy assumption from this association is that transport causes or in some way contributes to economic activity.[158] An alternative view is that transport activity is a *response* to economic activity rather than a cause of it; for example, travel increases because the economy is in good shape.[159] We believe both happen. Transport activity both causes and is caused by economic activity.

Transport activity both causes and is caused by economic activity.

Moreover, we would argue that transport activity is more of a cause of economic activity than the converse, not the least because of the chaos that can result when transport services are unavailable. Almost nothing causes governments to reach for their powers of persuasion and coercion more than disruption of transport services. In large cities, union or other activity that disrupts public transport service is rarely allowed to continue for more than 48 hours.[160]

Transport constitutes economic activity, as well as possibly stimulating it and being stimulated by it. In the US, transport's share of the Gross Domestic Product (GDP), a popular indicator of economic activity, was 10.3 percent in 2005, having declined from what may have been a peak of 11.6 percent in 1998 and 1999.[161] Transport's share of the EU's GDP in 2001 (EU15) was 6.4 percent, ranging from 3.3 percent for Greece to 9.1 percent for Belgium.[162] A decline in these shares could be achieved by reducing transport activity or by reducing the unit costs of transport activity. In either case, the result could be regarded as an adverse economic

TABLE 4.14 Local transport in Hong Kong and Toronto, 1995[163]

	Hong Kong	GTA
Population (millions)	6.31	4.63
GDP/person (US$)	22,968	19,456
Area (square kilometers)	1,096	7,075
Developed area (% of total area)	18	25
Density (persons/ha developed area)	320	26
Car ownership (cars/1000 persons)	47	464
Total trips (daily trips/person)	2.81	1.97
Motorized trips (daily trips/person)	1.85	1.73
Transit use (pkm/person)	3,675	1,050
Car use (pkm/person)	976	6,821
Car use (% of all motorized pkm)	25	86
Energy use for transport (GJ/person)	6.5	35.7
Annual cost of transport (US$/person)	964	2,490

Note: Abbreviations: GTA = Greater Toronto Area; ha = hectare; pkm = person-kilometer; GJ = gigajoule

impact, to the extent that whatever reduces GDP is so regarded. However, from other perspectives, a decline in transport's share of GDP could be regarded as positive.

Another way of looking at transport economics is through the prism of personal expenditures. In 1995, the only year for which there are readily comparable data, Hong Kong had a higher GDP per person than Toronto, and yet per capita Hong Kong residents spent only $964 on local travel while Toronto residents spent $2,490. These and other data are set out in Table 4.14, where it can be seen that Hong Kong residents also made more trips than Toronto residents, including more motorized trips. The largest differences of all concerned how the motorized trips were made. They were almost all by public transport in the case of Hong Kong residents and almost all by car in the case of Toronto residents.

Hong Kong residents, living in the richer world's densest development, enjoyed—and still enjoy—superlative public transport arrangements that mostly obviated car ownership. Toronto residents, living in what has been described as "Vienna, Austria, surrounded by Phoenix, Arizona,"[164] are highly car-dependent, despite having among the most comprehensive public transport systems in North America.

Toronto residents spent 2.6 times per capita what Hong Kong residents spent in order to meet their local transport needs. For this they had the comparative luxury, and frustrations, of traveling mostly by car. As well, by their transport spending, Toronto residents helped maintain employment in the automotive and other industries. Hong Kong residents had money to spend on other purchases, which may have supported other industries, or to save, which brings its own kind of economic advantages.

Thus, at the heart of considerations of transport's adverse impacts on the economy lies the question as to whether economic activity is good in itself, no matter what its products. Should more attention be given to the case that industrialized societies in particular have become locked into road modes of moving people and freight that are unduly expensive in terms of financial and energy resources? Or should spending money on transport be praised because it helps maintain employment?

We believe that an unavoidable feature of coming transport revolutions will be movement away from car travel and air travel. Cars as we know them could still exist decades from now, but there will likely be many fewer of them and much more travel by fairly familiar public transport modes and other collective means. Aircraft will still be used, but much more longer-distance travel will be by rail or water. New transport-related employment will be generated in richer countries, but it will not offset the employment that will be lost. In poorer countries there could also be net additions to employment, but not to the extent that would occur if present richer-country practices were to be adopted. Managing these economic and social transitions will be of vital importance.

Relative Impacts of Transport Modes and Means of Traction

First we want to note the interrelatedness of many of transport's impacts. This was well illustrated in a diagram in a UK Royal Commission report that showed "part of the web of connections between increased car ownership and use and environmental and social outcomes in urban areas." The diagram is reproduced as Figure 4.12. Similar webs could be drawn for transport modes other than the car, including freight transport modes, and for settings other than urban areas. In many cases, the webs would themselves be linked. For example, airports at the edge of urban areas can encourage access by car, whereas rail stations in the center of urban areas may be more compatible with access by public transport.

By way of summary of this chapter, Table 4.15 provides our conclusions as to the overall impacts of current transport modes, divided fur-

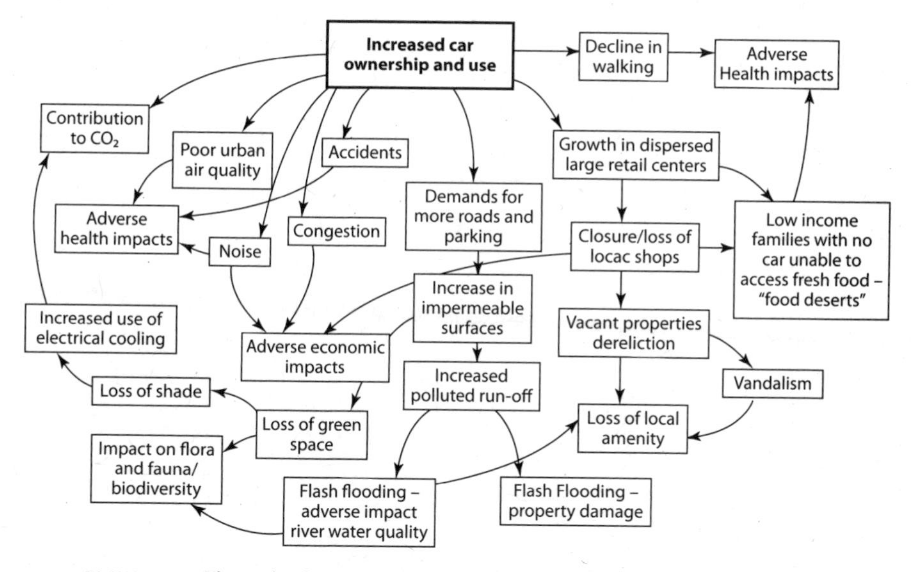

FIGURE 4.12 The web of connectedness of some of transport's impacts[165]

ther where appropriate by means of traction. These conclusions are based on the foregoing and are in the four right-most columns of Table 4.15. They should be regarded as impacts *per person-kilometer* (pkm) for moving people and *per tonne-kilometer* (tkm) for moving freight. Also provided are estimates of energy use *at the conveyance* in megajoules (MJ) per pkm or tkm. These estimates are presented as typical values, based more on North American than on other experience because of the ready availability of appropriate data. Elsewhere, because road vehicles are generally smaller, energy use per pkm by road could be lower and energy use per tkm by road could be higher.

As for all comparisons of energy use across modes, the values compared in Table 4.15 are sensitive to the stated assumptions about occupancies and loading. A particular point is the table's indication of higher energy use per kilometer by ICE-based public transport than by cars. This arises in part because of the relatively low occupancy of public transport vehicles in the US. In 2004, each bus—typically about 40 seats—carried an average of 8.8 passengers (22 percent occupancy). Each car—typically five seats—carried an average of 1.6 passengers (32 percent occupancy).[167]

The ratings of impacts in Table 4.15—high, medium and low—refer more to overall impacts worldwide than to the intensity of specific impacts. The selection of the five right-hand columns in Table 4.15 is

TABLE 4.15 Relative impacts of transport modes and means of traction[166]

Mode	Rate of fuel use (MJ/pkm, tkm)	Global impacts	Local and regional impacts	Material disturbance	Land take	Net socio-economic impacts
Moving people						
Personal vehicle (ICE)	2.6	High	High	Low	High	High
Local public transport (ICE)	2.8	Medium	Medium	Low	Medium	Low
Local public transport (electric)	0.6	Low	Low	Low	Medium	Low
Intercity bus (ICE)	0.7	Low	Medium	Low	High	Low
Intercity rail (ICE)	0.9	Low	Low	Medium	Medium	Low
Intercity rail (electric)	0.2	Low	Low	Medium	Medium	Low
High-speed rail (electric)	0.3	Low	Low	High	Medium	Low
Marine (dom. and int.)	1.0	Low	Medium	Low	Low	Low
Aircraft (domestic)	2.0	High	Low	Low	Low	Medium
Aircraft (international)	2.3	High	Low	Low	Low	Medium
Moving freight						
Truck (ICE)	2.6	High	High	Low	High	Medium
Rail (ICE)	0.2	Low	Low	Medium	Medium	Low
Rail (electric)	0.1	Low	Low	Medium	Medium	Low
Pipeline	0.5	Low	Low	Low	Low	Low
Marine (domestic)	0.5	Medium	Medium	Low	Low	Low
Marine (international)	0.2	Low	Low	Low	Low	Low
Air (domestic)	23.0	High	Low	Low	Low	Medium
Air (international)	9.9	High	Low	Low	Low	Medium

Note: ICE = internal combustion engine; pkm = person-kilometer; tkm = tonne-kilometer

somewhat arbitrary. Moreover, we are not sure if the final column, "Net socioeconomic impacts," has much meaning in view of the lack of or frailty of most available data.

Notwithstanding these major qualifications, and the questionable conclusions that intercity bus and pipeline have the lowest overall impacts, we believe Table 4.15 may be useful as an assessment of the present relative strengths of the modes. As we move to considering transport futures in the next chapter, we should stress again our belief that energy considerations will be paramount. Thus, the column "Rate of fuel use" is by far the most important. The other factors are for consideration when energy imperatives have been addressed.

5

The Next Transport Revolutions

Setting the Scene for 2025

Almost everything in this book so far has concerned the past or the present. In this chapter we venture forward, mindful of the advice that "prediction is very hard, especially about the future" (advice given—perhaps independently—by Neils Bohr, Danish atomic physicist, and then by Yogi Berra, US baseball player and team manager). In the spirit of this advice, what we present in this chapter about transport's future is more in the form of suggestions as to how transport *could* unfold than predictions as to how it *will* unfold.

In Chapter 1, we defined a transport revolution as a substantial change in a society's transport activity—moving people or freight, or both—that occurs in less than about 25 years. "Substantial change" means one or both of the following: an ongoing transport activity increases or decreases dramatically, say by 50 percent, or a new means of transport becomes prevalent to the extent that it becomes a part of the lives of 10 percent or more of the society's population.

We believe one of two types of transport revolution is likely to happen during the next few decades, brought on by the peaking of and subsequent decline in world oil supply described in Chapter 3. One revolution could be that transport activity will decline substantially, either because of continuing economic recession or because there will be little in the way of transport that is not dependent on oil, or for both reasons. The other type of revolution would involve maintaining and even raising overall levels of most transport activity but increasingly using means of transport that do not rely on oil.

Even before the dramatic events of 2008, noted in Chapter 3 and analyzed further in Chapter 6, there was a growing preoccupation with economic decline and societal collapse.[1] This is an outcome of limits to the supply of oil that we strive to transcend. We also want to avoid the challenges of trying to sustain oil-based transport systems in an era of what we described in Chapter 4 as *oil depletion*, i.e., falling oil production. This effort could well result in a vicious cycle of high oil prices, economic recession, lower oil prices, modest economic recovery, and high oil prices again. In Chapter 6 we will suggest that such a cycle is what began in late 2007 and could well continue beyond 2009.

We are much more interested in achieving the second type of revolution, whereby transport shifts to consuming less oil through use of transport technology that does not require oil for propulsion. We believe that such preparation for oil depletion is essential for enabling the continuation of humanity's gains in comfort, convenience, productivity and freedom from want. We cannot predict with any confidence whether this type of revolution will occur or how, if it occurs, it will unfold. We do feel we can help the prospects for this course of action by offering suggestions as to future outcomes that might be aspired to.

The purpose of this chapter is thus to show how a transition away from oil used for transport could be managed through anticipation. To do this we focus on two countries, the US and China. We chose these countries because they present the most challenging cases among what are now richer countries and poorer countries that are striving to become affluent. The US is the most challenging case among richer countries because it generates much more transport activity and uses much more oil for transport per capita than any other country, as was detailed in Chapters 2 and 3. China provides the most challenging case among poorer countries in pursuit of affluence because it is larger and more populous than any other and because its transport is expanding and motorizing more quickly than any other, as also detailed in Chapters 2 and 3. These two countries are first and second among consumers of oil in the world, and first and third among importers of oil.[2] (China has been behind Japan in net oil imports, but may have surpassed Japan in 2009.)

A reminder may be in order that China remains among the poorer countries in spite of her remarkable growth in industrial production and wealthy entrepreneurs. Although China's gross domestic product (GDP) is by some counts the second or third highest among the world's economies, after the European Union and the US, her GDP *per* capita ranks below halfway down the 220-plus list and is well under the world average.[3]

As well as being the most challenging cases—and the weightiest cases on account of their high levels of oil consumption—the US and China are important because they are the most influential countries in a wide range of socio-cultural and economic domains. In addition to their economic significance, they have importance as centers of the world's two most widely used languages, Mandarin and English.[4] The US in particular has a disproportionate cultural reach because of its dominance in film and other mass media. A US move toward moving people and freight without oil could have extraordinary influence in many ways.

The US and China are important because they are the most influential countries in a wide range of socio-cultural and economic domains.

We will go into some detail about *what* could happen in the transport revolutions to come in the US and in China, but we will not provide detailed prescriptions as to *how* to redesign every aspect of mobility during these transport revolutions. Such designs would easily fill another book, and several more will likely be written about the methods of changing mobility arrangements in response to oil depletion.

This chapter's prelude to transport system redesign is built around two pairs of scenarios for 2025: scenarios for the movement of people and freight for each of the US and China. Each scenario presents suggested targets for transport and energy use for each mode, set out in relation to our estimates of what was happening in 2007. We could have used 2008 or even 2009 as a baseline, but stayed with 2007—the baseline year in the first edition of the book—because of the considerable changes in 2008 and 2009 from previous years, which we detail further in Chapter 6.

We chose 2025 as the target year for several reasons:

- It is near enough to provide a meaningfully close target date that could motivate action—as opposed to simply planning for action—by incumbent government and corporate leaders. If we had chosen 2050, there could be a strong temptation to put off action until 2025 or later, or at least until the next generation of leaders will be in a position to deal with these challenges. For some people, 2025 may seem far in the future; but in relation to 2010, which may be the latest date to begin anticipatory redesign that can keep ahead of oil depletion, 2025 is only as far in the future as 1995 is in the past.

- This 15-year period is also a sufficient period within which to attain significant results from redesigning transport systems. Several of the dramatic changes in transport activity described in Chapter 1 occurred over shorter periods. Moreover, 2025 will be a decade or so after we expect the occurrence of the world peak in production of petroleum liquids to have become evident. (We suggested in Chapter 3 that the actual peak could occur by 2012, but its attainment will not be apparent for a few years.)
- If we had set the target year even five years later, in 2030, the required cuts in oil consumption would have been larger, and more likely to appear intimidating and engender defeatism.

The key factor in establishing our redesign scenarios is therefore the extent of the reductions in oil consumption that would be required by 2025. For this we look first to Figure 3.7 and the left-hand panel of Figure 3.8 in Chapter 3, which suggest that world oil production—and consumption—in 2025 will be in the order of 26.3 billion barrels (bb). This is only 15 percent below the likely production in 2007 and it is actually 7 percent *above* production in 1990. However, the most important difference is that anticipated oil production of 26.3 bb in 2025 would be 30 percent below the projected "business-as-usual" consumption of 37.6 bb in 2025. These production estimates and percentage changes are all set out in the bottom row of Table 5.1.

We should stress that world oil production and consumption in any year are essentially identical. What is produced is mostly used in the same year, and the amounts carried over from one year to another—in transit or in storage—are mostly similar and in any case small in relation to the totals. Thus, within limits the terms "world production" and "world consumption" may be used interchangeably. We sometimes use the term "potential consumption" to refer to what might be consumed if it were available at something like today's prices. The "business-as-usual" projection in Figure 3.8 represents potential consumption. We suggested in Chapter 3 that when production becomes constrained the difference be-

TABLE 5.1 Actual, likely and projected world oil consumption, and targets for 2025[5]

	Billions of barrels (bb)				Percentage by which 2025 target is higher (+) or lower (–) than:		
	Actual 1990	**Likely 2007**	**BAU projection 2025**	**Target 2025**	**Actual 1990**	**Likely 2007**	**BAU 2025**
Richer countries	18.1	18.6	18.3	10.7	–41%	–42%	–41%
Poorer countries	6.3	12.5	19.3	15.6	+146%	+25%	–19%
World	24.5	31.1	37.6	26.3	+7%	–15%	–30%

Note: "Oil" includes all petroleum liquids including natural gas liquids but not other liquid fuels such as ethanol, biodiesel and liquids from coal. "BAU" means business as usual. "Richer countries" comprise OECD member countries and countries in Eastern Europe and Eurasia. "Poorer countries" are the rest of the world. Variances are due to rounding

tween potential consumption and actual production determines the price of oil.

Table 5.1 also sets out how we think the shortfall between production in 2025 and potential consumption should be shared between richer and poorer countries, in this case counting among the richer countries OECD members and the "transition" countries (former USSR and Eastern Europe). We propose that about two thirds of the shortfall be borne by the richer countries, roughly corresponding to their share of total consumption in 1990.[6] The application of this proposal produces the results shown in the rows for richer and poorer countries in Table 5.1. To put the proposed changes in another way: If consumption in richer countries were to rise no more before 2010 and then decline by 42 percent to the 2025 target, this would require annual reductions in oil use across these 15 years by an average of about 3.6 percent. If consumption by poorer countries were to rise by 25 percent between 2007 and 2025, this would represent an average annual increase in consumption of 1.2 percent.

What this could mean in practice for richer countries is that their annual consumption could fall by less than 3.6 percent soon after 2010 and by more as 2025 approached. Poorer countries' consumption could rise more steeply than 1.2 percent a year at first—say, until about 2020—and then begin to fall. In both cases, the ease with which consumption could be reduced later in the period will depend in good measure on actions taken early in the period.

The cuts in oil used by transport should be proportionately larger than the overall reduction in oil use. This would allow for smaller cuts in, or even temporary maintenance of, consumption of currently essential uses of oil. These include the oil used as feedstock in chemical industries, particularly the production of fertilizers, pesticides and pharmaceuticals. In 2005 in the US, these and other non-fuel uses of oil comprised about 13.5 percent of total oil use.[7] Such uses should be protected or even allowed to grow in proportion to population.

On the other hand, there will also be some replacement of petroleum oil for transport by liquid biofuels and by liquid fuels derived from coal. As we discussed in Chapter 3, we do not expect more than a few percent of total transport fuels to be replaced by such liquids. Biofuels could be constrained by land availability and soil depletion. Coal to liquids could be constrained by concerns about atmospheric and other pollution caused by the conversion process. Production of both could be limited by their inherent energy requirements, which may well be unacceptable when global oil production is declining. We assume here that the yield of other liquid fuels will offset the greater cuts to be required of oil for transport. Thus, the net result would be a cut of about 30 percent in the use of oil and other liquid fuels.

Taking all these considerations into account, and allowing for a margin of error in the form of underestimation of the rate of production after the peak, we conclude the following. By 2025 richer countries should plan to reduce their use of liquid fuels for transport—almost all oil—by about 40 percent below 2007 levels. Poorer countries could plan to increase their use liquid fuels for transport by no more than 25 percent, i.e., more than they are using now but less than they are projected to use. Most of the rest of this chapter is taken up with suggesting the kinds of plans that could be put in place and how they might be implemented. We use the US and China as examples of how richer and poorer countries could redesign their transport systems to use much less oil than projected.

By 2025 richer countries should plan to reduce their use of liquid fuels for transport — almost all oil — by about 40 percent below 2007 levels.

The alternatives to developing such plans could well be high oil prices, much less availability of transport and all that goes with it, or economic collapse, or all of these in the kind of vicious cycle detailed in Chapter 6.

It's hard to say whether richer or poorer countries would be more affected by such outcomes. Richer countries are much more dependent on good transport, and have more to lose during economic hardship. But, they may in many ways be more resilient than poorer countries with large numbers of people living in poverty for whom economic collapse could mean devastating setbacks from already precarious positions. Even worse would be the prospect of military conflict in pursuit of increasingly scarce oil.

We are focusing on reducing oil use for transport as a means of preparing for oil depletion and high oil prices because we believe this is the most important and urgent reason for such reductions. However, the oil future we foresee may not happen. There may be no early peak in production or, if there is, the steep rise in oil prices experienced in the first half of 2008 may not recur. There would still be good reasons to reduce the oil consumed by transport. We discussed some of these reasons in Chapter 4, notably avoidance of further local and global pollution. Avoiding such pollution could provide sufficient justification for what we propose here.

Scenarios for 2025 for the US

Moving People in 2007 and 2025

What we propose for the motorized movement of people in the US in 2025 is in Table 5.2, which provides comparisons with the likely transport activity and energy use for each mode in 2007. A key underlying feature of Table 5.2—and also Table 5.3, which does the same thing for freight in the US—is that the US population is expected to grow by 16 percent across this period, from 306 million to 355 million.[8] Accordingly, in some cases we have shown both total and per capita values.

We suggested in connection with Table 5.1 that a decline in oil use for transport of about 40 percent by 2025 should be pursued by richer countries. We have set the proposed cut in US liquid fuel consumption for transport at a little higher than 40 percent (actually 44 percent, from 23.4 to 13.1 EJ) to respond to those within the US who might argue that there are more important uses of oil than for transport. The higher cut could also respond to those outside the US who might argue that the US, being by far the largest user of oil for transport, should attempt the largest cut in oil use for this purpose. Note that this is an absolute decline by 44 percent; the per capita decline is 51 percent.

TABLE 5.2 Motorized movement of US residents in 2007 (estimated) and 2025 (proposed)[9]

Values and totals in this table are rounded to aid comprehension	2007				2025						
Mode	pkm in billions (except per capita)	Fuel use per pkm, in MJ	Total liquid fuel use in EJ (GJ for per capita)	Total electricity use in EJ (GJ for per capita)	Local pkm in billions (except per capita)	Non-local pkm in billions (except per capita)	Fuel use per pkm in MJ	Total liquid fuel use in EJ (GJ for per capita)	Total electricity use in EJ (GJ for per capita)	Liquid fuel powered pkm	Electrically powered pkm
Personal vehicle (ICE)	7700	2.6	20.4		2500	2,500	2.1	10.5		5000	
Personal vehicle (electric)					1000		1.0		1.0		1000
Future transport					200		0.5		0.1		200
Local public transport (ICE)	50	2.8	0.1		100		2.0	0.2		100	
Local public transport (electric)	40	0.6		0.0	400		0.5		0.2		400
Bus (intercity, ICE)	200	0.7	0.1			500	0.5	0.3		500	
Bus (intercity, electric)						500	0.4		0.2		500
Rail (intercity, ICE)	6	0.9	0.0			100	0.6	0.1		100	
Rail (intercity, electric)	3	0.3		0.0		400	0.2		0.1		400
Aircraft (domestic)	950	2.0	1.9			600	1.8	1.1		600	
Aircraft (international)	330	2.3	0.8			400	2.1	0.8		400	
Airship (dom. and int.)						100	1.2	0.1		100	
Marine (dom. and int.)						100	0.7	0.1		100	
Totals	9300		23.4	0.0	4200	5200		13.1	1.6	6900	2500
Per capita	30,500		76.5	0.1	26,500			37.0	4.5		

Note: Abbreviations in this table: ICE = internal combustion engine; pkm = person-kilometer(s); MJ = megajoule; EJ = exajoule; GJ = gigajoule.
Electric modes are in italics

We are not proposing that the motorized movement of people decline in direct proportion to the reduction in oil used by transport. Indeed, Table 5.2 indicates that motorized person-kilometers would increase very slightly between 2007 and 2025 (by 1 percent) although fall per capita (by 13 percent). This would be achieved in two ways: by using oil more efficiently and by a substantial shift to electricity generated from other energy sources. Overall in 2025 compared with 2007, each liter of oil products would fuel about a third more person-kilometers. Electricity would fuel 27 percent of the motorized movement of people in 2025 compared with about 0.5 percent in 2007.

There are several ways in which we could have plausibly constructed a pattern of motorized movement within the framework imposed by the 44 percent reduction in liquid fuel use.[10] For example, we could have allowed for more aviation (consistent with trends through 2007) and less movement by car, perhaps compensating for the lower car use with more public transport. For each mode, we have used our best judgment as to what would be the most feasible and acceptable, informed by some understanding of the cost implications.

Some particular features of Table 5.2 are worth noting:

- ICE-based personal vehicles (cars) will be providing just over half of the movement of people in the US in 2025, but much less than today's share, which is over 80 percent. Their average energy consumption in megajoules per person-kilometer (MJ/pkm) will be about 20 percent lower because of technical improvements and higher occupancies.
- Electric personal vehicles in 2025 will include cars and two-wheelers that have only electric motors and a declining number of "plug-in hybrids," which can use a built-in internal combustion engine (ICE) but do so rarely. The average MJ/pkm is a conservative suggestion based on information in Table 3.3 in Chapter 3.
- Future transport signifies the availability of new local transport options that may encompass electric jitneys and various kinds of on-demand transport including what is known as personal rapid transport (PRT), discussed in Chapter 3. Average MJ/pkm for this category is that suggested for PRT.[11]
- By 2025, local public transport would have expanded substantially and become largely electrified. Electrified service would comprise a mix of modes: trolley buses, light rail (trams) and heavy rail (metro).

Considerable reduction in MJ/pkm is anticipated in ICE-based transport, chiefly through higher occupancies, but less in electrified transport, which already operates with high efficiency and, as it expands, will encounter capacity challenges.

- Intercity bus would also expand. Roughly half of it would be powered by more efficient versions of today's diesel engines and half would be electric buses, drawing power for overhead wires that are installed for bus and truck use along some major highways. Modest reduction in energy use is anticipated for the already efficient intercity ICE buses. Electric intercity buses will use a little less electricity per pkm than their urban counterparts, chiefly because of longer distances between stops.
- Intercity rail undergoes one of the largest increases by 2025, talking up some of the reduction in travel by car and replacing many short-haul flights under 500 kilometers. Most of the expansion would be by electrified rail, much of it providing high-speed service—more than 200 kilometers per hour (125 miles per hour). Unit energy use by ICE and electrified trains is expected to fall by a third, largely on account of higher occupancies.
- In the period until 2025, there would also be expansion of diesel-powered passenger rail service, over tracks not yet electrified, although eventually just about all rail, passenger and freight, will be powered by electricity.
- Domestic aviation would have contracted substantially by 2025. A trend will be under way to use larger aircraft—because they are more fuel-efficient—flying over fewer routes, that is, those that can generate the high occupancies that will also be required to attain low levels of fuel use per person-kilometer. Modest reductions in unit fuel use are expected by 2025, chiefly on account of higher occupancies. Further reductions can be expected after 2025 as more of the fleet will be large, efficient aircraft.
- International aviation would have expanded a little by 2025, although remaining about the same per capita. Its character would be changing toward movement of larger, more fuel-efficient aircraft flying over fewer routes, to be a much more prominent feature after 2025, when even international aviation will decline.

- By 2025 there would be some use of fuel-efficient, partially solar-powered airships for moving people over distances, particularly to remote locations for which service by rail or road is impracticable.[12]
- There would be more use of water-based modes by 2025, particularly domestically in the form of coastal, lake and river ferries, but also by trans-oceanic vessels. The last especially could make considerable use of wind power through use of kites or solid sails.[13]

The particular values for transport activity and unit energy consumption in 2025 in Table 5.2—and in the three tables concerning 2025 that follow—are less important than the process of developing and making use of such a table. Readers may well want to use different values, which may in turn require different balances among the modes. Our point is not that we have found the most appropriate values (although we have tried to identify them) but that development and use of tables such as Table 5.2 is an essential part of serious transport redesign. The five key steps in the process are these:

1. Set the table's parameters—in the case of Table 5.2, a 44 percent cut in use of liquid fuels between 2007 and 2025 and essentially no change in motorized transport activity (actually a one percent increase in person-kilometers).
2. Estimate current transport activity and energy use.
3. Anticipate available modes in 2025 and their unit energy use.
4. Develop a plausible balance of 2025 modes that fits the 2025 transport activity and energy use to the set parameters.
5. Engage in continuous improvement of energy use estimates and proposals for transport activity.

In constructing Table 5.2 and the similar tables that follow, it would have been possible—and indeed more plausible—to give *ranges* of values rather than specific values, particularly for unit energy consumption. For all modes, consumption can vary widely according to occupancy, efficiency of the propulsion system, quality of the fuel, circumstances of the motion (e.g., wind speed) and very many other factors. We felt that providing ranges would have added undue complexity to tables that already contain what some readers may regard as too much information. Moreover, we stress again that at this stage the particular values are less important than the process of putting the values together within the kind of framework we use here and propose for general use.

Moving Freight in 2007 and 2025

Similar considerations apply to the movement of freight in, to and from the US in 2025, which is shown in Table 5.3 in comparison with freight movement and fuel use in 2007. Again, there are several ways in which we could have constructed a pattern of freight movement, and we have used our best judgment as to what could be the most feasible and acceptable, informed by cost implications.

Our overarching consideration in configuring freight movement in the US in 2025 was the same as for the movement of people: that use of liquid fuels for transport—chiefly petroleum fuels—will decline by a little over 40 percent. Almost three times as much fuel was used for moving people in, to and from the US in 2007 as for moving freight. Moving people could be considered to be less essential and thus worthy of a larger proportional cut. On the other hand, much of the current movement of freight can be considered at least as inessential the current movement of people. We decided to avoid such comparisons by applying similar declines in liquid fuel use to the movement of both people and freight. The actual proposed cut for freight movement is 42 percent (49 percent per capita).

We are however suggesting that by 2025 slightly more of the movement of freight be electrified than of the movement of people: 33 vs. 27 percent.[15] This reflects the relative ease with which electrification can be applied to the already high share of freight movement by rail in the US. The busier and thus more efficient routes are likely to be electrified first. The result would be that smaller efficiency gains in the use of liquid fuels for moving freight might be expected than for moving people (17 vs. 32 percent).

Overall, we are proposing essentially no change in total tonne-kilometers (actually, a 2 percent increase), although a 12 percent decline in per capita tkm. Other features of Table 5.3 are these:

- Regular trucks—shown as "Truck (ICE)"—would still be in evidence, operating within and between urban regions, but performing in total only about 75 percent of the freight movement performed today.

TABLE 5.3 Motorized movement of US freight in 2007 (estimated) and 2025 (proposed)[14]

Values and totals in this table are rounded to aid comprehension	2007				2025					
Mode	tkm in billions (except per capita)	Fuel use per tkm, in MJ	Total liquid fuel use in EJ (GJ for per capita)	Total electricity use in EJ (GJ for per capita)	tkm in billions (except per capita)	Fuel use per tkm in MJ	Total liquid fuel use in EJ (GJ for per capita)	Total electricity use in EJ (GJ for per capita)	Liquid fuel powered tkm	Electrically powered tkm
Truck (ICE)	2050	2.6	5.4		1500	2.0	2.9		1500	
Truck (battery)					500	1.0		0.5		500
Truck (trolley)					500	0.8		0.4		500
Rail (ICE)	2650	0.2	0.6		900	0.2	0.0		900	
Rail (electric)					2700	0.1		0.2		2700
Pipeline	1250	0.5	0.6		800	0.3	0.3		800	
Air (domestic)	15	23.0	0.3		10	17.3	0.2		10	
Air (international)	25	9.9	0.3		25	7.4	0.2		25	
Airship (dom. and int.)					50	5.0	0.3		50	
Marine (domestic)	700	0.5	0.4		1100	0.4	0.4		1100	
Marine (international)	4200	0.2	0.8		3000	0.2	0.5		3000	
Totals	10,900		8.3		11,100		4.8	1.1	7350	3700
Per capita	35,600		27.3		31,200		13.3	3.0		

Note: Abbreviations in this table: ICE = internal combustion engine; pkm = person-kilometer(s); MJ = megajoule; EJ = exajoule; GJ = gigajoule.
Electric modes are in italics.

These trucks would be more efficient than now, but their use would nevertheless be in decline.

- "Truck (battery)" refers to the use of battery-electric trucks and vans, chiefly in urban areas, often making trips between distribution centers and destinations. We expect their energy use to be in about the same relation to ICE trucks as electric cars to ICE cars.
- "Truck (trolley)" refers to the use of road vehicles that can draw electric power from overhead wires along highways between places without rail service. These would use less energy than battery trucks because there would be no charging and discharging losses.
- Rail freight activity would increase by about 35 percent. The US already has one of the highest shares of movement of freight by rail, and the scope for further increases is limited. Moreover, with the proposed massive expansion in movement of people by rail (see Table 5.2), there could be some sharing of facilities that could further limit the expansion of rail freight. There may not be much scope for further reductions in energy use by ICE locomotives. Freight movement by electrified rail would not have quite the energy-use advantage of electricity as would be available for the movement of people.
- Pipeline activity would decline by almost 40 percent, chiefly because it is now mostly fossil fuels that move by this mode and consumption of these fuels would decline.
- Movement of freight by conventional aircraft within the US would decline, because of its energy and other costs, although there could be some use of airships, as we proposed for moving people. Some reduction in aircraft energy use can be expected, partly through better loading. Airship energy use for freight movement is comparable to that for moving people.
- Ocean freight activity to and from the US would decline by almost 30 percent, partly because of higher fuel costs for ships and their connecting transport but more because of a general movement toward local production stimulated by energy constraints.
- By contrast, domestic movement of freight by water—including short-sea shipping and movement by river and canal—could increase substantially as fuel costs enhanced its relative advantage in comparison with movement by road.

Again, the particular values chosen for 2025 are less important at this stage than the process of planning for particular levels of energy use and transport activity, as long as the process includes provision for continuous improvement of estimates and proposals.

Implementing Transport Redesign in the US

National Security Will Justify US Government Leadership

This section explores how Americans[16] could launch a shift away from being the world's most voracious consumer of oil used in transport and embrace energy efficient mobility that relies increasingly on electric traction. While achieving such a transition will require profound changes in the way that mobility is provided, it is worth emphasizing that these adjustments need to be embraced because they could well involve less disruption and turmoil than any alternative course of action. Little geopolitical expertise is required to see that in a world of declining oil production, societies lacking the technology and organization to meet basic human needs with less oil will be at greater risk of international and internecine conflict over access to petroleum. A US transport revolution that develops the capacity to use as little oil for the transport sector as possible would appear a more attractive prospect than further use of military power to secure a steady flow of oil from foreign sources.[17]

Who in May 2008 could have predicted that one year later General Motors would be under government ownership?

Transport revolutions inspired by anticipating redesign for oil depletion would also appear easier than trying to manage decline by allocating a steady supply of oil to "essential" functions in US society. Many of the measures discussed below may seem impolitic or even inconceivable under the "normal" economic logic of the second half of the 20th century. But, the US will not be experiencing normal conditions once oil depletion becomes manifest. The events of 2008–2009 provided an indication of how the economic world has changed and may well change more. Who in May 2008 could have predicted that one year later General Motors would be under government ownership and Chrysler would be owned by United Auto Workers, or that both companies would have entered and exited from bankruptcy protection in the space of several months?

Each step in the redesign we envision could be accepted, even embraced, if the perceived risk and pain of the particular change were understood to be less daunting than the risk and pain of alternative actions that do not advance a transport revolution. Thus, existing transport arrangements could be set aside because the technology and organizational innovations that reshape mobility will be seen to provide improvement over the increasingly dark prospects for living with less oil through managed decline or global conflict.

Despite frequent celebrations of the free market's role in leading America's transport innovations, US mobility has also been advanced by a series of planning and sponsorship arrangements established by governments of all political stripes since the republic's earliest days. Transport planning in the US began with state aids to canals and roads in the 18th century,[18] and continued through land grants and financial incentives to building a continental railway network in the 19th century.[19] The federal bureaucracy embraced a direct role in planning, designing and financing highways, beginning in 1916.[20] It was then instrumental in launching an ambitious master plan for a "National System of Interstate and Defense Highways" in 1956.[21] The challenges faced in redesigning America's transport sector to rely upon electric traction will not be diminished by public sector involvement. Indeed, government's engagement will be an essential component of success.

When government demonstrates a commitment and a capacity to support the redesign of transport in the US, the private sector can be expected to bring much dynamism and innovation to implementing these efforts. We highlight both what the front end of such a transformation could look like and the likely dynamics of change once America's transport revolutions are firmly in progress in 2025. The start of a transport revolution needs to have a high enough profile that those who are (or will become) involved in the transformation can recognize when fundamental changes are under way. Successful plans also require a compelling depiction of the end stage, where the goals are brought to life in an inspirational vision. A powerful example of such a vision for contemporary American transport can be found in the book *Magic Motorways,* written to complement the General Motors "Futurama" exhibit at the 1939 World's Fair.[22] Both the exhibit and the book depicted an appealing American *autopia* in which the car offered unprecedented ease of mobility for local and long distance travel.

Launching transport revolutions in the US could require three elements to steer change away from chaos and conflict. First would be establishment of an agency that can develop a detailed plan and ensure its effective implementation through financial arrangements that encourage state and local governments and the private sector to embrace redesign efforts. Second would be termination of existing programs and plans for expanding airport and highway capacity for oil-fueled mobility, and redirection of human and financial resources toward developing electric traction capabilities. Third would be a new tax on oil products used for road transport, enough initially to raise the national average pump price by about 13 cents per litre (50 cents per gallon). This tax would also be applied to fuel for domestic aviation. Future tax increases would be needed to fully fund the redesign of America's transport infrastructure. These could be gauged subsequently, once the costs and economic impacts of America's energy transition became apparent. The proceeds would be used in part to induce individuals and businesses to retire what could soon be "stranded assets." These would include jet aircraft and motor vehicles that can be fueled only by petroleum products. The proceeds of the tax would also be used to stimulate private, state and local investment in electric traction infrastructure in much the way that fuel and airline ticket taxes are used now for expanding aviation and road infrastructure.

The three measures would clearly commit the US to a new course with respect to transport energy. These first steps away from the status quo would be among the most difficult during the course of America's impending transport revolutions. Interests vested in today's oil-powered mobility would highlight and fiercely oppose the costs and disruption of change. The more numerous eventual beneficiaries of electric traction would be less motivated to support redesign, not least because its advantages could arrive further into the future. Most Americans would be unfamiliar with the transport alternatives being developed and thus uneasy at the prospect of such radical change. Communicating the risks of inaction in the face of oil depletion will require strong, effective leadership, a matter we return to in Chapter 6.

Launching Revolutions: 1. Create a Transport Redevelopment Agency

Creating a new public agency to plan, facilitate and monitor transport redesign will be a prerequisite for engaging the US in coming transport revolutions. We see no good candidate for such a role from among the myriad government agencies and private organizations that contribute to America's current transport planning. They include industry associations and several government departments, not only those concerned mostly with transport but also, for example, the armed forces, which have often been concerned with the capacity and technology of America's civilian transport and have recently focused on oil depletion as a strategic challenge.[23] They will all need to be involved in developing the plan for a redesigned transport system. However, this restructuring will best be facilitated by a new entity that leaves aside much of the baggage of existing interests in—and approaches to—moving people and freight in the US. The new agency should be constituted as a transparent, inclusive and fair enabler of the considerable efforts that lie ahead.

The mandate for an agency to guide the redesign will need to come from the top, meaning the US president. We propose a name such as the "Transportation Redevelopment Administration" (TRA), which signals clearly what the organization will be about and provides a relevant and unambiguous acronym. TRA would have a board chaired by the US vice-president whose members include the secretaries of Defense, Energy, Treasury and Transportation. As well, state, county and city governments should be represented on TRA's board. TRA could draw upon the expertise associated with the Transportation Research Board (TRB), an affiliate of the National Academy of Sciences that assembles America's analytical and technological expertise in over 200 standing committees involved in examining every aspect of mobility.

The mandate for an agency to guide the redesign will need to come from the top, meaning the US president.

TRA would draw on past US experience in creating public agencies to develop innovative solutions to serious challenges. Earlier organizations overseeing considerable social and economic redesign efforts include the Reconstruction Finance Corporation established by President Hoover in 1932, the Tennessee Valley Authority inaugurated by President Roosevelt in 1933, the United States Railway Association created by President Nixon in 1973, and the Air Transportation Stabilization Board signed into law by President G. W. Bush on September 22, 2001. The last organization had the job of using loan guarantees to sustain the US airline industry, hurt by the terrorist attacks 11 days earlier and the attacks' aftermath. TRA would combine elements of each of these agencies' structures to serve four key functions.

First, TRA would provide a forum for consultation with industry, organized labor and interested citizens on changes that would create considerable new benefits, as well as impose real burdens. Second, TRA would become a repository of managerial and technical expertise in energy-efficient transport redesign. Third, TRA would serve as a banker and broker for financing deployment of the technology and infrastructure needed to make electric traction the prime mover in the US. Fourth, TRA would become an assessor and evaluator of the work in progress to redesign American mobility. TRA could thus be seen as a "superagency" along the lines of the Department of Homeland Security (DHS), which grew quickly and assumed wide-ranging responsibilities in its mission to keep Americans secure on the home front.

DHS grew through the rapid transfer and adaptation of pre-existing government programs, from the US Coast Guard to the Immigration and Naturalization Service. So could TRA gain momentum rapidly through staff and resource transfers from agencies within the US Department of Transportation (US DOT), which has some 55,000 employees and received a $70.3-billion budget in 2008.[24] Among agencies within the US DOT umbrella are the Federal Aviation Administration, the Federal Highway Administration, the Federal Transit Administration and the Federal Railroad Administration. Many employees of US DOT and its agencies, and many from among the tens of thousands working on transport issues in state and local governments, could quickly migrate to TRA's planning, design and finance departments. All these departments and agencies would have a considerable surplus of program delivery capacity as a result of the second, more challenging requirement to successfully launch America's next transport revolutions.

Launching Revolutions: 2. Terminate Highway and Airport Expansion Programs

Existing aviation and road development policies and programs that are not compatible with a shift away from oil-powered transport will have to be terminated, as will policies and programs that support associated land uses. The skill and effort needed to remove existing policies and dismantle established programs is far from trivial. Lack of a focus on *policy termination* has undermined many efforts by leaders—across the spectrum of political orientation—to change the direction of US policy.

These efforts include, for example, the Carter administration's agenda of government leadership in energy conservation of the late 1970s and the Reagan administration's goal of replacing Social Security pensions with private alternatives in the 1980s. Failures to terminate existing policies have undermined the key priorities of more than these two presidents. One analyst noted that the political dynamics of terminating established public policies differ fundamentally from those involved in creating new policies because "...distinctive coalitions generally form on both sides... [and] termination contests are usually more bitter and harder to win than most policy adoption contests."[25]

A clear principle would thus be useful for justifying the end of programs that support oil-powered mobility in the US. The logic is that of no longer digging once the determination to get out of a hole has been reached. Many of the previous efforts in the US to cultivate energy-efficient transport alternatives—including local public transport and intercity rail passenger improvements—have been undermined by simultaneous additions to road and airport capacity, usually paid for with earmarked trust funds from fuel and other taxes. Such an approach to transport development, which some portray as "balanced" spending, is analogous to applying a car's accelerator and brake at the same time. The result undermines the performance of both systems and eventually destroys the engine. The onset of the next transport revolutions should be most noticeable for what *stops* happening, namely the expansion of highways and airports. A scan of the US federal budget suggests the magnitude of resources that could then become available following such redeployment.

Lack of a focus on *policy termination* has undermined many efforts by leaders — across the spectrum of political orientation — to change the direction of US policy.

US DOT's budget proposal allocated spending of $47.3 billion in 2008 to reduce the congestion of America's roads and airways.[26] Such funding would build upon decades of similar spending aimed at adding more air and road infrastructure to move ever-expanding volumes of cars and planes. Once oil depletion is recognized as a constraint on future growth of driving and flying, spending such large amounts on expanding capacity could be seen as wasteful and counterproductive, even amid the pleas against termination by those who gain great benefits from continuing with such efforts.

State and local governments spend about four times the federal funding of transport infrastructure, mostly on roads and airports, which respectively comprise about 84 and 11 percent of the total. The total was $160 billion in 2005–2006—the latest fiscal year for which complete data are available—some of which supported maintenance and rehabilitation of existing infrastructure.[27] Several states' constitutions prohibit the spending of gasoline tax revenues on anything but roads. Thus, state and local termination of spending on new road infrastructure could be expected to change more slowly than federal highway investment. Nevertheless, if only the unrestricted air and road spending by state and local governments were initially shifted away from airports and road building, one could expect at least $50 billion a year to be reallocated to developing the infrastructure and accelerating the diffusion of electric traction in America's road and rail network. This could include conversion of many airports into "travelports" that connect the remaining international and long-distance flights to electric road and rail feeders, as discussed below.[28] Other airports and sections of urban highway infrastructure could be decommissioned, much as military installations are still being decommissioned following the Cold War.

This spending transition would create intense opposition from interests that see themselves as economic losers from such changes. Government will have to anticipate such opposition, defusing it with incentives for change and compensation for losses. The incentives will be generated by the $50 billion in annual spending that will go into electrification and expansion of rail corridors, major road networks in metropolitan areas, and massive expansion of traditional and "advanced" public transit systems. Existing producers of electric-powered transport will reap a windfall from this bonanza, as will firms that rush to join this sector by adding traction and related technology to their product lines. The ability to compensate losers—including, for example, airlines and the lessors of what would be their surplus aircraft—will require the most controversial, third element of the launch of transport revolutions in the US: a new tax on oil products used to fuel transport.

Launching Revolutions: 3. Tax Oil-Based Transport Fuels

Plans for taxing oil used as motor fuel to spur conservation are not new. John B. Anderson received 7 percent of the vote in the 1980 presidential election during the campaign for which he proposed an additional gasoline tax of 50 cents per gallon as an energy conservation measure. This is equivalent to about $1.30 per US gallon or 35 cents per liter in 2009 dollars. Anderson proposed allocating the revenue to meet projected deficits in federal old-age and disability (Social Security) pensions.[29] This may well have been the best transport policy proposal to have been ignored by the US government during the decades when growing oil consumption could have been easily restrained. While increasing taxes on motor fuel has been politically traumatic in the US, the UK shows that such leadership can occur—on occasion. In 1993, a Conservative government introduced the "fuel price escalator": an annual increase in the tax on gasoline of 3 percent above the inflation rate, later raised to 5 percent and then, by a Labour government, to 6 percent. The escalator was repealed by the same Labour government in 2000, soon after oil prices began their present rise.[30]

Any government that sets out to pass a major increase in transport fuel tax and then secure re-election will need to make (and keep) commitments to spend the proceeds on tangible and popular programs. Rail infrastructure within and between cities would be a prime candidate for such spending. One should not underestimate the popularity of intercity rail and rail transit as potential travel options, even in car-dependent America. Despite the minuscule marketing efforts by Amtrak and public transport agencies—especially in comparison to the relentless promotion of cars, airlines and even ocean cruises—there exists considerable support for rail, notably new rail systems that are faster, more frequent and more extensive than those managed by Amtrak, the federal government agency that runs intercity passenger trains in the US.[31]

Plans for taxing oil used as motor fuel to spur conservation are not new.

Creating effective alternatives to flying and driving will make an oil tax seem less confiscatory. In the interim, while new mobility alternatives are under development, fuel tax revenues could be used to compensate some of the losses that accompany this transition. An example would be providing credits to owners who trade in cars and trucks with ICEs for electric-powered vehicles. Another would be to give owners of ICE gas guzzlers (e.g., Hummers, Navigators, Escalades) decade-long public transport passes in return for their increasingly worthless vehicles. This kind of compensation could convince voters that the fuel tax was not another fiscal grab, but a mechanism for making the transition to new energy sources widely affordable.[32]

The political acceptability of motor fuel taxes collected by state and federal governments has been attributed to their clear connection to road building. These earmarked taxes are judged fairer by many because they connect the "pain" of the tax to a "gain" in mobility. Compensating businesses and households for the lost value of their assets during the shift to new energy sources could provide the winning formula for this politically risky policy initiative.[33]

Once the adjustment costs of individuals and firms become history, the revenues from this tax could shift to financing electric traction infrastructure in the same way that today's American motor fuel taxes largely go to fund road infrastructure. Earmarking the motor fuel tax for transition assistance and for development of traction infrastructure is not a "magic bullet" that would eliminate the conflict over what will inevitably be a contentious policy, but it could tip the balance toward early action in a world where time is of the essence.

Thus primed for change, the US could turn its considerable public and private initiative to reshaping the organizations that provide or contribute to mobility, and to introducing long established, but unfamiliar, electric traction technology into its rail and road infrastructure. Such transformation will involve major departures from existing organizational and ownership arrangements in local public transport, railways and airports. It will also see the restructuring of airlines, road haulage firms, car and aircraft manufacturers and many other transport sector businesses in ways that can only be guessed at from this pre-revolutionary vantage point. The pace of change will need to be faster than almost anything that took place during the 20th century. The closest equivalent would be the Second World War pause in mass motorization described in Chapter 1. The extent of change will see some mighty organizations fall into decline, just as some of America's colossal 20th-century railway companies and steamship lines disappeared while others became focused niche carriers in their search for a new mobility role.

The effects of this first wave of change will be uneven as some parts of the US already have more in the way of oil-free mobility alternatives than others. To yield a substantial reduction by 2025 in the amount of oil used to fuel the movement of people, daily life in Manhattan will change less in its current rhythms and routines than will life in Atlanta or Houston. Contentious political decisions will have to be made regarding how fast to move in rebuilding transport and other infrastructure in newer "sunbelt" cities whose physical development was shaped entirely around automotive transport. Older "rustbelt" cities—including Baltimore, Philadelphia, Pittsburgh and Cleveland—offer housing and public transport capacity that makes them less car-dependent today than Atlanta or Houston might be able to become by 2025. How much should society consider spending to retrofit highly car-dependent urban regions with transit-oriented development when affordable housing stock and vacant lots along electric tram and train lines in Northeast and Midwest urban areas are available for redevelopment? Such questions will need early answers if the US is to move smoothly into the coming transport revolutions.

Whatever the outcomes of the debates over such questions, the next transport revolutions in the US will see a considerable oil-depletion influence on its already substantial amount of internal migration, with the emerging trend becoming an exodus from relatively new areas of settlement to older places. Part of this migration will be influenced by a relocalization of extended families. When cheap flights are no longer the norm, living more than 1,000 kilometers apart will mean a level of family separation that many Americans have never experienced. When people find that they cannot participate in family celebrations or contribute to the care of children and elderly relatives by flying a few hours each way on weekends and holidays, many will likely relocate closer to their relatives. Such a trend has already become noticeable among people who had retired from northern to southern states that are thousands of kilometers from their families.[34]

Mass migration such as occurred after the devastation of hurricanes Katrina and Rita in 2005 suggests that part of the natural adaptation to major change is to resettle in more hospitable locations. Thus, some relocation of population should be expected during the upheavals that will occur during oil depletion. Not every community across the US will need to duplicate its current transport capacity with electric traction. Some places should plan for less transport capacity, as their populations will decline even as overall population increases. While some urban areas shrink as others grow, the links between population centers will all need extensive redesign. This is because most Americans now move between cities by flying and driving, the two transport modes that will have to stop expanding in order to meet oil-reduction targets.

Anticipating declines in air travel and motoring will be hard to imagine in a nation where physical mobility is equated with social mobility. The most persuasive approach to convincing Americans to step back from travel by car and plane is to provide a transport alternative that does not depend on oil. Railways offer great promise for a rapid shift to electric traction and are the focus of our next discussion.

Developing Rail in the US

The scenarios for 2025 sketched out in Table 5.2 and Table 5.3 imply an unprecedented expansion of rail transport in the US. The projections for moving people by train would comprise a true revolution in rail travel in the US that would likely include development of dedicated high-speed passenger lines such as are found in Europe, Japan, Korea and Taiwan, and were discussed in Chapter 1. The freight transport projections anticipate both the expansion and electrification of existing tracks, many of which are already at or near capacity.

US railways are presently designed and managed for the efficient movement of bulk freight, with little sharing of tracks with passenger trains. Moreover, the proposed expansion in rail freight means that sharing tracks could become more challenging. Thus, most of the proposed 50-fold increase in rail passenger movement could occur on new infrastructure, much of it designed for relatively high speeds (200 km/h and higher).

To achieve a sense of the scale of the infrastructure requirements for this revolution, we can first assume that that 200 billion of the anticipated 500 billion person-kilometers (pkm) of rail travel in 2025 will be accomplished over dedicated high-speed passenger lines. Using estimates for provision of high-speed service in California as a basis, this will require throughout the US about 25,000 km of double-tracked line (i.e., twin tracks, each normally for one direction only).[35] To put this in perspective, there are presently about 170,000 km of railway in the US, much of it single-tracked.[36]

The infrastructure and rolling stock cost for the 25,000 km of track could be about $1 trillion.[37] This seems a large amount but over 30 years would amount to about $0.22 per pkm.[38] Operating costs of high-speed rail are generally a small fraction of total costs, rarely more than a quarter.[39] Thus the total cost per pkm would be under $0.30, or less than $165 for a 550 km trip, the Los Angeles–San Francisco distance. Currently, this is more than the lowest one-way cost by air (about $70 inclusive, airport to airport) but less than the total cost of moving a single-occupant car between the two cities.[40]

As oil prices rise, ticket prices for a high-speed rail system could become more competitive with travel by air and car, and government subsidies such as the provision of interest-free capital could help this process along. As well, with a massive program to build 25,000 km of high-speed double-track lines, there could be cost reductions from economies of scale and much "learning by doing."[41] These could have the effect of reducing costs but are not allowed for here. There could also be shortages of material and skilled personnel that would drive up construction costs, especially if the railway manufacturing sector cannot keep pace with the pace of infrastructure expansion. Growing this sector ahead of oil depletion—particularly the skilled workforce capable of building new trains and tracks—would be a wise policy.

As challenging as introducing dedicated high-speed passenger lines, would be provision for the remaining 300 billion pkm of rail travel per year. About 200 billion pkm of this would be electrified but would not be high-speed service—that is, top speeds would be less than 200 km/h—for reasons of track configuration, stop frequency or fuel economy. The remaining 100 billion pkm of rail travel would be on track that had not been electrified by 2025. Some might never be electrified, for reasons of cost or power availability, but most would eventually become grid connected.

It's difficult to suggest how much actual track would be electrified overall by 2025 and what the infrastructure costs might be. Although the cost per kilometer would be lower, the overall costs of lower-speed rail infrastructure could be similar to that for the construction of a 25,000 km network of dedicated high-speed passenger lines. If this were to be the case, the total required investment in passenger rail could be in the order of $2 trillion, almost all of which would be for infrastructure as opposed to rolling stock. If this investment were made over 15 years, say from 2010 to 2025, the average annual requirement would be about $140 billion. This is similar to the amount of annual private investment in transport in the US in 2007 and about half the amount of government investment.[42]

Electrification of the freight lines could require a similar order of investment. Some of this would be covered in the above provision for non-high-speed passenger lines, which could be shared with freight service. Most of the existing track would remain freight only. Much of it would require electrification and replacement of locomotives. Providing a more precise estimate of the required investment would be an early step in the development of a detailed strategy for the next two decades of US transport.

The scope of what is proposed would necessitate diversion of resources from other parts of the US economy. While the transport revolutions were unfolding, transport's role in the economy could well increase. Presently, it accounts for about a tenth of US GDP, or about $1.4 trillion of about $14 trillion of annual total expenditure.[43] If the required annual investments in rail were to be double the total indicated for passenger rail (i.e., a total of $280 billion) this would amount to roughly 2 percent of total GDP.

New approaches to planning and development will be required to realize America's extensive railway redesign. Given the extent of what needs to be done, there is no chance that existing railway finance, design and construction efforts could simply be scaled up. Nor could most past approaches to developing America's rail infrastructure be revived. One has to reach back to the 19th century to find a comparable phase of railway building in America.[44]

During the years in which they opened up large sections of the American continent to trade and settlement, railway companies raised much of their investment in Europe. Then, investing in the US was viewed akin to investing in China's emerging economy today. The companies imported labor to build the tracks, and settlers to live along the lines, in a vertically integrated development model that came to be vilified as an abusive big-business monopoly.

Private railways had a relatively free hand in building most of the infrastructure they continue using to this day. The labor practices and environmental impacts that were the norm during this development period would no longer be considered acceptable. Conversely, the construction methods and regulatory restrictions that have been applied to more recent rail infrastructure additions, including the modest expansion of electrification from New Haven to Boston and upgrading of tracks for high-speed passenger trains—which took over a decade to complete—are also unsuited to massively expanding rail's capacity. A new model for rail development will need to be devised to meet the challenges of oil depletion and volatile prices.

The current expertise in the US in respect of transport infrastructure mostly concerns highway engineering and planning. It offers limited possibility for direct transfer to the needs of railway redesign. Termination of highway expansion programs will make available many thousands of public-sector highway engineers, planners and project managers. A way will need to be found to redeploy their experience to rebuilding rail infrastructure that is mostly privately owned. A good part of the 25,000 km of high-speed passenger lines could be built along new public rights-of-way, or fill in some of the highway infrastructure that will become surplus during oil depletion. However, most of the new passengers and nearly all of the freight will travel along conventional electrified tracks, and the best place to develop them is on the existing rail rights-of-way.

The current expertise in the US in respect of transport infrastructure mostly concerns highway engineering and planning. It offers limited possibility for direct transfer to the needs of railway redesign.

Most of the America's "main line" rail rights-of-way have excess capacity for tracks that were previously removed in response to declining rail traffic between the 1950s and 1990s. New signal systems that use global positioning system (GPS) technology could also increase the capacity of existing tracks. In some places, such as the rail tunnels beneath the Hudson River serving New York City, existing rights-of-way are fully used and incremental redevelopment of rail would not be possible. In regions where population growth occurred mostly after the early 20th-century boom in railways and public transport, such as California and the Pacific Northwest, new rail infrastructure would feature prominently in transport system redesign. Finding a way to facilitate public investment in, and construction of, new tracks on existing rights-of-way will be essential to meeting the challenges of the proposed vast rail infrastructure expansion.

With time being of the essence, producing new rail capacity will draw upon previous plans and designs where possible, rather than starting from scratch. Numerous plans for rail infrastructure redevelopment have been produced since the 1990s. Such efforts focused mostly on expand-

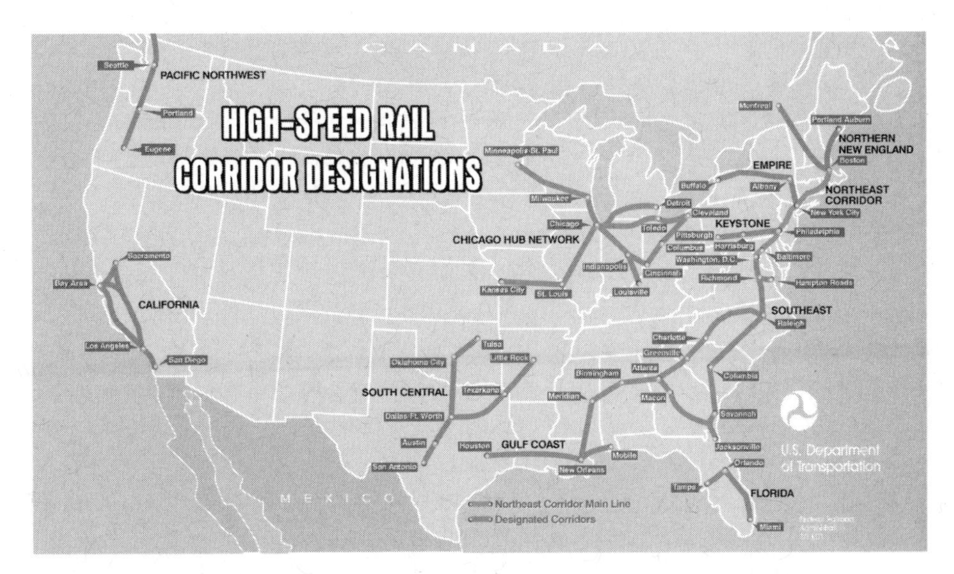

FIGURE 5.1 Corridors designated for high-speed rail in the US during the 1990s.[46]

ing, and in some cases introducing, rail passenger capacity. They were undertaken by the Federal Railroad Administration and state transport agencies including those of California, Florida, Texas and a consortium of Midwestern states.[45] The results have included numerous useful insights into what it would take to develop well used intercity passenger rail corridors. In 1991 and again in 1998, the US Congress authorized the US Department of Transportation to designate high-speed rail corridors that could then become eligible for federal planning and financial assistance. Eleven such corridors were approved. They are illustrated in Figure 5.1. These corridors could well become the backbone of the 25,000 km dedicated high-speed rail network that would support 200 billion pkm of travel by 2025. Recent plans for them are discussed in Chapter 6.

Several of the designated corridors have seen modest improvements of conventional train performance, including additional frequencies and modest speed increases, all below 150 km/h. However, only the Boston–Washington Northeast Corridor (NEC) experienced a significant infrastructure upgrade that extended electrification over its full length, during the late 1990s, as well as introduction of high-speed service—known as Acela—in 2000.[47] This difference is mostly explained by the public ownership of NEC infrastructure and the private ownership of the other rail corridors, whose owners earned most of their income from carrying freight in high volumes. Introducing higher-speed passenger trains into the mix of bulk freight (e.g., coal, chemicals and grain) and time-sensitive goods trains (e.g., auto parts, assembled vehicles and containerized shipments) was seen by the owners to offer more risk than reward and thus for the most part has not been welcomed.

The NEC offers lessons for those seeking to develop electric rail infrastructure in the near term. The right-of-way from Boston to Washington is North America's busiest inter-city passenger rail line. Moreover, it accommodates a wide range of train service speeds. The NEC illustrates what can be accomplished when an infrastructure owner invests in electrification, which could power both passenger and freight trains but today serves only passenger operations. Contrasting the NEC with the other corridors designated for passenger rail improvement highlights organizational innovations that would facilitate introduction of electric-powered mobility on privately owned railway infrastructure.

The most striking feature of the 734 km NEC is the diverse uses to which it is put. Whereas the French and Japanese railways hosting the well known TGV and Shinkansen were built for high-speed trains, and are used sparingly for anything else, the NEC accommodates an astonishing range of train technology among its 1,700 weekday train movements. These include the Acela express trains derived from the TGV, electric-locomotive-hauled intercity trains operated by Amtrak, diesel- and electric-powered commuter trains operated by several public transport agencies, and diesel-electric freight trains running mostly between New Jersey and Washington. Some 700,000 passengers travel over some part of the NEC each weekday.[48] Of these, about 32,000 make intercity trips aboard one of Amtrak's 100 daily train services, comprising roughly half of all the intercity rail travel in the US. Most of the remaining travel is on commuter trains, some of which travel only a short distance over the NEC, starting or ending their journeys on numerous branch lines. Freight train operations on the NEC are modest.

The NEC's experience demonstrates that it is possible to upgrade and electrify a bustling US railway corridor to serve both regular and high-speed passenger travel while still leaving room for freight operation. In 2025, much of the proposed 200 billion pkm of electric train travel not using fully dedicated tracks and 100 billion pkm of non-electric train travel could be accommodated by upgrading rail rights-of-way incrementally to look more like the NEC, with much more intensive electric freight service. The question remains how best to develop such multifunctional railway corridors from existing privately owned rights-of-way.

An organizational innovation that could enable a large-scale partnership between public and private rail redevelopers is the *infrastructure condominium*. This legal device separates the ownership of land along a transport right-of-way from what is built upon it. In a railway infrastructure condominium, the right-of-way would be owned separately from the tracks, communications and signaling systems and electric power distribution equipment.[49] This ownership structure would enable a privately owned rail right-of-way to be shared by freight carriers, local transit operators, long distance passenger carriers and high-speed corridor operators (who might well be carriers that are transitioning from airline operations). The tracks occupying a railway condominium could remain quite segregated operationally, or some could be integrated into a common network. Rents could be paid to the ground-lease holder, which in many cases would be a private railway company that had previously owned the integrated tracks, signals and right-of-way. These rents could be used to finance electrification of America's freight rail network, which could remain in private hands. Third parties such as electric power companies could also partner with either (or both) private and public railway infrastructure owners to use these corridors for power transmission as well.

Once legislation is in place to recognize the infrastructure condominium as a new mode of transport asset ownership, the door will have been opened for innovative mechanisms that direct public and private investment into railway electrification and expansion, along with the redesign of other infrastructure assets. Infrastructure condominiums could also help in the redesign of other modes' infrastructure, discussed below. TRA would become a major player in packaging investment deals that make the most of new options for infrastructure ownership.

By leasing space along private rail rights-of-way to locate new passenger train tracks, TRA would channel investment into expansion and electrification of track capacity without threatening the private ownership of existing assets and rights-of-way. Rail industry resistance to such perceived threats have stymied many previous efforts to develop passenger rail corridors. As a device for facilitating public-private partnership, the infrastructure condominium could avoid the conflicts that arise when spending public funds on expanding private infrastructure or when nationalizing railway infrastructure to enable public investment. In some cases, this approach could produce public ownership of new tracks used by high-speed trains and concomitant electrification of existing tracks. In other cases, government and private freight carriers could manage the new and improved railway jointly, with both providing rent payments to obtain operating slots assigned to freight and passenger train movements on the corridor.

TRA road redesign efforts could develop partnerships among energy, trucking and logistics corporations (e.g., BP, UPS and J. B. Hunt, among others) to own elements of the system that would distribute power for electric traction along America's highway network. Electrifying road infrastructure for use by trolley buses, trolley trucks and grid-connected cars would be a largely new departure from existing highway design and engineering. This task would be simpler than designing high-speed rail corridors, because the roads themselves would not require extensive upgrading. The learning curve could be steeper, but not last as long as for railway redesign. Unlike for railways, the demand for highway capacity would likely shrink. Higher energy prices would push some movement of people and freight to rail.

Another major transformation of infrastructure involving TRA would be conversion of some airports into already-noted travelports, where intercity trains, intercity electric buses and local public transport interconnect with one another and with aviation. In 2005, America had 575 airports authorized to receive commercial flights of aircraft with nine or more passengers.[50] Perhaps less than two dozen of these will be heavily used in 2025, and fewer still beyond that date. Some will be suitable for establishing passenger-oriented travelports. Others could serve as freight logistics centers where electric trucks and trains connect with local pick-up and delivery services. Almost all travelports will need remodeling to introduce electric rail and road infrastructure. The airports in Philadelphia and South Bend, Indiana, are the only ones presently served by electric rail infrastructure in North America, although 13 more have tram, metro or diesel commuter train stations on their premises.[51] Local governments and airport authorities should be eager to work with TRA on such conversions. They have borrowed many billions of dollars to expand airport capacity, based upon projections of steady growth in flying that are unlikely to hold true. The federal government may well need to allocate some of the new fuel tax revenues to municipalities that convert or decommission airports to help pay down some of their debt.

The Dynamics of Change

Along with new kinds of infrastructure development will come new carriers to make use of them. There will be both consolidation and exit among firms in the most oil-intensive modes of transport, including those concerned with aviation and motor vehicle production and operation. Amid the volatility of such industrial restructuring, carriers with novel organizational arrangements are likely to emerge. New links among energy production, manufacturing and vehicle operation will be seen, as will configurations that operate transport services over new electric rail and road infrastructure. Joint ventures among public and private companies could also emerge, including freight carriers, bus companies and marine operators. These carriers could exploit new mobility niches, and develop services that enable travelers to make the most of the transition away from conventional cars for short trips, aircraft for medium distance travel and trucks for long-haul shipping.

This path to the redesign of America's mobility will not be smooth, although it will be smoother than any unplanned and unmediated transport revolution. As the pain of lost jobs and investments—and the much higher costs of driving and flying—becomes widespread, government's role in providing transition assistance will make a major difference in avoiding panic, conflict and defeatism over the moves away from oil and toward electric traction. The fiscal mechanisms of taxing oil used in transport and investing in infrastructure condominiums will need to be supple enough to address unintended consequences and unmet needs that become apparent during the course of the transitions. What will also become apparent well before the 44 percent reduction in transport's oil consumption by 2025 is a vision of a future that differs from, and is better than, the great uncertainty about America's prospects that hovers over discussions about energy and economic crisis in 2009. The US government's engagement in transport redesign will produce evidence that a planned transition can enable Americans to "do better with less"—with the resulting betterment measured in terms of healthier communities, economic recovery, more attractive built and natural spaces, and a reduction in violent conflict.

The US transport system in 2025 could provide very different mobility experiences from those of today. Turn-of-the-century travel options will not vanish, but air travel and, to a lesser extent, travel by personal vehicle will be recognized as luxury transport options, priced accordingly and used more sparingly than before. In place of the narrow focus on vehicle speed and occupant flexibility of the "fly and drive" syndrome that characterizes pre-revolutionary America, attractive features will be discovered among the new options for moving people and goods. Electric trains and buses will replace much intercity driving and most flights under 1,000 km. These vehicles will move more slowly than jet aircraft and require more coordination than driving one's own vehicle. However, the alternative they will provide to the "door to door" drive and the two-to-three-hour flight need not provoke pangs of deprivation, except among those nostalgic for the mobility myths of the open road and the wild blue yonder.

For intercity trips up to 500 km (about 300 miles), the new mobility alternatives should require little or no more total travel time than today's aviation system and highways. Planned and run well, new intercity mobility options could do away with much of the "hurry up and wait" that adds time and stress to flying and driving. This includes waiting for increasingly invasive security procedures at airports,[52] waiting in traffic jams of road vehicles (gridlock) and aircraft queues for takeoff (winglock), waiting for connections through air hubs, and waiting to claim luggage that must be checked to keep a growing number of items that could be weaponized out of passenger cabins (e.g., liquids and gels). A well-run train and bus system will reduce the time and the discomfort of these travel experiences. Passengers on Japan's Shinkansen routinely arrive a minute or two before departure because there are no ticketing, luggage and security formalities to contend with. The Eurostar service linking London with Brussels, Paris and elsewhere has more elaborate pre-departure screening, with identity and luggage checks, but even there passengers are accepted up to 20 minutes before departure.

Trains and buses can distribute travel time differently, with the longer journey offset by shorter waits to board, connect and collect one's belongings. The speed of the vehicle in motion will be slower than an aircraft, but travelers will reclaim minutes or hours that are consumed by today's unpleasant aspects of travel. More time spent aboard the vehicle will create the greater opportunity for undisturbed work and rest, as well as for taking meals and for the social interactions that are largely missing from contemporary flying and driving.

Even when the total journey time increases by many hours for land travel in the thousands of kilometers, and by days for voyages by sea, the experience will offer compensation. Wireless network connectivity, comfortable accommodations and appealing food and drink could offer the chance to work and play in motion that is today reserved for those who can afford first-class flying and luxury cruise ships. Flying could remain an option for truly urgent travels, but otherwise people may not regret the time they will be able to spend aboard tomorrow's upgraded vehicles and vessels. Being in transit will become more commonplace, but also more appealing for many travelers.

The opportunity that can be created by trading speed for comfort is well illustrated in the transformation of the outmoded ocean liner of the 1960s into today's floating resort, producing a cruise sector that is more profitable than most airlines. Just as gourmets celebrate the emphasis on quality found in the "slow food" movement, so too could post-peak-oil travelers appreciate the quality gained by slower motion. Taking the time to savor a relaxing, productive and enjoyable trip while it is happening could seem natural to future travelers. When they think about what came before, they may well look back on the twilight of the "jet set" era—when passengers brought along their own food and drink, pillows, blankets and other creature comforts to survive no-frills flights—with the kind of bemusement with which their parents regarded photos of passengers packed into steerage accommodation on early 20th-century ocean liners. How Americans of the 2000s embraced the waste and discomfort of cramming onto planes and jamming the roads with huge motor vehicles occupied by a lone driver may seem as odd to Americans of the 2030s as the way in which 18th-century Americans "bundled" in the beds of strangers at inns and roadhouses during overnight journeys.

Flying could remain an option for truly urgent travels.

People could be more likely to enjoy their travels in 2025 than before the transport revolutions, but they will be making fewer long-distance trips, because such journeys will require more time and could be more expensive. All but the most urgent transcontinental travels will take more than a day, and will no longer be made weekly or even monthly. Business travel will still happen—transatlantic and even transpacific trade was a major feature of the US and European economies well before the advent of jet aircraft. The convention and trade show business still has a bright future. Businesspeople will be eager to have a periodic opportunity to encounter colleagues and competitors from afar whom they will meet personally much less often between such events than today. Attending one or two such gatherings a year, taking a week or more out of business traveler's schedule, will become much more common than flying across time zones for short meetings or brief visits with clients. These post-peak-oil commercial gatherings may echo—at a global scale—the dynamic of the medieval trade fairs held in Champagne, source of that French region's fame long before it produced a sparkling wine. Then and there, producers, merchants and financiers gathered to make deals that would set the terms of trade among distant parts of Europe.

Recreational travel could change even more dramatically in comparison with the jet-fueled global tourism of the period before about 2010. People will still take short breaks, but these will be to tourist attractions in their own region, or perhaps to an adjoining region less than a day's travel by fast train or bus. Air travel for leisure trips will be the exception. With prices similar to what was once charged for the Concorde, intercontinental "economy" flights may tempt people to take just once-in-a-lifetime holidays involving a grand tour of another continent. Those who have time available—students, pensioners and Europeans taking their six-week holidays—will sail the oceans between continents aboard hybridized ships that blend the diversions and comfort of today's cruise vessels with the point-to-point mobility offered by yesteryear's ocean liners, much as the way the *Queen Mary 2* maintains the tradition across the North Atlantic today. These ships will be the principal means of tourism to familiar and exotic getaways that are remote or inaccessible by electric trains and buses, including Caribbean islands, Hawaii and the South Pacific, the southern Mediterranean, Alaska and northern Norway.

Some of the greatest gains from their transport revolutions will become apparent when Americans are not traveling. This will happen because the need to restrain energy use, oil in particular, will favor transport organization and technology that are consistent with building and maintaining attractive and convivial communities. Electric vehicles are much quieter and more easily integrated with human settlements than vehicles using ICEs. Rail lines require much less space to move comparable volumes of people and freight than do roads. Travelports can be much more congenial to community development when the number of flights using them drops and the number of electric trains and buses passing through rises.

Urban communities designed around pedestrians—for example, Society Hill in Philadelphia, Greenwich Village in New York and Beacon Hill in Boston—have long been some of America's most desirable city locations. The same attractiveness can be found in "streetcar suburbs" such as Shaker Heights in Cleveland and Brookline, Massachusetts that were not designed around the car. Even suburban towns developed around commuter rail lines such as Maplewood, New Jersey; Lake Forest, Illinois; and Palo Alto, California have notably more appealing and more valued physical attributes than nearby suburbs that have been designed for cars as the sole mobility option.

Urban communities designed around pedestrians have long been some of America's most desirable city locations.

These community attributes will be much easier to recreate, and improve upon, in settlements that no longer depend upon the car. Immense value will be created by the ability to develop neighborhoods without the space required for motor vehicles as the ubiquitous mobility mode. This value could be offset by decline in the worth of today's car-dependent suburbs and exurbs. Such losses may well be offset by the new premium placed on land for periurban agriculture in a future when the energy costs of food production encourage relocalization of much agricultural activity.

The coming transport revolutions in the US will set the stage for a profound spatial transformation of American life, one that will take several generations to unfold. The sprawl and isolation of communities that grew up around the car will take longer to adapt than the mobility mode that gave birth to them. When trams, trolley buses and advanced forms of collective transport become the norm in US cities and towns, their physical redesign will no longer be constrained by the need to accommodate so many motor vehicles. Freed from the constraint of car-dependence, American architects, planners and urban designers will create many places that inspire a desire to stay put rather than an urge to speed away for shopping, work or entertainment at a distance. When these dividends become apparent, much nostalgia for the open road will become tempered by the realization that previous mobility patterns had imposed substantial physical and social costs on American communities.

Scenarios for 2025 for China

We are much less familiar with China than with the US and many fewer data about China are available to us. Also, China is undergoing several transformations of breathtaking scale and speed. Thus, the informed tentativeness we expressed about anticipating transport futures for the US comes close to pure speculation when we turn to China. Nevertheless, we hope that our attempt to suggest a transport future for China will be found useful and will stimulate more substantial analyses. Again, we stress that the particular features we propose for transport in China in 2025 are less important than the exercise of envisioning a future within set parameters of energy use and transport activity.

China's projected growth in transport could yield energy and pollution problems that more than offset improvements made elsewhere in the world.

What we can say with certainty is that without planning for reducing oil used in mobility, China's projected growth in transport could yield energy and pollution problems that more than offset improvements made elsewhere in the world, if current trends prevail. China's size draws an immense flow of raw materials and resources from across the globe. The key staples that support human civilization are increasingly flowing to and within China. This extraordinary country is today the world's largest consumer of grain, meat, coal and steel,[53] and is behind only the US in oil consumption.[54]

China's vast appetite for raw materials yields and is stimulated by a corresponding industrial output of global significance.[55] China's near-leading role in manufacturing is physically reflected in its prime share of the world's freight movement.[56] China's economy and society are being driven by an industrial revolution that is unprecedented in the size of the global market that is being served and the number of workers who are producing the goods sold in that market.

As with previous industrial revolutions in the UK and the US, the wealth created by such economic transformation is having profound social and spatial effects. China is seeing a shift from rural to urban settlement that resembles the European and North American populations' reorientation from farming to industrial labor, but dwarfs both the speed and scale of previous human migrations. In 1985, 23 percent of China's population lived in urban areas. That share had climbed to 41 percent by 2005. Between 1985 and 2005, China's urban population grew by 248 million people, mostly through migration from rural areas. Between 2005 and 2025, this urban population is expected to grow by a further 294 million people, again chiefly through migration, bringing the urbanization rate to 57 percent and more than tripling the urban population of 1985.[57] This movement of people—chiefly from agricultural subsistence to wage earning in urban industry—must rate among the greatest migrations in human history. Tang Jun, a sociologist at the Chinese Academy of Social Sciences, views such urban growth as inevitable. He wrote, "to ask whether China wants urbanization is like asking whether a person needs to eat."[58]

If these urbanization projections are realized, hundreds of millions more Chinese will migrate into physical and social spaces that do not yet exist and will likely differ from current urban arrangements. No other nation faces the prospect of developing new communities on such a vast scale. This is in addition to the huge investment in commercial and industrial buildings.[59]

The environmental implications are profound. One assessment concluded,

> China's achievement of developed-world consumption standards will approximately double the world's human resource use and environmental impact. But it is doubtful whether even the current human resource use and impact on the world can be sustained. Something has to give, or change. This is why China's environmental problems are the world's.[60]

An assessment of "China's burgeoning love affair with the automobile" captured China's sustainability challenge to affluent nations succinctly: "The day their future looks like our present, we're done for."[61] This troubling resemblance is growing, from the congested ring roads of Beijing to the South China Mall in Dongguan, reported to be the largest shopping center in the world.[62]

The motivation to launch transport revolutions in China—particularly in conjunction with her massive urban development—is thus quite high. The stakes are large, and the global consequences of not reducing the oil consumed by transport are sufficiently serious that substantial encouragement for China's transport revolutions should be forthcoming from the international community. Opportunities for international support of Chinese transport innovation, akin to the World Bank loan that enabled Japanese deployment of the Shinkansen, are noted below.

As for the US, we highlight the three key points that a successful plan should address in order to inspire and guide change successfully. These include specifying the launching actions that signal a clear break with current mobility approaches and depicting the expected consequences from achieving the plan's mobility and energy targets in 2025. Identifying the wider impacts arising from these transport revolutions will enable leaders and citizens to see the value created by an organized transition away from oil-fueled mobility in contrast to the chaotic and uncertain future that would follow a major global energy crisis. First, however, we present our scenarios for China for 2025, for moving people and then for moving freight.

Moving People in 2007 and 2025

What we propose for the motorized movement of people in the China in 2025 is in Table 5.4, which provides comparisons with the likely transport activity and energy use for each mode in 2007. An underlying feature of Table 5.4—and also Table 5.5, which does the same thing for freight in China—is that the population of China is expected to grow by 9 percent across this period, from 1.33 to 1.45 billion.[63] Thus, in some cases we have shown both total and per capita values. China's population is expected to grow at a slower pace than that of the US. Both rates of population growth could be lower than anticipated if food supply becomes constrained as a result of high oil prices that restrict transport of food and production of fertilizers and pesticides.

Our overarching consideration in constructing Table 5.4 was that use of liquid fuels for moving people—chiefly petroleum fuels—would increase by no more than about 23 percent between 2007 and 2025, according to the target set for poorer countries in Table 5.1. We proposed an increase of 22 percent in use of liquid fuels for transport between 2007

TABLE 5.4 Motorized movement of residents of China in 2007 (estimated) and 2025 (proposed)[64]

Values and totals in this table are rounded to aid comprehension	2007				2025						
Mode	**pkm in billions (except per capita)**	**Fuel use per pkm, in MJ**	**Total liquid fuel use in EJ (GJ for per capita)**	**Total electricity use in EJ (GJ for per capita)**	**Local pkm in billions (except per capita)**	**Non-local pkm in billions (except per capita)**	**Fuel use per pkm in MJ**	**Total liquid fuel use in EJ (GJ for per capita)**	**Total electricity use in EJ (GJ for per capita)**	**Liquid fuel powered pkm**	**Electrically powered pkm**
Personal vehicle (ICE)	500	2.0	1.0		750	500	1.7	2.1		1250	
Personal vehicle (electric)	200	0.0		0.0	1,500		0.5		0.8		1,500
Future transport					500		0.4		0.2		500
Local public transport (ICE)	300	2.1	0.6								
Local public transport (electric)	30	0.4		0.0	1000		0.4		0.4		1,000
Bus (intercity, ICE)	500	0.5	0.3			500	0.4	0.2		500	
Bus (intercity, electric)						500	0.3		0.2		500
Rail (intercity, ICE)	300	0.7	0.2			100	0.5	0.0		100	
Rail (intercity, electric)	300	0.2		0.1		900	0.2		0.1		900
Aircraft (domestic)	150	2.0	0.3			150	1.8	0.3		150	
Aircraft (international)	50	2.3	0.1			100	2.1	0.2		100	
Airship (dom. and int.)	0					100	0.9	0.1		100	
Marine (dom. and int.)	5	1.0	0.0			100	0.6	0.1		100	
Totals	2,350		2.5	0.1	4,250	3,000		3.0	1.6	2,300	4,400
Per capita	1,750		1.9	0.1	5,000			2.1	1.1		

Note: Abbreviations in this table: ICE = internal combustion engine; pkm = person-kilometer(s); MJ = megajoule; EJ = exajoule; GJ = gigajoule. *Electric modes are in italics.*

and 2025. This corresponds to a per capita increase of 13 percent. Use of these fuels could have risen to a higher level in the years before 2025, but by 2025 consumption of them would be falling and would continue to fall thereafter.

Another consideration was that China should experience substantially more growth in the movement of people by 2025 than in the movement of freight. Overall we propose more than a threefold increase in the movement of people (just under threefold per capita) compared with "only" a 60 percent increase in the movement of freight (50 percent per capita).

The emphasis on moving people in China is in sharp contrast to our approach to the US, for which we proposed similar reductions in the movement of people and freight (respectively by 6 and 2 percent). The emphasis is justified because compared with richer countries there is much less movement of people in China than movement of freight in China, and between China and other places. Table 5.3 and Table 5.5 suggest that Americans are associated with four times as much freight movement, in tonne-kilometers per capita, as the people of China. Table 5.2 and Table 5.4 suggest that residents of the US engage in 17 *times* as much motorized movement, in person-kilometers per capita, as residents of China. Put another way, in China in 2007 only 20 percent of transport energy was used for moving people; in the US this share is 72 percent. The US and China use about the same total amount of energy for moving freight, but the US uses almost ten times more energy for moving people.

Even with the proposed major increase in motorized travel by 2025, residents of China will still be traveling much less per capita than residents of the US. The gap will have narrowed, from 17 times as much to 5 times as much.

As for the US, there are several ways in which we could have plausibly constructed a pattern of motorized movement in China in 2025 within the framework imposed by the proposed limits on oil use. Again, we have used our best judgment as to what could be the most feasible and acceptable, aware that for China our judgment may be even more fallible than for the US.

Some particular features of Table 5.4 are worth noting:

- We see a near fourfold increase in personal vehicle use by 2025, most to come from substantial growth in electric—chiefly battery-electric—

vehicles. The present growth in ICE vehicles could continue for a while and then fall back to about 2.5 times the current level. Meanwhile, the number of personal vehicles propelled by electric motors will grow. The higher unit fuel use for this category in 2025 compared with 2007 represents the inclusion of substantial numbers of electric cars as well as electric scooters and bicycles.

- As for the US, "Future transport" in Table 5.4 signifies the availability of new local transport options that may encompass electric jitneys and various kinds of on-demand transport including what is known as personal rapid transport (PRT). The average megajoules per person-kilometer (MJ/pkm) for this category is that suggested for PRT. The lower unit energy use than the US reflects, as it does for other categories, smaller vehicles and higher occupancies. PRT, if deployed, may not be as appropriate for China as it could be for the many low-density settlements in the US and, to a lesser extent, Europe. However, China could develop domestic PRT systems as part of an effort to compete in a world market for such systems.
- Local public transport will have become entirely electrified, partly in response to the concerns about local air quality noted in Chapter 4.
- Intercity rail will have become almost entirely electrified, much of it providing high-speed service. The remaining ICE-based rail will serve low-capacity routes and thus have higher unit fuel use than in 2007.
- Domestic and international aviation will be at or below 2007 levels in 2025, having risen in the years between and then fallen. Rail—and to some degree airship and marine—will accommodate much of the potential growth in intercity travel.
- As for the US, domestic and international aviation will increasingly comprise larger aircraft flying over fewer routes.
- Also as for the US, ocean ships, increasingly wind-assisted, will be beginning to accommodate international travel across water.

Overall in 2025, the motorized movement of residents of China will be much more electrified than that of residents of the US. About 66 percent of travel by residents of China will be electrically propelled, compared with about 27 percent of travel by US residents. This reflects the greater scope for innovation when a transport system is less developed and technology is not so "locked in."

TABLE 5.5 Motorized movement of China's freight in 2007 (estimated) and 2025 (proposed)[65]

Values and totals in this table are rounded to aid comprehension	2007				2025					
Mode	tkm in billions (except per capita)	Fuel use per tkm, in MJ	Total liquid fuel use in EJ (GJ for per capita)	Total electricity use in EJ (GJ for per capita)	tkm in billions (except per capita)	Fuel use per tkm in MJ	Total liquid fuel use in EJ (GJ for per capita)	Total electricity use in EJ (GJ for per capita)	Liquid fuel powered tkm	Electrically powered tkm
Truck (ICE)	900	2.6	2.3		1,600	2.0	3.1		1,600	
Truck (battery)					1,000	1.0		1.0		1,000
Truck (trolley)					500	0.8		0.4		500
Rail (ICE)	1,050	0.2	0.2		1,500	0.2	0.3			
Rail (electric)	1,050	0.1		0.1	3,000	0.1		0.2		3,000
Pipeline	100	0.5	0.0		200	0.3	0.1		200	
Air (domestic)	5	23.0	0.1		5	17.3	0.1		5	
Air (international)	5	9.9	0.0		15	7.4	0.1		15	
Airship (dom. and int.)					100	5	0.5		100	
Marine (domestic)	4,750	0.5	2.4		5,500	0.4	2.1		5,500	
Marine (international)	3,750	0.2	0.8		5000	0.2	0.8		5,000	
Totals	11,600		5.8	0.1	18,400		7.0	1.6	12,400	4,500
Per capita	8,700		4.4	0.0	12,700		4.8	1.1		

Note: Abbreviations in this table: ICE = internal combustion engine; pkm = person-kilometer(s); MJ = megajoule; EJ = exajoule; GJ = gigajoule. *Electric modes are in italics.*

Moving Freight in 2007 and 2025

Our suggestions as to China's freight movement are shown in Table 5.5 in comparison with freight movement and fuel use in 2007. Again, there are several ways in which we could have constructed a pattern of freight movement, and we have used our best judgment as to what could be the most feasible and acceptable.

Our overarching consideration for configuring freight movement in China in 2025 was the same as that for the movement of people. Use of liquid fuels for transport—chiefly petroleum fuels—will have increased by no more than about 23 percent, as proposed in Table 5.1. The actual proposed increase is 19 percent, or about 10 percent per capita. However, the number of tonne-kilometers of freight movement would increase by about 60 percent. Other features of Table 5.5 are these:

- We see a more than threefold increase in road freight traffic, but with less than a third of the increase to be in ICE trucks and vans; the remainder will be in battery-electric vehicles or grid-connected vehicles (GCVs). This large increase would be part of a major shift in emphasis of freight movement from international to domestic, as material prosperity in China increases and importers of Chinese products begin to rely more on local production.
- Rail freight movement will more than double, and the focus on electrification will be even stronger than for road freight movement.
- Movement by pipeline will increase as pipes are laid and some rail and road transport of oil and coal is moved to pipeline.
- Growth that might have occurred in domestic movement of freight by air will be taken up by rail, and in some cases by airship. International movement by air will continue to grow. As for other aviation, aircraft moving freight will be larger and fly over fewer routes.
- Marine transport will not grow at the same rate as much other transport. International marine activity will be limited by a shift to local production. Local marine activity will be limited by its necessary reliance on liquid fuels—in contrast to surface modes, which can switch to electricity.

Rail electrification could be less of a challenge in China than in the US. In late 2006, almost a third of China's rail track was electrified, accommodating about half of all of the movement of freight by rail and perhaps a similar share of the movement of people.[66] A larger challenge could be provision of sufficient electricity, discussed below.

Launching China's Transport Revolutions

The world's future very much depends on China's developing a capacity for sustainable urban living that will create the place for the huge population that will move to urban areas before 2025. Thus, the revolutions at the heart of China's many mobility innovations will need to be mostly urban. Sustainable new "ecocities" have become China's highly publicized approach to reconciling economic development with a clean environment since 1986, when China's State Environmental Protection Administration launched a pilot project in Dafeng City in Jiangsu Province.[67] The highest profile step of this initiative is Dongtan—a community with an area of 86 km² being planned on Chongming Island off the coast of Shanghai. Dongtan is to be one of ten satellite cities around Shanghai. Four of these, including Dongtan, are to have a population of 500,000 or more. Dongtan is being conceived of as "...a zero-pollution, largely car-free, renewable-energy powered, sewage-recycling, green-fringed Utopia."[68] However, most of Shanghai's other satellite cities, and many more cities now under development across China, are entrenching a car-oriented and oil-intensive future.

The revolutions at the heart of China's many mobility innovations will need to be mostly urban.

Even Dongtan's plan lacks a vision of sustainable transport beyond the city itself. The new fixed link that will connect Dongtan to the mainland is being designed for conventional cars and trucks, rather than rail and new forms of advanced public transport. The expectation appears to be that travelers to and from Shanghai will drive across the bridge and park on Dongtan's outskirts. The reality could well be that Dongtan residents and visitors will demand to drive motor vehicles inside the city because cars would be the principal means of accessing the city. Development on adjacent parts of Chongming Island is already oriented toward auto travel. Indeed, a 2008 report suggested that the plan for Dongtan has been orphaned after its local champion, Shanghai Communist Party leader Chen Liangyu, was imprisoned on corruption charges. It noted that, "...a rash of building has begun around the Dongtan site to cash in on a road tunnel that is scheduled to connect the island to the mainland next year.... Many of the new high-rise apartment blocks are marketing themselves as "ecocities," borrowing the language of Dongtan but few of its environmentally friendly principles. Instead, at "Yingding eco-village" tourists can take boat rides, go fishing, and enjoy "cockfighting and goatfighting displays."

If Dongtan's original plan represented the "yin" of Chinese efforts to attain an urban ecological utopia, the "yang" can be found in the nearby satellite city already developed in Jiading district, a case of environmental and social sacrifice for the economic gains from developing an autopia of motor vehicle manufacturing.

Just 20 km from Shanghai's center, $5 billion in development has transformed Jiading's farmland into the "Shanghai International Automobile City," quite consciously creating the Detroit of Asia.[69] The office towers, car parts manufacturing and assembly plants have been complemented with a $300 million Formula 1 racetrack that hosts the Chinese Grand Prix.[70] Energy and environmental risks associated with the car becoming a pillar of China's economic development have been judged to be manageable, or at least creating no worse challenges than what other nations will have to face in the future. In answering the question, "Should Chinese households have cars?" one author set out China's elite consensus that the car is a key generator of prosperity that justifies ignoring concerns about energy and environment,

> ...energy...should never be an excuse for restricting the development of the auto industry and denying people's access to a car culture. Otherwise, Japan and the Republic of Korea, which produce no oil and rely totally on imports, would not be entitled to the development of the auto industry at all.[71]

A less extreme, but equally unsustainable, example of building today's automotive transport into China's future urbanization can be found in Beijing's development plan. In 2004, a multi-centric urban region was proposed, with 11 new satellite cities planned for development by 2020. Citizens of these new cities were expected to make half their trips by public transport, compared to Beijing's 27 percent share of travel by this mode in 2000.[72]

When the detailed plans for the satellite cities of Shunyi, Tongzhou and Yizhuang were presented to China's Municipal Development and Reform Commission in 2006, a vast increase in road capacity was a featured part of the urban design. News reports noted that "the standard for each city is six kilometers of road per square kilometer, almost three times as many as the current standard in Beijing."[73] Such a high automotive capacity suggests that planners expect almost all the travel in these satellites that does not occur on public transport will be in motor vehicles, in contrast to the traditional predominance of walking and bicycling, which accounted for 65 percent of China's urban travel in 1995.[74]

Car-oriented development has thus been the emergent norm in China, despite highly publicized eco-city alternatives. The massive investment in China's road infrastructure has been only partly recovered from transport users. China's road development program has accumulated a public debt of 600 billion yuan renminbi, roughly $79 billion, since highway construction began in 1988.[75] Until recently, taxes on transport fuel have been low by world standards: the equivalent of about 2.6 US cents per liter, compared with typical values of about 10 cents per liter in the US and 108 cents ($1.08) per liter in Europe. Vehicle purchases have supported less than 50 percent of road building costs, providing a major stimulus to motorization from general revenues that echoes the US experience.[76]

China's large "stimulus" investment, initiated late in 2008 to counter the effects of the global recession, supported car-oriented development as a means of stimulating domestic consumption. As we discuss in some detail in Chapter 6, the "stimulus" investment has also provided substantial support for public transport, particularly electrified inter-city transport. We note there too that China is also moving strongly toward electrification of road vehicles, with a focus on battery technology. What is uncertain is how much, if at all, China is moving away for the norm of car-oriented development we identify above.

We believe China has an opportunity to become the world leader in producing grid-connected transport technology, with benefits that far outweigh the profits from manufacturing dominance in a soon-to-be-declining ICE-powered automotive sector. Taking the lead in producing the next generation of advanced transport technology would be much more likely if China's domestic market were to create the core demand for such output. This would be much as the UK's early 19th-century boom in rail travelers did for production of the steam engine and the automotive boom in the US did for the 20th-century production of cars powered by ICEs.

China's strategy of creating the market for the next generation of advanced transport technology would begin with signals that car-oriented urban expansion will yield diminishing returns. Complementary signals would indicate that production of grid-connected vehicles and the technology for integrating them into new cities will gain rapid rewards. As in the case of our proposal for the US to tax oil used in transport and direct part of that revenue stream to compensate transport asset holders who make the shift to grid connection, the Chinese government's fiscal capacity should be used to raise revenue from higher taxes on gasoline and ICE-powered vehicles. These funds would support the conversion of today's car manufacturing facilities to produce trains, trams and grid-connected trucks and cars. New airport and highway construction programs would be terminated, and advanced urban transit and high-speed train programs launched in their stead. A new eco-city transport program could ensure that personal rapid transport (PRT) would be built into the design of new cities, including Dongtan, Shunyi, Tongzhou and Yizhuang. This would be an expansion of the "Community Energy Management" approach, based on the premise that "a significant share of future urban energy consumption is predetermined when land-use and urban form are designated."[77]

If Dongtan's eco-city plan were to be revived, an important change would set the car aside and integrate a PRT system into this community. The key difference from Dongtan's original plan would come in redesigning the bridge linking it to the mainland so that it would be used by PRT vehicles only. This would greatly reduce the bridge's cost, since the weights of PRT vehicles would be much lower and the PRT guideways would require a much narrower structure to move the same traffic volume. The resulting savings, along with further savings from greatly slimming down Dongtan's road infrastructure to accommodate only pedestrians and bicycles, would be allocated to financing a comprehensive PRT system that could serve Dongtan's residents better than a road network. The remaining PRT development costs would be supported by funds from China's new tax revenues on gasoline and ICE-powered vehicles.

A regional rail hub would need to be established on the mainland side of the bridge, facilitating easy connection to Shanghai and other satellite cities. With PRT added to its menu of leading-edge sustainability designs, Dongtan would offer an illustration of how a new generation of grid-connected urban transport can meet future mobility needs. China would then have an alternative urban development model that demonstrated the economic, energy and environmental payoffs from building in grid-connected transport as the principal means of motorized local travel. The time may have passed to recover Dongtan's eco-city ambitions, but such an approach could well be tried in one of the many dozens of new cities to be created across China in the next ten years.

Opposition can be expected to China's turning away from car-oriented urban development, airport expansion and the production of vehicles with ICEs. Large investments have already been made to support car-dependent local and intercity travel and aviation for longer distances, creating assets for both private- and public-sector organizations. At the regional level, the investment relationship that supports much existing transport infrastructure resembles a joint venture between government and private developers. In explaining the tensions in China between regional economic development and environmental policy, one author wrote, "Although government and enterprise appear separate it is more realistic to regard the two as a joint local territorial corporation, with the township government serving as the corporate headquarters and the enterprises serving as the various business arms."[78] Chinese transport revolutions will thus pose a governance challenge for local, provincial and national political leaders.

Local and regional governments are poorly placed to lead a transport revolution that would integrate PRT into China's future urban development, but they are well positioned to follow a lead that originates from above. China's transition from a state-led to a market economy has changed local governments "...into local states with a strong interest in [economic] development... No longer passive agents of the central government, China's [municipal] governments are responsible for local prosperity."[79] The current urban planning context resembles the 1970s American "growth machine" model in which "...the desire for growth provides the key operative motivation toward consensus for members of politically mobilized local elites, however split they might be on other issues."[80] Thus, if grid-connected transport development can be positioned as China's "next big thing" in urban economic development, local and regional governments can be expected to become enthusiastic agents of transport revolutions.

China's "...cities today are first and foremost hierarchically ordered administrative centers even though their primary function is often economic."[81] Much political effort is devoted to negotiating and maintaining local autonomy from provincial and national superiors. Nevertheless, the national government has set the direction for urban priorities such as the one-child policy, which local governments have implemented with considerable effectiveness. When three conditions are present, central government can obtain high levels of local compliance for policy priorities [if]...

- all top leaders agree on the issue.
- all top leaders are willing to give the issue priority.
- the degree of compliance of lower levels is measurable.[82]

In the US, the national highway program, the mortgage interest subsidy and the urban renewal program reflected an elite consensus from the 1950s through the 1990s that the country's future was well served by mass motorization and suburban sprawl. These measures guided state and local governments in transforming urban fringes into a "geography of nowhere."[83] China's leaders will need to produce the fiscal and administrative incentives that turn provinces and local governments into transport revolutionaries.

The national government has the means to reorient transport and urban development away from today's energy intensive and environmentally unsustainable trends. It will discover the motive to do so on realizing the opportunity to launch an economic bonanza from creating 21st-century urban spaces designed around GCVs. China's gains could well rival those that accrued to the US from having been the leading producer of ICE-powered vehicles throughout the 20th century. China is among the best-positioned nations to embrace such a future. The cost of building grid-connected transport infrastructure—PRT, trolley buses, trams and other vehicles—in many new large cities will be lower than the cost of providing the road capacity, parking spaces and other facilities that would otherwise be needed to support ICE vehicles. The capital cost of urban development will be lowered as well as the mobility costs of the residents of these new cities. Retrofitting existing cities will be another challenge, perhaps to be embraced fully only when the deployment of revolutionary transport in new cities is the norm.

Emulating Dongtan's potential integration of the grid into urban mobility in the many dozens of new satellite cities that will arise across China would create a critical mass of demand for advanced transport propulsion, vehicles and guideways. Producing for this market would provide Chinese industry with powerful cost and performance advantages that would boost exports of this technology. The initial costs of repositioning China's urban transport trajectory might even serve to help stimulate China's domestic economy at a time when falling demand for exports has yielded manufacturing overcapacity. China's national leaders thus have every reason to use their influence to stimulate the design and production of technology that will support future transport revolutions.

Electricity Generation in the US and China

The above scenarios for 2025 would represent substantial shifts toward electric traction, and corresponding increases in electricity generation—unless there were compensating reductions in other sectors. We began a discussion of how an increase in electricity consumption for transport might be accommodated in Chapter 3. There we made several relevant points:

- Switching *all* the current oil-fueled personal transport to electricity in a richer country could require 15–40 percent of current electricity consumption, plus another few percentage points for switching road freight. Generating *capacity* may not have to be increased as much.
- Until about 2025, such an increase in electricity consumption *could* be provided by coal-fired generation, although after that time world coal availability could begin to decline. Whether it *should* be produced from coal is another matter.
- At least as controversially, although with potentially less impact on the environment, nuclear energy could be used to provide some increased generation; however, what is now the main nuclear fuel, uranium, could be in short supply before 2025.
- Wholesale conversion to renewable means of electricity generation—from wind, thermal and photovoltaic solar, tidal and geothermal sources—is desirable for several reasons, and is possible, although not likely by 2025.
- In many richer countries, there is huge scope for reducing electricity use while maintaining comfort, convenience and productivity.

We continue the discussion of electricity generation and consumption chiefly to examine what the above proposed increases in electricity consumption for transport might mean for the US and China and to suggest how they might be accommodated.

Our discussion is based on Table 5.6. The upper part of this table shows for the US and China, for 2007 and 2025, installed generating capacity, electricity generation and the generation required for the above scenarios for 2025. Values for 2007 are actual or likely values, extrapolated from recent trends. Values for 2025 are "business-as-usual" projections for each country.

Both countries are expected to experience substantial growth in generating capacity and generation, much more for China than for the US. The key feature of the upper part of Table 5.6 is this: the electrification of transport proposed until 2025 would require use of 6 and 8 percent respectively of the expected electricity consumption in the US and China in that year. If there were to be a lesser increase in consumption the maximum shares would be greater, but no more than 7 percent in the US and 17 percent in China.

The additional amount required for transport electrification would not increase total generation if there were commensurate savings in other sectors. Investment in electricity conservation in both the US and China is at a low level compared with investment in generating capacity,[85] yet investment in conservation is often more cost-effective.[86]

The lower part of Table 5.6 shows the primary energy used in electricity generation in the US and China, both now and what is expected in 2025 if present trends continue. In both countries, more than 70 percent of generation is and will be from fossil fuels,[87] and in both countries the share of renewable energy sources in the mix is expected to decline by 2025. The growth in generation from coal—44 percent in the US and 125 percent in China—could be particularly alarming on account of the increased emissions of globally and locally acting gases. A proposal to add to uses of electricity, such as this book's central message, could thus also be alarming—unless it comes with an indication as to other ways in which the required electricity could be generated.

The potential for generation of electricity from renewable energy is considerable in both the US and China, much more than is suggested in the lower part of Table 5.6. The American Council for Renewable Energy

TABLE 5.6 Likely electricity generation in the US and China in 2007, and business-as-usual projections for 2025[84]

	US	China
Installed capacity 2007 (GW)	948	448
Installed capacity 2025 (GW)	1076	886
Growth in installed capacity	14%	98%
Generation 2007 (TWh)	4003	2470
Generation 2025 (TWh)	5125	5446
Growth in generation	28%	120%
Generation required for 2025 transport scenario (TWh)	299	428
Generation for transport as a share of all 2007 generation	7%	17%
Generation for transport as a share of all 2025 generation	6%	8%
Shares of 2007 generation		
Oil	3%	2%
Natural gas	20%	0%
Coal	48%	82%
Nuclear	19%	2%
Renewable	10%	14%
Shares of 2025 generation		
Oil	2%	1%
Natural gas	19%	2%
Coal	53%	84%
Nuclear	16%	5%
Renewable	9%	9%
Growth in generation by source		
Oil	-3%	28%
Natural gas	19%	662%
Coal	44%	125%
Nuclear	12%	398%
Renewable	21%	42%

(ACORE) estimates that renewables could feasibly be used to provide 635 gigawatts of generating capacity in the US by 2025. The ACORE report concluded:

> Based on the assessments in this report, it is critically clear that if our nation has the courage to commit to change, to make real change in policies that drive market economies and decision making, it will be possible to achieve as much as 25% of our nation's energy supply from renewable energy sources by 2025.[88]

This is well above the requirement of 6 or 7 percent of generation suggested by Table 5.6.

The potential for electricity generation from renewable sources in China appears to be similarly large. One estimate suggested that the feasible potential is 815 gigawatts, although no time frame for its development was given. The report concluded:

> There exists a window of opportunity for renewable energy in China to be considered, and its potential to be maximised through policy decisions. The Chinese government has given high priority to renewable energy as part of future sustainable electricity system.[89]

Our conclusion from this brief examination of the implications of our proposed electrification of transport for electricity generation is that it is feasible to accommodate the required generation through accelerated investment in renewable production of electricity. Because of the importance of transport and doubts about future availability of coal and uranium as well as oil (see Chapter 3), accommodating the proposed transport revolutions in this way would be prudent and expeditious. Electrification of transport would highlight the importance of electricity in future energy arrangements and could serve as a stimulus to the kinds of policy decisions called for in both the US and China.

Moving into the Future

The future is a lot closer for some people than for others. Visionaries and innovators can identify something new that lies just over the horizon and inspire others to make it happen. These changes can make a great contribution to civilization (e.g., the steam trains described in Chapter 1). They can also be fairly trivial (e.g., pet rocks and robot dogs). Such changes usually touch many lives and enrich many more than the people who foresaw them. As the 21st century unfolds, we see a pressing need for society to move into a future in which mobility is no longer dependent on oil. The transport revolutions that can achieve this require little in the way of technological breakthroughs, at least in the short term. What launching the transition to transport without oil will require is considerable innovation in implementing change as compared to inventing new technology.

While there remains plenty of room to improve electric propulsion, battery capacity and grid connections over the long term, the transport revolutions we have explored in this chapter could begin today by deploying "off-the-shelf" technology for railways, trolley buses, trams and metros. Governments around the world, and even some private carriers, are purchasing and installing these grid-connected transport options as we write. What needs to change is the scale of such deployment. What is most needed is innovation in answering the "how to" questions.

Rather than pursuing new technology in the first instance, we see the need for fresh ideas and approaches that make the most of existing technology to reduce the oil consumed by transport. This means moving resources away from research into transport technology that can deliver only longer-term solutions (i.e., after 2025) and redeploying them into the deployment of available technology that can deliver short-term reductions in oil consumption (i.e., before 2015). The key to keeping ahead of oil depletion is to begin transforming our current mobility arrangements sooner rather than later.

Until now, this process of inventing a new future for moving people and freight has been mostly focused on exploring options for change that would be deployed a decade or longer into the future . Much of that attention has been given to changing energy sources in ways that would allow present traction systems to continue for as long as possible. For example, there has been considerable focus on fuel cells as portable electric generators that could keep vehicles moving independently of the grid, albeit with a high energy cost for that autonomy. Other initiatives target biofuel production in order to keep the ICE running well into the future.[90] Little is being done to seize the immediate opportunities for revolutionary change in mobility that arise from connecting vehicular propulsion systems directly to the grid, mostly while the vehicle is in motion, but also with better batteries to allow for autonomous movement.[91] The analysis we have provided in this chapter and the preceding ones suggests that grid connection may offer the best prospects for moving land vehicles away from oil on account of its energy efficiency and readiness for deployment. Massive investment in infrastructure for grid connection may well be what is required to accommodate oil depletion and to protect against the dark prospects for conflict over dwindling oil supplies.

In the next and final chapter of this book, we consider how the people best positioned to make the future of grid-connected transport a reality could help bring about this and other transport revolutions. Given the magnitude of the changes envisioned, strong and decisive political leadership will be required to support the visionaries and innovators, particularly in the two nations facing the biggest challenge from oil depletion, the US and China. We try to imagine circumstances in which leaders will embrace opportunities to launch the transport revolutions that will sustain the comfort and convenience of motorized mobility, forestall global conflict and, not least, leave a better environment for unborn generations.

6

Leading the Way Forward

The closing chapter of our first edition was written in mid 2007, just before the world began to change dramatically—from a transport and energy perspective and in many other ways. In that chapter, we looked forward two or three years—as we will in this closing chapter. We hinged that chapter on the possibility of a "tipping point" that would rouse policy-makers to act on the realization that a peak in oil production was imminent. We said the tipping point would "...alert richer and poorer countries alike to the approach of shortfalls in oil production and the prospect of very high prices [and provide] the strongest possible spur to the kind of appropriate action that was suggested for the US and China in Chapter 5."

Our first candidate for a tipping point was that there would be emergence of widespread understanding of what at the time were mounting concerns that Saudi Arabia's oil production had begun to decline inexorably. Saudi Arabia, the world's major producer, was the source of about 13 percent of all oil consumed in 2006, and its share of anticipated growth in production was expected to be higher. Its production had indeed been falling. However, rather than continue to fall, it picked up again after the first quarter of 2007 and then rose until the third quarter of 2008, by a total of about 10 percent.[1] We and many others were wrong to suggest that Saudi Arabia might not be able to raise its production. Now, with inauguration of production from the giant Khurais field,[2] Saudi Arabia's production appears capable of rising further, although perhaps still not enough to offset falling production elsewhere.

Our second candidate for a tipping point was "...early steep price increases in retail petroleum products, steeper than we have had during the last few years." We did not feel this was a leading candidate. We expected steep price increases only after oil production had clearly entered decline, i.e., after what we expected to be a 2012 peak in production, and only if there had been insufficient preparation for oil depletion.

We were right to suggest that early steep price increases could prove to be a tipping point, but wrong to expect them only after the peak in global oil production. As we discuss in the present edition's Chapter 3, oil prices, which had been rising for several years, began to take off late in 2007. They surged far ahead of what we expected until reaching a peak in July 2008. They then collapsed in the most extraordinary way, and rose again during 2009.

An interesting point for us is that a barrel of crude oil cost about $70 when we completed the first version of this chapter in July 2007, and it cost about the same amount in mid 2009. During the intervening two years, the price more than doubled to $147 in July 2008, and fell to $32 in December 2008. This huge and unprecedented convulsion has been mostly matched by more modest changes in the price of gasoline. In the US—where changes in oil price cause gasoline prices to change more than in most other countries because taxes on it are low—a gallon averaged $3.00 ($0.77/liter) in July 2007, $4.10 in July 2008, $1.75 in December 2008, and $2.58 in July 2009.

In this closing chapter, we do not dwell on potential tipping points because we feel the oil price shock of 2008 provided a wake-up call about the challenges of peak oil production. Instead, we build on the present edition's Chapter 3 to consider the dynamics of events before and after the July 2008 peak in oil prices. We then review the relevant responses of the governments of the US and China to these events, most of which are ongoing. We conclude that the 2008 oil price shock has served to begin moving China toward the scenarios set out in Chapter 5, although we are less sure that the US has embarked on this kind of path. Actions are noted that could put both China and the US on track to deal with their energy and transport challenges while advancing economic recovery.

The Wake-Up Call Provided by an Oil Price Shock

The oil price rise of 2007–2008 was as close to a tipping point for the matters we address in this book as has been seen since the 1970s and perhaps earlier, a tipping point in the sense elaborated by Malcolm Gladwell, who popularized the term. It didn't quite change "everything…all at once,"[3] but the price rise's impact on transport, energy and associated matters has been profound.

One major change has already been touched on in Chapter 3: the sharp decline in the sale of new light-duty vehicles in the US. There, Figure 3.11 shows that a fall in sales of light trucks (i.e., SUVs, passenger vans and pick-up trucks) began late in 2007. It was followed by a decline in sales of regular automobiles in mid 2008. Figure 6.1 shows total sales of all light-duty vehicles in more detail, adjusted for normal seasonal variation and on a per-capita basis to allow for the quite rapid growth of the US population. Sales per capita had been falling slowly for several years. From 2006, they fell a little more steeply. Then, between late 2007 and early 2009, they plunged by more than 40 percent.

It didn't quite change "everything…all at once," but the price rise's impact on transport, energy and associated matters has been profound.

The *New York Times* occasioned this plunge with the front-page headline, "Industry Fears Americans May Quit New Car Habit." It was over an article noting that total sales had fallen from an annual rate of over 17 million (for 2005) to less than 10 million in early 2009.[4] Explanations given for the decline—and for why sales might not recover—included transition of the baby boom generation into retirement, a period associated with lower rates of car purchase. We believe another factor at play could

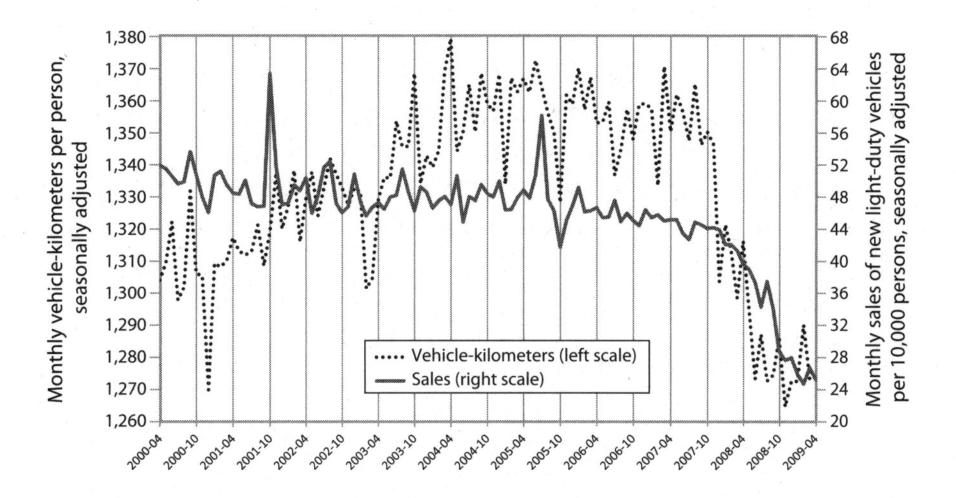

FIGURE 6.1 Vehicle movement and sales in the US, 2000–2009[7]

be the striking decline in driving by young people in North America.[5] In south-central Ontario, for example, weekday trips by 16- to 20-year-olds as a car driver fell by about 40 percent between 1986 and 2006.[6] Contributory factors appear to be higher insurance rates and the introduction of graduated licensing schemes.

The remarkable fall in new-vehicle sales illustrated in Figure 6.1, if continued for a few more months, might be viewed as rapid and substantial enough to qualify as a transport revolution according to the definition proposed in Chapter 1. However, the fall does not constitute such a revolution because vehicle sales and ownership are not transport activity but important precursors to transport activity, as discussed in Chapter 2. If sustained, the notable drop in sales could well give rise to a revolutionary decline in transport activity, specifically car driving.

Figure 6.1 shows, too, that car driving had been following something of a similar trajectory to car sales but with smaller relative changes. Vehicle-kilometers traveled per person had been increasing until 2006, although the rate of increase was moderating. This indicator of amount traveled then fell a little until late 2007 when it began to fall much more steeply.[8] Between November 2007 and March 2009, vehicle-kilometers per person fell by "only" 6 percent. This is much less than the 43 percent change in sales noted above. Nevertheless, it was the largest decline across such a period since 1970, the first year for which comparable data are available. Moreover, it was reversal of a long-term upward trend. If sustained for 16 years (less if the previous upward trend is the baseline) the accumulated reduction in driving would qualify as revolutionary according to our definition in Chapter 1.[9]

What these two rather different indicators—of vehicle sales and travel—may have in common is sensitivity to the price of oil and oil products, particularly gasoline. Both indicators have fallen together and substantially *only* during and after oil price shocks, notably those of 1979 and 1990.[10]

What was it about the price of gasoline in the US in late 2007 that might have caused the declines in car sales and travel? Oil prices during the summer of 2007 were hardly higher than during the previous two summers, with weekly national averages rising above $3.00 per gallon in each year. One thing that was different in 2007 was that gasoline prices

What was it about the price of gasoline in the US in late 2007 that might have caused the declines in car sales and travel?

did not fall as much after the summer as they had done in 2005 and 2006. In the fall of 2007, prices stayed near $3.00 per gallon and then began rising again at the end of the year rather than falling to about $2.50 per gallon as they had done in the previous two years.[11] Also, public awareness and acceptance that a peak in world oil production was approaching appeared to be growing.[12]

Similar changes happened in Europe, although there the steep decline in vehicle sales began a month or two later and was not so precipitous. As in the US, the price of transport fuels did not decline by much after the summer of 2007, and soon began rising again. In the UK, for example, the average pump price of gasoline had risen to £0.95 per liter (about $7.00 per US gallon) in 2007, as it had during the previous summer.[13] Then, instead of falling, it rose further, passing the milestone of £1.00 per liter during November 2007. European car sales per capita, which had been steady for some years, subsequently fell by 20 percent.[14]

Table 6.1 summarizes what happened in the US and Europe and also provides information about Japan. There, the steep fall in car sales occurred later still, after the peak in oil and gasoline prices, perhaps because they were already falling more steeply, and perhaps too related in some way to the extraordinary decline in Japan's automotive export market, also shown in Table 6.1. Yet another factor in the later timing of Japan's steep fall in domestic sales could have been a short-term reduction in the tax on gasoline early in 2008 that interrupted the oil-price-induced rise in

TABLE 6.1 Features of the sharp falls in car sales in the US and Europe, and car sales in and exports from Japan[16]

			Monthly sales *or exports* per 10,000 population		
	Per-capita trend before the steep fall	**Period of the steep fall**	**Before the fall**	**After the fall**	**Change**
US (domestic sales)	Sales falling by 2%/year	Dec-07 to Jan-09	44	25	–43%
Europe-18 (domestic sales)	Sales constant	Feb-08 to Dec-08	31	25	–19%
Japan (domestic sales)	Sales falling by 3%/year	Nov-08 to Apr-09	31	25	–19%
Japan (exports)	*Exports rising by 15%/year*	*Feb-08 to Feb-09*	42	14	–67%

gasoline prices.[15] Of interest in Table 6.1 is the indication that the changes may have reduced car sales per capita in the US to the same level as in Europe and Japan.

Examining Strands of the Recent Economic Unraveling

In Chapter 3 we endorsed the opinions of several economists that the oil price rise of 2007–2008 was an immediate cause of the global economic disarray that became intense at the end of 2008 and continued to unfold in 2009. Support for this position appears to be growing,[17] although the conventional wisdom remains that the current recession is mostly the result of risky lending behavior and even corruption on the part of US financial institutions.[18]

What other factors might have contributed, particularly to the decline in vehicle sales in the US? One factor often invoked as a trigger of the downturn is the collapse of the US housing bubble. A surge in home prices had begun in 2001. In late 2004, the rate of increase in prices began to fall. During 2005, mortgage delinquencies and foreclosures began to rise, and home sales and housing starts began to fall. In early 2006, home prices began to fall,[19] as did the amount of borrowing in the form of mortgages.[20]

Employment in residential construction also began to fall in 2006, shown in Figure 6.2. The figure shows too that at about the same time the employment decline in vehicle and parts production quickened, presumably a response by manufacturers to the steepening decline in vehicle sales noted in Figure 6.1. The more-or-less simultaneous falls in home and vehicle sales are suggestive of links between the two, or between each of them and another factor.

One link between housing and transport is discussed briefly in Chapter 3: that rising prices of transport fuels in the middle of the decade may have contributed to problems in the US housing market, especially by making living in auto-dependent suburbs less affordable and thus depressing the demand for, and value of, much recently built housing. In other words, some of the problems in the housing market and the fall in vehicle sales and travel may all have been consequences of the oil price increase.

Another link posited between the housing and automobile markets has been that when house prices were rising, homeowners were "using

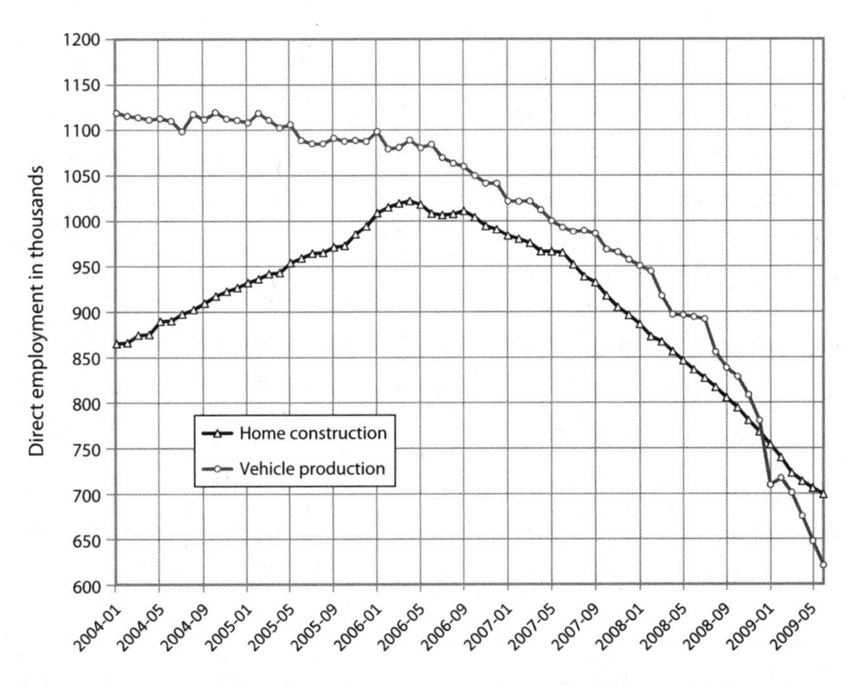

FIGURE 6.2 Direct employment in home construction and vehicle production, seasonally adjusted, US, 2004–2009.[21]

their homes as cash machines."[22] By this is meant that as properties increased in value, homeowners were able to spend the rising equity in their properties through remortgaging their homes or borrowing against them in other ways. Thus, when the housing market collapsed, big ticket purchases such as automobiles could have fallen as a consequence.

Thus, there are reasons for viewing the interacting problems in the US financial, housing and auto sectors as multiple factors contributing to the economic recession. While the US economy was increasingly burdened by the weight of these economic problems during 2007, evident signs of recession did not emerge until after the truly steep falls in vehicle sales beginning in late 2007 (Figure 6.1) and in employment in the vehicle and parts industries (Figure 6.2). Even with this steep fall in employment, the overall unemployment rate did not begin to rise until May 2008.[23] Gross domestic product (GDP) did not begin to fall steeply until later in 2008.[24]

The economic impact of a financial sector meltdown alongside downturns in the housing and auto sectors will take many more years to untangle. Retrenchment in the auto sector, spurred by increases in oil price, played an important role. The main argument for viewing the auto industry's collapse in late 2007 and early 2008 as causative is that the timing of the recession was more closely linked to changes in oil price and the auto sector than to changes in the housing sector. Also, although direct employment in both sectors was similar before the recession (see Figure 6.2), job losses in the auto sector could have had a larger economic impact because the jobs paid more and may have had a larger multiplier effect.[25]

The conventional wisdom remains that distress in global financial markets was the main cause of the recession, resulting from risky lending behavior for home mortgages and other practices. Perturbations in these markets began in 2007, prompted in part by US mortgage defaults.[26] They could have contributed to the fall in car sales by making borrowing for car purchases more difficult. However, US consumer borrowing did not begin to fall dramatically until late in 2008, well after car sales began their steep decline in 2007. Indeed, the decline in consumer borrowing from levels of the previous years may well have been a *consequence* of the fall in car sales rather than a cause, in that car purchases are a major reason for consumer borrowing.[27]

For commentators writing in early 2008, the evidence of dysfunction in the US and other financial systems in 2007 seemed hardly different from most previous episodes of disruption.[28] The financial industry's meltdown did not occur until after the July 2008 oil price peak, although the factors that triggered it could have been at work for some time in advance. The first among an extraordinary series of financial crises in September 2008 was repossession by the US federal government on September 7 of two then-private agencies, the Federal National Mortgage Association and the Federal Home Loan Mortgage Corporation. This incurred a potential cost of $200 billion, chiefly through mortgage delinquencies.

Then, on September 15, 2008, Lehman Brothers, a global financial services firm, filed for what became the largest bankruptcy in US history. It was followed by massive bailouts of financial institutions by governments in the US and Europe. The bailouts began with the purchase of the insurance company American International Group (AIG) by the US government for $85 billion on September 16 (an investment subsequently increased to more than $180 million). These happenings were followed by extreme distress in money markets reflected in near collapse in early October in the ability of banks to borrow funds from other banks.[29]

Global Economic Crisis

As we proposed in Chapter 3: rising oil prices, which were behind the dramatic decline in car sales, have been an underlying cause of the current recession in the US and perhaps elsewhere. Oil price increases also weakened the housing sector and played a role in undermining an already fragile financial sector. Other effects of rising oil prices, such as mounting losses in the aviation sector, for example, likely also played a role. However, in our discussion of how the next few years might unfold, we will give prime importance to oil prices and to their relationship with automobile manufacture, sales and use.

Our focus on autos rather than housing, finance or aviation, is justified by the pivotal position of the auto sector in transitioning to the kind of energy diversification that will be needed to successfully adapt to oil depletion. In the US, for example, manufacture and use of road vehicles employs about five times as many people as manufacture and use of aircraft (about 15 times as many if paid road vehicle operators are included) and contributes about six times as much to GDP.[30] There is roughly six times as much movement of people by car as by air, and roughly 50 times as much movement of freight by truck as by air (see Tables 5.2 and 5.3).

World oil prices rose from 2002–2008, we argued in Chapter 3, because demand was beginning to be constrained by perceptions that supply could not continue to rise. Prices became high enough to constitute an oil shock that triggered a recession in the US through significant and interconnected impacts on the auto, housing and financial sectors. Other economies were affected because they were tightly linked to the US through trade or because of direct impacts of oil prices on them, or for both reasons.

The effect on world trade was remarkable. After increasing by 75 percent from the first quarter of 2005 to the third quarter of 2008, the value of world trade fell by 37 percent during the next two quarters.[31] The decline in the amount of goods movement represented by this change could well constitute the beginning of another potential transport revolution, although confirmation will have to wait until data on freight movement become available.

The recession eased the demand pressures on oil and oil products. Prices tumbled, but began to recover early in 2009. Oil prices rose again more because supply constraints continued—and may even have been tightened by lack of investment[32]—than because demand was picking up again. In May 2009, *The Economist* suggested that it is hard to find anyone in the oil industry who does not believe that prices could rise even higher than the "giddy leap" to above $147 a barrel during July 2008.[33]

In mid 2009, the global economic situation seems as hard to predict as ever. Unemployment is not falling, and there is debate over the need for more "stimulus." The price of a barrel of oil has declined a little in response to intimations of continued economic gloom and then risen to above $70 again. Wise people do not venture a strong opinion as to whether a year or two from now the global economy will be in better or worse shape or as to whether oil prices will be higher or lower than they are at present. But, we do see enough evidence to suggest that changes in the transport sector, particularly in the auto industry's restructuring, could deliver an essential contribution to restoring economic capacity to function in a future with less oil.

Transport's Position in the Vicious Cycle of Limited Oil Supply and Economic Distress

Figure 6.3 sets out our simple model of the vicious cycle that was set off by the 2008 oil price spike. These destructive forces were propagated by the pivotal role that oil-fueled motor vehicles play in the economic activity of richer countries, particularly those with a substantial auto industry.

Stage 1 of the cycle reflects the condition during the run up to the oil price peak in July 2008. Then, the key factor in the convergence of demand and supply was the growth in oil consumption in and oil imports by industrializing countries, chiefly China. In subsequent iterations of the cycle, tightening supply, exacerbated by falling investment, could make a more important contribution to the resulting price increases represented as Stage 2.

If personal transport is fueled by oil products, as it is almost entirely now, and the price of the fuels rises steeply or unexpectedly or crosses salient levels in relation to income—e.g., $4.00/gallon in the US in mid 2008—Stage 2 leads to Stage 3, which comprises falls in vehicles sales and travel.

The extent to which Stage 3 leads to Stage 4, an economic slump, and the depth of the slump, depends, among other things, on the importance of the automotive industry in the national economy. Countries with significant auto production can be hit hard as production and employment

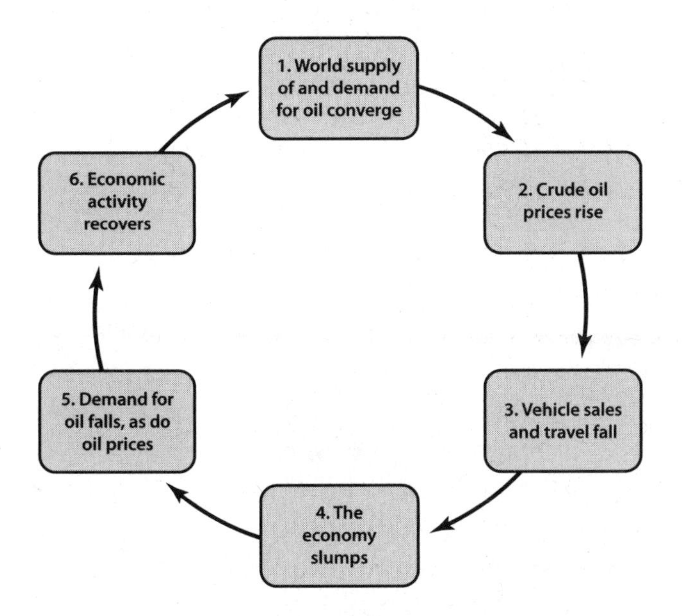

FIGURE 6.3 Vicious cycle of oil price rises and recessions

in this sector drop off. Countries without an automotive manufacturing industry can also experience economic decline through one or more of: (i) reductions in trade with affected countries; (ii) reductions in non-manufacturing aspects of automobile use (e.g., consumer sales, repairs, etc.); and (iii) disruptions to the global financial system resulting from reduced economic activity. The interaction of declines in the auto industry with housing, finance and other sectors can ramify these economic effects. Stage 5, falling demand for oil and falling oil price, is a usual consequence of such economic deterioration.

The recovery of economic activity in Stage 6 could occur in part because oil prices have fallen, as in Stage 5, but also because recovery is a usual sequel to an economic slump. One factor could be "pent-up demand" for cars and other vehicles, expressed after purchases forgone in Stage 3. A result could be an increase in oil consumption, or at least a halt to a fall in consumption. Attrition of oil production, exacerbated if it occurs by rising demand, would return the cycle to the dynamics of Stage 1.

With each iteration of the cycle, the economy could rebound to a lower level. This could mean that subsequent iterations would have a more modest impact: an economy cannot be hit so hard if it is already down. Nevertheless, the prospects for an economy caught in this vicious cycle would be ominous if such a pattern could not be broken.

The main way to break out of the downward spiral would be to sever the link between Stages 2 and 3, so that key transport and transport-related activities become less affected by subsequent rises in crude oil prices. The main aim of this book is to demonstrate the advantages of reducing transport's dependence on oil, chiefly by shifting land transport to electric traction. Among the many benefits of moving people and freight without oil, escaping the vicious cycle of economic decline that became evident in 2008 may provide the most powerful stimulus for initiating transport revolutions.

There are other ways of breaking out of the vicious cycle. One is to increase oil production, and thus prevent convergence of demand and supply. As elaborated in Chapter 3, we believe this will not be possible for much longer. Production in some places—e.g., from Canada's oil sands and off the coast of Brazil—can continue to grow. However, such production of "unconventional oil" is much costlier than production of the "conventional oil" that comprises most current output (as illustrated in Figure 3.7).

Another way to break the vicious cycle would be to reduce the dependence of the economy on transport. With the exception of aviation, we do not think that such a strategy would be desirable. Modern society and its numerous advantages over former ways of living depend on abundant and efficient mobility, and we see every reason to maintain these advantages except when no alternative to increasing oil consumption can be found. If aircraft could be fueled renewably, or if there were no reasonable alternative to much aviation, then we would not make an exception for aviation.

The vicious cycle portrayed in Figure 6.3 is a simplification to provide some clarity to a complex set of interacting factors. These include interactions among economies through changes in trade and changes in the availability of credit. Such interactions can become vectors for the transmission of economic well-being or malaise. These factors can prevail in richer countries without auto manufacturing capacity and in poorer countries generally. China is the exception—a poorer country (in per-capita terms) that has major automotive manufacturing capacity. It is discussed below in considering China's response to the present recession.

The Neglect of Transport in Searching for Cures for the Recession

Before considering the adequacy of the actions of the US and China in combating the current recession and in moving toward the scenarios set out in Chapter 5, we take a broader look at the recession itself. Two questions among many that have been asked shed light on the forces of change that could trigger transport revolutions. The first concerns how bad the recession is, particularly in relation to the Great Depression of the 1930s. The second is how long the recession might last.

Barry Eichengreen of University of California Berkeley and Kevin O'Rourke of Trinity College Dublin have examined several features of the first year of the current recession, which they say began in April 2008. They compared them with comparable features of the Great Depression, which for Eichengreen and O'Rourke began in June 1929.[34] They concluded the following:

- World industrial output has fallen roughly in step with how it fell in 1929–1930, as has US output. Industrial output of some countries, notably Japan, has fallen much more steeply (as evidenced in Table 6.2, particularly the indication of the large fall in automobile exports).
- On average, stock market indices and trade have both fallen proportionately more than in 1929–1930.
- The responses of governments and central banks have been much speedier and stronger than in 1929–1930. Money supplies have been increased. Interest rates have been brought down more. Government spending to stimulate economic activity has been proportionately much greater.

Eichengreen and O'Rourke's main argument is that "today's crisis is at least as bad as the Great Depression," especially outside the US. The World Bank, which focuses on the fate of poorer countries, has been not quite as apocalyptic. It described the present recession as "the biggest financial crisis since the Great Depression," and projected that "the global economy [will] begin expanding again in the second half of 2009," although not at full potential until 2011.[35] The Organization for Economic Cooperation and Development (OECD), which focuses on richer countries, has offered a similar message. The present crisis is "the deepest decline in post-war history. The ensuing recovery is likely to be both weak and fragile for some time. And the negative economic and social consequences of the crisis will be long-lasting. Yet, it could have been worse."[36]

None of these assessments pays much attention to oil prices as a factor in the recession. The World Bank echoed the conventional wisdom on the crisis in this way: "Almost two years after problems in the US mortgage market set in motion the biggest financial crisis since the Great Depression, global financial markets remain unsettled, and prospects for capital flows to the developing world are dim." The OECD added: "The downturn has been global in scope, even though its financial epicenter was in the OECD area. Indeed, trade and financial linkages prompted a synchronized collapse in activity and trade after financial markets froze in the second half of 2008." For the most part, the role of an ongoing vicious cycle of the kind we have sketched out above has not been considered in these and other mainstream economic attempts to characterize either the dynamics or the magnitude of the current downturn.

We are as skeptical of the guarded optimism offered by the World Bank and the OECD as we are of their emphasis on remedying defective financial arrangements as the key to curing our current problems. Such a prescription has caused governments to compensate for the depredations of the current economic recession chiefly by using borrowed funds to buttress financial institutions and stimulate economic activity, almost any economic activity, and then inflating their way out of debt.[37] Such strategies may well produce short-term benefit, but with the longer-term costs of high inflation, insufficient preparation for oil depletion, and resulting economic stagnation.[38] Our view is that similar short-term benefit could be realized through reducing modern society's dependence on oil for mobility. Unless this objective becomes the centerpiece of government strategy, and we move toward something resembling the scenarios of Chapter 5, we can expect further setbacks of worsening impact.

In the rest of this chapter, we will apply this perspective—advancing economic recovery through moving transport off oil—to examine briefly some of the responses of the governments of China and the US to recent events, and we will look at some of the ways oil prices and economic activity could unfold during the next few years. Then we'll assess the government responses in terms of their potential effectiveness making progress toward the kind of scenarios for 2025 outlined in Chapter 5.

To put this another way, building on an analogy used in Chapter 5: We are in a hole and need to get out. Do the government responses to recent economic distress amount to making the hole deeper? Or do they help us to get out by filling in the hole or by constructing a ladder to get out of the hole?

Chastened by recent events, we recognize even more uncertainty in looking ahead than we expressed at the beginning of Chapter 5. There, we noted that prediction is hard "especially about the future." We could add some words by the rediscovered economist, John Maynard Keynes. He wrote, in reference to such matters as the future of oil prices and responses to them, "...our knowledge of the future is fluctuating, vague and uncertain."[39] We might add too that our certainty about the future extends only a little beyond knowing that the sun will rise tomorrow and that most of us will be able to benefit from it. Despite our uncertainty, we believe it is essential to try and puzzle through the prospects for reducing our use of oil in transport systems.

Rescuing Chrysler and General Motors

The most significant transport restructuring initiative of the US government during the last year has been the provision of emergency financial support to the three major automobile manufacturers. For the second largest, Ford Motor Co., this has amounted only to extension of a line of credit, so far unused. For the third largest, Chrysler Corp., the support comprised grants, loans and eventual assumption of eight percent of the equity of a reorganized company, now known as Chrysler Group LLC. For the largest, General Motors Corp. (GM), the support comprised grants, loans and eventual assumption of 60 percent of the company's equity.

Much of the funding has been provided through the Troubled Assets Relief Program (TARP), established in October 2008 to provide support for financial institutions in difficulty and extended to the auto industry in December 2008. By June 2009, the total amount made available to the auto industry from TARP appeared to be $85.3 billion.[40] More may have been provided from other sources.

The two auto manufacturers are undergoing radical restructuring, mostly guided by the analyses of the Presidential Task Force on the Auto Industry. In March 2009, the Task Force concluded that the restructuring plans proposed by the managements of Chrysler and General Motors "did not establish a credible path to viability...[and]...were not sufficient to justify a substantial new investment of taxpayer resources."[41] Following this declaration of non-confidence, GM's CEO resigned, and GM and Chrysler pursued bankruptcy protection to accelerate the pace and expand the scope of their reorganizations.

Italian automaker, Fiat S.p.A.—which now owns a 20 percent share of Chrysler Group LLC, with the option of increasing this to 51 percent—has essentially taken control of Chrysler. Fiat has installed its chief executive, Italian-Canadian Sergio Marchionne, as Chrysler's chief executive. He replaced Robert Nardelli, who was positioning Chrysler to become a leader in electric vehicles but said nothing about this when he handed control to Marchionne.[42] Fiat, however, is already a partner in the production of electric vehicles in Europe and has at least one electric vehicle of its own under development.[43]

GM seems intent on becoming a smaller company. Several brands and operations have been sold or terminated. More significantly, GM is restructuring so as to be able to break even in a US market in which there are 10 million annual sales of light-duty vehicles.[44] Average annual sales from 1988 to 2007 were 16.7 million. Figure 6.1 indicates an early-2009 plateau in monthly sales at about 25 per 10,000 population, equivalent to annual sales of about 9.2 million.[45] Thus, GM appears to be planning for recovery to a level of US market activity closer to the recent depressed state than to the level before the recession, say about 11.5 million sales per year, or about 70 percent of annual sales during 1998 to 2007.

Toyota Motor Corp. replaced GM as the world's largest auto manufacturer in 2007, a position GM had held since 1931. In 2008, Toyota sustained its first annual loss on operations since it was founded in 1937, about $4.4 billion. (GM lost over $30 billion in 2008.) In June 2009, the new chief executive, Akio Toyoda, grandson of the founder, Kiichiro Toyoda, said that Toyota will be reorganizing so that it can be profitable while using only 70 percent of its factory capacity.[46]

Thus, the world's two largest auto manufacturers may be anticipating a future in which 30 percent fewer cars are sold in richer countries than has been the case for most of the last decade. As the years pass, such a lower level of sales would result in fewer cars on the road, perhaps 20–25 rather than 30 percent fewer than in the years before 2008 because owners may well hold on to their vehicles longer. Such a reduction in the number of vehicles could be consistent with our scenario for the US set out in Chapter 5, where the first three data rows of Table 5.2 anticipate a 22 percent fall in the amount of travel by personal vehicles (including future transport) by 2025.

A future with fewer cars in richer countries is also consistent with the suggestion near the beginning of this chapter that people in the US may be becoming less interested in owning cars, particularly older and younger adults. Similar concerns have been expressed by automobile industries in Europe and Japan, which have recognized the challenge of predictable "evolution towards an aging and a 'no-car' society."[47]

A challenge for the US government, which now has much at stake in the viability of these two companies, particularly GM, will be reconciling the goals of ensuring good returns on its investments and of steering the auto industry toward production of vehicles that use less petroleum-based fuel, including electric vehicles. Whether reconciliation will be easier or harder as the price of oil rises may depend on how adept the companies become at providing affordable electric and other vehicles. The Presidential Task Force on the Auto Industry was not enamored with GM's prime move in this direction: "...while the Chevy Volt holds promise, it will likely be too expensive to be commercially successful in the short term."[48] One alternative for the industry is to persuade consumers to pay more for small electric cars than they are prepared to pay now. An observer noted, "The challenge is not more difficult than it was to convince US residents who live in one of the 30 states south of the snow belt mostly graced with excellent paved roads that what they really need is a truck with all-weather and off-road capability."[49]

Another challenge for the US government will be defending its actions in the event of failure, i.e., the companies continue to lose large amounts of money. Criticism of the rescue of GM in particular has been stronger in Canada than in the US to date. The governments of Canada and Ontario contributed approximately a sixth of the funds for the rescue, reflecting Canada's share of the production of the integrated, but US-owned, Canada-US auto industry. A critic in Canada noted that Canada has paid $8 billion for an equity stake in GM that at the time had a market value of just under $58 million, with an obligation to sell this stake within nine years. If the value of GM were to rise fivefold during this period, Canadian governments would recover about two cents on each dollar invested. Looked at another way, the investment in the two companies means in effect that each of the vehicles that will likely be sold in Canada by GM and Chrysler during 2009 will be subsidized by about $30,000, i.e., about the same as the average retail value of these vehicles. A third observation was that saving the jobs of the GM employees in Canada is being achieved at the cost of about $2 million per job saved.[50]

In Canada and the US, the most grievous aspect of the auto sector rescue could be the cost of lost opportunities. A more productive strategy might have been to use the funds to foster companies intent on providing for the electric mobility required for the scenarios set out in Chapter 5. In the meantime, the moves by GM (and Toyota) toward accommodating declining personal vehicle use can be welcomed because of their consistency with the scenarios and because downsizing the production of ICE-powered vehicles by major manufacturers could provide market openings for innovative transport businesses to flourish.

Other US Steps Toward Creating New Mobility Options

Apart from the above actions concerning the viability of the major auto companies, there are five significant enacted or proposed legislative measures of the US federal government that are relevant to attaining the scenarios for the US developed in Chapter 5.[51]

The first two of these antedate the present administration. One is the Passenger Rail Investment and Improvement Act of 2008, which provides for the continuation of Amtrak, the national passenger rail service, and for a modest strengthening of the intercity passenger rail system. Among other things, it authorizes the Department of Transportation to support development of high-speed rail service along the ten corridors indicated in Figure 5.1, to the extent that funds are available. The only substantive action by mid-2009 was the issuance of a request for expressions of interest in developing high-speed rail (defined in the legislation as having operating speeds of 177 km/h or greater), with responses to be submitted by September 2009.[52]

The other product of the previous administration is the $25-billion Advanced Technology Vehicles Manufacturing Loan Program (ATVMLP), a feature of the Energy Independence and Security Act of 2007. The first three loans under this program were made in June 2009. Nissan North America Inc. secured a loan of $1.6 billion to support annual production of 100,000 battery-electric automobiles at its Smyrna, Tennessee plant by

2013. Production of this vehicle for the US market has already begun in Japan. It is said by Nissan to seat five, travel for 160 kilometers before recharging, and be fit for use on all highways.[53] Ford was lent $5.9 billion to re-tool several plants to produce 13 fuel-efficient models, including 5,000 to 10,000 battery-electric vehicles a year beginning in 2011. Tesla Motors received approval of a loan of $465 million toward development of an affordable family sedan based on its battery-electric Roadster Sport, which is in production and costs $105,000 to purchase. Legislation is before Congress that would double the annual allocation to ATVMLP. The differently powered vehicles whose mass production is fostered by ATVMLP could replace the models and brands that are now being phased out of Chrysler's and GM's business plans.

The American Recovery and Reinvestment Act of 2009, passed by Congress in February 2009, provides for "stimulus" spending of up to $787 billion. Most of this is for tax reductions, support for states and municipalities, health care and education. The remainder includes $126 billion for investment in infrastructure and science and $65 billion for investment in energy.[54] Within these allocations are four items that could advance attainment of the scenarios in Chapter 5:

- $9.0 billion for intercity rail, with priority for high-speed rail
- $8.5 billion for local and regional public transport
- $2.4 billion for development of electric vehicle technologies with a focus on batteries
- $18.8 billion for improvement of the electrical grid.

As well, there are funds for federal agencies to purchase electric vehicles and funds to subsidize consumer purchases of electric vehicles. Investment in established transportation is mostly confined to $27.5 billion for highway and bridge construction.

The measures proposed in the 2010 US federal budget include provision of a $5.0 billion contribution toward development of high-speed rail, which could be the first of five such annual installments.[55] The 2010 budget would also terminate work on fuel cells for mobile applications. When explaining this withdrawal of support for fuel-cell work, Energy Secretary Steven Chu said this is a technology "for the distant future." He added, "In order to get significant deployment you need four significant technological breakthroughs. That makes it unlikely."[56] He did not note the objection to this technology we elaborated in Chapter 2: using renewably generated electricity to make hydrogen and then using the hydrogen to make electricity is an extremely inefficient use of the renewable electricity. By the time the budget is approved, the cuts for fuel cell work may well have been restored in which case the debate would continue.

The above four actions and proposals are encouraging in that they will and could serve to move the US toward the scenarios set out in Chapter 5. Another encouraging line of policy development is actual and proposed requirements to improve the average fuel economy of light-duty vehicles. Presently the standards for 2010 model vehicles are 27.5 and 23.5 miles per gallon (mpg) for regular automobiles and light trucks (SUVs, passenger vans and pick-ups), respectively. For the 2016 model year, they are to be 39 mpg for cars and 30 mpg for light trucks.[57] (39 mpg is 6.0 L/100 km.) The US government believes that requiring manufacturers to improve the average fuel economy of the vehicles they produce could encourage the development and production of electric vehicles of many kinds.[58]

At variance with the above indications of progress could be moves to delay replacing the legislation that articulates federal surface transport policy in the US. The present law is the *Safe, Accountable, Flexible, Efficient Transportation Equity Act of 2005*, most of the provisions of which expire in September 2009. An ambitious successor bill has been drafted by the House of Representatives' Transportation and Infrastructure Committee. It would raise spending levels substantially above those mandated by the present *Act*, especially on intercity rail, public transport and facilities for non-motorized transport.

Federal funding for projects that flow from this legislation now comes mostly from the proceeds of the federal gasoline tax (18.4 cents per gallon), which have become increasingly inadequate because of inflation and declining fuel consumption. Options suggested for funding the proposed enhanced spending include raising this tax, tolling many more roads, and taxing oil speculators. The president and the secretary of transportation want the present legislation extended until after the elections in November 2010. Members of the Transportation and Infrastructure Committee want their bill to proceed now. They remind the president that he once gave "the fierce urgency of now"—words of Martin Luther King—as his reason for entering politics.[59] Unlike most presidencies, the first strong criticism of President Obama's regime by otherwise sympathetic commentators has arisen from concerns related to the transport sector.[60]

Approval of the proposed 2010 budget and successful attempts to advance the new surface transport legislation could mean that the US will be making progress toward the scenarios we set out in Chapter 5, particularly with respect to intercity rail. Annual federal funding for passenger rail could rise to $25 billion. State and local governments might add a similar amount. However, the total of $50 billion would be well short of the $140 billion per year we suggested in Chapter 5 would be required for upgrading passenger rail sufficiently to implement the scenarios for the US. Thus, even with respect to what may be the US government's most urgent transport priority—inter-city rail, particularly high-speed rail—initiatives to date will need to go much further to prepare the US for oil depletion.

One action at the state level is particularly relevant to attaining the Chapter 5 scenarios. In November 2008, California voters approved construction of a high-speed rail line between San Francisco and Anaheim (50 km beyond Los Angeles) and authorized issuance of a state bond for $9.95 billion chiefly for this purpose. The bond would fund the first stage of what has been described as the "the most expensive single infrastructure project in United States history."[61] Eventually, there would be branches to Sacramento and to San Diego. The lengths of the initial stage and the total system would be 835 and 1,290 kilometers respectively. Infrastructure and trainsets are projected to cost some $34 billion for the initial stage and $11 billion for remainder (totaling about $40 million per kilometer).[62]

Assuming 2008 costs of construction, driving and air travel, revenues from the first stage (to be completed by 2020) are estimated to be sufficient to pay for the second stage (to be completed by 2030). The full system is expected to replace most air travel and about 15 percent of intercity road travel within the corridors. The net financial benefit of the built-out system by 2050 is expected to be close to $100 billion, in 2008 terms.[63] If, as we suggest, oil becomes scarce or expensive or both, the direct benefits could be higher and the indirect benefits could be incalculable. The latter would chiefly be having a means to travel between California's major population centers mostly fueled by renewably produced electricity (perhaps from concentrating solar power plants in California's deserts).

The California High-Speed Rail Authority expects about $14 billion of the $34 billion cost of the first stage to come from the US federal government, and it is not presently clear how that contribution will be made. The balance of $11 billion (i.e., beyond the $9 billion to be borrowed by the state) is to come from local governments and private investors. A further wrinkle in the prospects for implementation may be the perilous conditions of California's state and local finances and governance.[64]

These initiatives fall short of the revolutionary changes needed to shift US mobility away from today's near total dependence on oil. They do, however, introduce the potential for US mobility to keep ahead of oil depletion. Much more will be needed to provide the pace of change we suggest. Actions elsewhere, particularly in China, could spur awareness that such transport revolutions hold the key to a sustainable economic recovery.

Actions by the Government of China Suggest How to Break the Vicious Cycle

While the US is taking the first steps in a generation to upgrade its passenger rail system, China has embarked on what has been described as the "largest railway expansion in history," committing more than $1 trillion to expand the rail network from 78,000 to 120,000 kilometers of which 18,000 km is to be capable of supporting high-speed passenger service (200–350 km/h). Much of this is to be completed by 2012. Almost all the new service will be electrified. The longer term plan is to provide high-speed rail service to all 300 or so cities with a population of more than 200,000.[65] The numerous foreign interests attracted by this astonishing burst of activity include the information technology company IBM. In June 2008, in partnership with numerous, mostly US partners—including the California High-Speed Rail Authority—IBM opened its Global Rail Innovation Center in Beijing.[66]

China's intercity rail expansion plans antedated the current recession but have been given impetus by it. Rail expansion is a major beneficiary of the RMB 4 trillion ($585 billion), two-year fiscal stimulus package approved by the government of China in November 2008. In all, a total of more than $500 billion is to be invested in rail infrastructure and rolling stock over the next three years.[67] The massive expansion of rail service in China, particularly electrified rail service, is entirely consistent with the scenarios for China we proposed in Chapter 5.

Urban transit is another focus of China's investment that has been given further impetus by the stimulus package, with much of this investment being made by provincial and municipal governments and agencies responding to pressures from the central government. New subway lines or major extensions to existing lines are under construction in at least 15 cities, and 12 more are planning them.[68]

The stimulus package, which also includes investment in road and aviation infrastructure, is aimed at spurring domestic spending in response to the decline in international trade that began in the second half of 2008. According to one estimate, "growth from China's fiscal and credit stimulus is bigger than the negative shock of the global crisis."[69]

One outcome of China's stimulus package has been a surge in sales of motor vehicles, chiefly automobiles, to a level that may be higher than the trend before 2008. Sales data are not as readily available as the closely related production data, shown in Figure 6.4. Vehicle production mostly fell during the last eight months of 2008, but then rose steeply enough to resume or even exceed the pre-2008 trend.

China is taking many steps to provide the electric power needed for transport electrification.

Why vehicle production in China fell between April and December 2008 requires further analysis. For the moment it can be noted that the fall was not evidently related to a decline in consumer confidence. The National Statistics Bureau of China's Index of Consumer Confidence did not begin to fall until August 2008, when it went below 94.0 for the first time, and then fell every month until April 2009.[71] Manufacturers generally could have been getting an early signal from a decline in business, particularly export business, but trade with the US, the main importer of China's goods, did not begin to fall until November 2008 (and it has fallen every month since then).[72] There was an 18 percent increase in retail gasoline prices in June 2008, but the decline in vehicle production started before that. Perhaps the only remaining plausible explanation is that vehicle manufacturers were aware of the rapid increase in world oil prices in the first half of 2008 and cut back on production in anticipation of a resulting fall in sales.

Factors contributing to the surge in vehicle production and sales in early 2009 include subsidies for rural purchasers and reductions in taxes on car purchases, both measures to encourage consumer spending that were in the stimulus package.[73] Two increases in gasoline prices early in 2009 did not appear to be much of a deterrent to car purchases.[74]

The rapid growth in vehicle sales in China may be inconsistent with the scenarios in Chapter 5. There we anticipate an almost fourfold growth

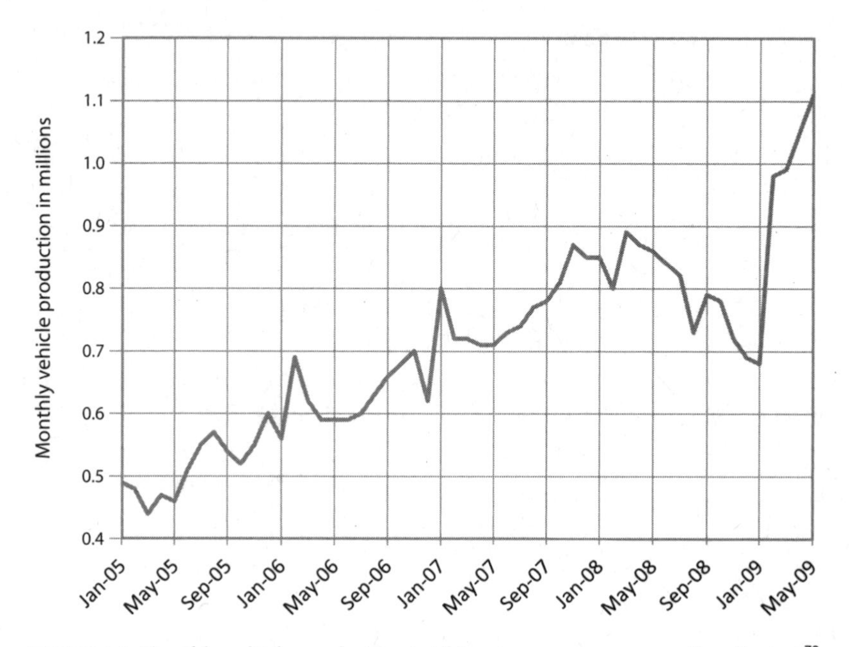

FIGURE 6.4 Monthly vehicle production in China, 2005–2009, seasonally adjusted.[70]

in travel by personal vehicles between 2007 and 2025 (see Table 5.4). This may correspond to a similar growth in vehicle production, i.e., at an annual rate of about 8 percent per year. According to Figure 6.4, the rate of growth during the period May 2007 to May 2009 has been much higher: almost 25 percent per year. Comparisons are complicated by the inclusion of electric two- and three-wheelers in Figure 5.4 and not in Figure 6.4. Another factor making assessment difficult is lack of information about the longevity of vehicles produced for the Chinese domestic market. Nevertheless, the actual rate of growth in 2007–2009 would appear to be much beyond what might be required to stay within availability of fossil fuels.

What could make a critical difference in the consequences of travel by personal vehicles in China is the pace of their electrification. In Chapter 5, we suggest that by 2025 most travel by personal vehicles in China should be powered by electric traction. The planned current rate of electrification of automobiles is high. The share of electric versus ICE-based propulsion is to rise from essentially zero in 2009 to about 3.5 percent by 2011. If the share grows at that rate until 2025, perhaps a third of travel by automobile will be powered by electric traction. Factoring in a larger share of electric motorcycles, tricycles and scooters could put the total share above 50 percent. Such growth in electric traction is a major element in a strategy to make China the world leader in electric vehicle technologies of all kinds.[75]

China is taking many steps to provide the electric power needed for transport electrification. Eventually this is to be produced from renewable sources. Toward that end, the plan adopted late in 2007 requires that renewables' share of 7.5 percent of electricity generation in 2005 be raised to 10 percent by 2010 and 15 percent by 2020. Sources include wind power, solar thermal and voltaic, hydropower and biomass. Nuclear energy, also targeted for expansion, is not included in this total. The plan also speaks to development by 2020 of "a relatively complete renewable energy technology and industry system" that would enable further growth in the share of renewably sourced electricity and position China well to export skills and products related to electrification.[76]

Critical to the exploitation of many renewable sources of electrical energy and conventional sources, is enhancement of what has been described in one report as China's "rickety and fragmented grid." Investment in the grid of some $44 billion is planned by 2012, and a further $44 billion by 2020. Almost all of this would go toward installation of ultra-high voltage (UHV) lines. Early in 2009, China completed the world's first 1,000 kilovolt AC line, 640 km in length, linking Shanxi and Hebei provinces, described as a "breakthrough in the technology of long-distance, large-capacity and low-loss UHV power transmission."[77]

China is also boosting its access to oil. In May 2009, PetroChina, with 75 projects in 29 countries, became the world's largest traded company by capitalization. Sinopec, China's other major internationally operating energy company, has larger revenues than PetroChina. Significant elements of Sinopec's development are construction of oil and gas pipelines from the Bay of Bengal through Myanmar to southwest China, and the development of the giant Yadavaran oil field in Iran. Meanwhile, China continues to expand its capacity for military action beyond its borders. According to one analyst, China's efforts to access expanding volumes of future oil production means that "the probability of a military confrontation...in which oil may play a role has been dramatically heightened."[78]

China Appears Poised between Virtuous and Vicious Cycles

In general, we find what is happening within China, compared with what is happening in the US, to be more compatible with the scenarios we propose for 2025. Although many of China's ongoing policy moves were in place before 2008, they have been greatly accelerated in response to the events of 2008. We are tempted, although with much caution, to suggest that China may be heading into a *virtuous* cycle of economic restructuring rather than the vicious cycle illustrated in Figure 6.3. The key difference from the US and other richer countries is the response at Stage 4, the economic slump that originated from rising oil prices. China's response has been a massive increase in investment in domestic transport that does not depend on oil. As a result, China is developing a flourishing transport manufacturing sector that can lead the world in producing much needed electric road vehicles and rail systems. When the economies of richer countries recover, China will be well positioned to export electric mobility to them, including to the US. If the vicious cycle in Figure 6.3 repeats, China can again at Stage 4 extend its pursuit of such a virtuous path toward economic development that does not rely on oil.

China would thus be contributing not only to attainment of the Chapter 5 scenarios for itself but also for the US. This could occur through a sizable drop in the price of electric rail and road vehicles, which can be anticipated given China's intended scale of production and the cost-cutting prowess that has secured China's reputation in other manufacturing areas. The overall result could go a long way toward enabling other countries to meet the imperative of reducing world oil demand to within the envelope of declining supply. China could thereby become the economic superpower of the post-petroleum era, much as the US led in developing and reaping the economic benefits from automotive and aerospace technology deployed during the oil age.

The reality, however, is that China's dependence on oil-fueled transport is also growing, as shown in Figure 6.4 and by the efforts noted above to secure future oil supplies. These trends could offset the trends toward transport electrification and lead China more into a vicious than a virtuous cycle. China is poised between these two paths, and it is impossible to predict which will be taken. However, there is every reason for hope that China's revolutionary tendencies will prevail at this critical juncture of energy and transport redesign.

Concluding Remarks

Oil prices have mostly risen in the first half of 2009, fueled chiefly by China's rapid growth in oil imports, in turn fueled at least in part by the extraordinary growth in vehicle sales in China shown in Figure 6.4.[79] Will we again see a repeat of the steep run up in oil prices that occurred in 2008? Will there be another iteration of the vicious cycle portrayed in Figure 6.3? Or have things changed sufficiently that the cycle has already been broken for one or more of the following reasons?

- The current recession has destroyed demand for oil enough to delay a steep run up in prices for several years.
- If oil prices do rise, they will not reduce vehicle use and sales in the US much more because both are as low as they can go without radically transforming the way Americans live.
- If oil prices rise, and vehicle use and sales fall, the economy will be less affected because the transport sector is a smaller part of the economy than in 2008.

We are inclined to anticipate a repeat of the cycle depicted in Figure 6.3. Each of the reasons noted above could contribute to a lessening of the viciousness of the cycle, but recent actions have not been enough to break the cycle altogether. This is particularly true of the US, where the opportunities for change offered by the turmoil of 2008 are still being puzzled through. It is also true of China, where progress on electrification has been immense, but where there has also been more digging of the hole of increased dependence on oil in road and air transport.

The vicious cycle could be intensified by a factor touched on only briefly in this chapter: collapse of the aviation industry. The industry was heading in this direction during 2008 on account of the rapidly rising cost of jet kerosene. The price of this fuel then plunged with the crude oil price, but so did the demand for air transport. In mid 2009, as fuel prices but not ridership have increased again, the outlook seems more precarious than ever. Giovanni Bisignani, director general of the International Air Transport Association, said on June 25, "Even if we look beyond the crisis, it is difficult to see a return to business as usual. This crisis is re-shaping the industry."[80] In the US, cuts by domestic carriers planned for the fall of 2009 mean that capacity will fall to the lowest September level since 1984.[81]

A civil aviation crisis is more likely to roil the US than China, at first.

This is because of the greater role of aviation in the transport sector, the high exposure of US "legacy" carriers to massive debt and pension obligations, and the greater vulnerability of many low-cost carriers to discretionary travel demand. Because there is often little opportunity in the US to travel by rail or bus instead of by air, the pressure to subsidize "essential" air services could be strong, bringing the risk of diverting resources to sustain an unsustainable level of mobility by air.

Maritime transport appears to have been affected even more by the global economic downturn than aviation, but may be showing signs of recovery. The Baltic Dry Index, which reflects the cost of shipping bulk goods such as iron ore and wheat, fell by more than 90 percent between June and December 2008, but by June 2009 had regained a third of the fall.[82] The Hamburg Index, which reflects prices of shipping by container, fell less but by May 2009 was not showing signs of recovery.[83] Some of the surplus oil tankers are being used to store oil. Other vessels are moored, waiting for trading activity to return. Even though the numbers suggest more of a fall in business than that for aviation, the industry does not appear to be in crisis. However, much less is known about marine transport than about other modes—including its vulnerabilities—and superficial appearances could be misleading.

Another oil-price-induced recession, or deepening of the present recession, however unwelcome, could provide a second, unmistakable, wake-up call that may be necessary to fully launch the transport revolutions considered in previous chapters. The revolutions include electrification of land transport, use of wind power for marine transport, and substantial contraction of air transport in favor of land-based modes, chiefly higher-speed rail. The wake-up call could truly force a halt to hole digging—no more expansion of oil-fueled transport—and sharply focus investment of human and financial resources on deploying alternatives.

The final chapter of the first edition of this book had the title "Leading the Way Forward," as does this chapter, because leadership makes the difference between recognition of a need and meeting the need. Our focus then was on the US, because of its importance as the world's most mobile society and because a change in US leadership was imminent. The new US president seems as disposed toward attainment of something like the scenarios of Chapter 5 as any of his predecessors since President Carter, and perhaps more capable of inspiring people to move toward them. President Obama also inherited the opportunities for change afforded by the events of 2008. "Rule one: Never allow a crisis to go to waste," said Rahm Emanuel, his chief of staff, a few days after the election.[84]

But the reality in 2009 is that the US "remains stuck in a protracted slump."[85] "Once again a Democratic president has pushed through job-creation policies that will mitigate the slump but aren't aggressive enough to produce a full recovery."[86] Obama has been compared with Herbert Hoover, the unloved mining engineer who had the misfortune to be US president when the Great Depression began: "...just like another very good man, Barack Obama is moving prudently, carefully, reasonably towards disaster."[87]

The slow start on America's road to recovery could result from the truism that for leadership to succeed there have to be followers who are ready and able to implement desired changes. The US will have to overcome a major deficit in its capacity to implement the kind of transport revolutions we have proposed. This seems to be increasingly recognized as necessary. The example of high-speed intercity trains, identified as a transport priority at the highest level of government, will illustrate this challenge.

After setting out a new vision for high-speed rail in the US in April 2009, President Obama turned to the Department of Transportation (USDOT) to implement the vision.[88] Had his call to action concerned expanding mobility by air or road, many thousands of federal bureaucrats could have been mobilized in response. They are staff responsible for, and experienced in, working with state and local governments and the private sector to develop bigger and better airports and highways.

But the task of realizing the high-speed rail initiative fell to perhaps half a dozen specialists in the USDOT's Federal Railroad Administration (FRA). Fewer than a dozen states were prepared to join these efforts because they lacked staff capable of intercity passenger rail planning. And unlike airports and roads, which are the responsibility of municipal governments and port authorities across the US, there is not a single engineer or planner employed by local government who has responsibility for intercity rail infrastructure. A further challenge is that almost all US rail infrastructure is owned by freight carriers who are openly skeptical about, and at times actively hostile to, passenger train operations along their properties.[89]

One result has been that the Government Accountability Office (GAO) has found the USDOT's efforts to fall short of a workable plan to implement high-speed rail transport. The GAO judged the FRA's high-speed rail plan to be "more of a vision than a strategic plan." While recognizing the challenge of preparing such a plan in 60 days as mandated by the American Recovery and Reinvestment Act of 2009, GAO advised Congress that the plan "does not define a clear federal role for involvement in high speed rail projects other than providing Recovery Act funds."[90] What neither the GAO nor anybody else in government has yet admitted publicly is that the capacity to achieve the Obama administration's vision will require the knowledge and effort of many more people than are presently involved in developing high-speed rail solutions for the US.

A key part of the challenge that still lies ahead for the US will involve redeploying human resources to establish a capacity to follow the lead toward transport revolutions. In Chapter 5, we proposed that a Transportation Redevelopment Administration (TRA) be created to facilitate the shift of thousands of public sector engineers and planners who have experience developing air and road infrastructure to new positions where they could turn those skills to developing electric mobility infrastructure, such as the high-speed rail initiative that is so far moving too slowly to keep ahead of oil depletion. Aiming for radically different results without major organizational change is not a promising strategy for advancing transport revolutions, based on past experience. We still see the need for a new transport agency that can redeploy both human and fiscal resources from the kinds of mobility that deepen America's decline in the vicious cycle to alternatives that can foster the shift to a virtuous cycle.

Not everybody who has been trained to build and manage air and road infrastructure may be willing and able to shift to fostering electric mobility. Many who prove incapable of making the switch may need to be moved out of their current roles, because the demand for air and road infrastructure will drop appreciably. Effective leadership of US transport revolutions will thus put a premium on reorganizing established public agencies at each level of government. Transport "know-how" will need to be placed where it can do the most good.

Meanwhile, China is booming. According to the International Monetary Fund, China's GDP growth will be 7.5 percent in 2009.[91] (That of the US will be −2.6 percent, and that of almost every other richer country will be even lower.) Credit was given to China's "substantial macroeconomic stimulus." China's relatively rosy position presents a different kind of leadership challenge in the transition to post-carbon transport systems. Part of the distinct leadership needs for advancing transport revolutions in China stem from the particularities of Chinese leadership.

It's hard to compare the diffuse and circumspect Chinese leadership with the more familiar, transparent, and enduring arrangements of US political leadership. One apparent difference is that a surprising number of senior Chinese government officials appear to have practiced as engineers, while a disproportionate number of senior US government officials have practiced as lawyers.[92] Both countries seem at first sight to be ungovernable on account of their size, complexity and massive social challenges, with inequality in each case being among the most prominent of these challenges. For the US, further examination sustains the impression of ungovernability, partly because conflict within government is an explicit part of the constitutional structure,[93] and partly because every aspect of government seems in permanent election mode. Governments in China have the luxury of freedom from such electoral considerations and can focus more keenly on providing effective leadership.

The risk for China is that wrong decisions may be more likely to be made and sustained without the accountability provided by political opposition. The continued focus on massive deployment of gasoline-fueled automobiles, as noted above, may be an example of a wrong decision that has not been recognized as such because of insufficient pressure to reconsider it.

In the US, the bigger risk is that the governing process becomes mired in institutionalized conflict and no decisions are made, or that actions are insufficient. Yet, on the particular matter of changing the emphasis on automobile production, the US appears in its own way to have been more effective.

The US seems likely to have another chance at implementing the elements of one or more transport revolutions. The first attempt at stimulating the economy—the American Recovery and Reinvestment Act of 2009, discussed above—appears to have fallen short of what is needed to spur a turnabout. The tax reductions provided by the Act appear to have been converted into savings rather than consumer spending. The grants to state and local governments have been swamped by those governments' revenue shortfalls. Moreover, banks remain unwilling or unable to lend, whether to consumers or to businesses that might create jobs more directly.[94] Thus, there is considerable discussion about the need for a second stimulus package,[95] and perhaps an opportunity to be more effective in the sense of breaking the vicious cycle in Figure 6.3 that may characterize the US economy.

A new stimulus package could move the US a long way toward the scenarios of Chapter 5, meanwhile creating millions of shorter- and longer-term jobs and reviving the US economy. Based on our analyses in Chapters 3 and 5, we suggest massive investments in five areas:

- Electricity generation, chiefly concentrating solar power plants in the US southwest, as discussed in Chapter 3
- Further allocations to reinforce interstate and local electrical grids
- Further allocations to intercity passenger rail
- Production of tens of thousands of trolley buses to boost local transit services
- Provision of electrical infrastructure and reserved lanes for these buses along tens of thousands of street kilometers.

To ensure success, these investments would need to be accompanied by a significant reorganization of federal, state and local transport bureaucracies. This would be to develop the capacity for implementing transport revolutions while winding down the capacity that supports outmoded air and road infrastructure programs. Given the precarious finances of most state and local governments, future federal relief will likely be sought. It should be granted with an obligation to undertake the necessary administrative reorganizations.

Such a further-reaching stimulus program could truly turn crisis into opportunity. It could cut the unemployment rate by half and boost manufacturing capacity in critical areas. It would move the US toward the transport revolutions needed to maintain the remarkable benefits provided by motorized mobility. This stimulus package would help the US share a measure of leadership in transport technology with China, rather than being eclipsed by Asian vehicles, infrastructure design and mobility management. Because the US continues to be an exemplar for many developing countries, transport revolutions in the US could be especially influential in ensuring that the challenges of oil depletion are met in ways that provide benefits to all humankind.

Appendix A: List of Acronyms and Abbreviations

ACORE	American Council for Renewable Energy
ATA	Air Transport Association
ATVMLP	Advanced Technology Vehicles Manufacturing Loan Program
bb/y	barrels per year
BEV	battery electric vehicle
BOAC	British Overseas Airways Corporation
BTU	British Thermal Unit
CAFE	Corporate Average Fuel Economy
CCS	carbon capture and storage
CFC	chlorofluorocarbon
CNG	compressed natural gas
dB	decibel
DHS	Department of Homeland Security
EC	European Commission
EJ	exajoule
EM	electric motor
EPA	Environmental Protection Agency
EU	European Union
FRA	Federal Railroad Administration
GAO	Government Accountability Office
GCV	grid-connected vehicle
GDP	Gross Domestic Product
GHG	greenhouse gas
GLA	gross leasable area
GPS	global positioning system
GTL	gas-to-liquids
g/vkm	grams per vehicle-kilometer
h/cap/d	hours per person per day (in Figure 2.9)
HCFC	hydrochlorofluorocarbon

ICAO	International Civil Aviation Organization
ICE	internal combustion engine
IEA	International Energy Agency
IPCC	Intergovernmental Panel on Climate Change
JIT	just-in-time
JNR	Japanese National Railway
kJ	kilojoule
km/h	kilometer per hour
L/100km	liters per 100 kilometers
LEV	low emission vehicle
LNG	liquefied natural gas
LPG	liquefied petroleum gas
LULUCF	land use, land use changes and forests
mb/d	million barrels a day
$\mu g/m^3$	micrograms per cubic meter
MFA	Material Flow Analysis
MIPS	material input per unit of service
mg/kg	milligrams per kilogram
MJ/kg	megajoules per kilogram
MJ/pkm	megajoules per person kilometer
MW(h)	megawatt (hour)
NAFTA	North American Free Trade Agreement
NDAC	National Defense Advisory Commission
NEC	(Boston–Washington) Northeast Corridor
NiMH	nickel metal hydride
NIST	National Institute of Standards and Technology (US)
NOx	nitrogen oxides
NYMEX	New York Mercantile Exchange
ODS	ozone-depleting substance
OECD	Organization for Economic Cooperation and Development
OPACS	Office of Price Administration and Civilian Supply
OPEC	Organization of Petroleum Exporting Countries
OPM	Office of Production Management
Pan Am	Pan American Airways
PHEV	plug-in hybrid electric vehicle
pkm	person-kilometer / passenger-kilometer
PM10	particulate matter less than 10 microns in diameter
PM2.5	particulate matter less than 2.5 microns in diameter
POP	persistent organic pollutant
pp/ha	persons per hectare
ppm	parts per million
$ppmCO_2e$	parts per million CO_2 equivalent
PRT	personal rapid transport

RTRI	Railway Technical Research Institute
SMP	Sustainable Mobility Project
SNCF	Société Nationale des Chemins de Fer Français (French National Railways)
SULEV	super low emission vehicle
SUV	sport utility vehicle
TARP	Troubled Assets Relief Program
TEU	twenty-foot equivalent unit
TGV	Train à Grand Vitesse
tkm	tonne-kilometer
TRA	Transportation Redevelopment Administration
TRB	Transportation Research Board
TW(h)	terawatt (hour)
UCG	underground coal gasification
UHV	ultra-high voltage
URL	uniform resource locator
USDOT	US Department of Transportation
US EIA	US Energy Information Administration
V2G	vehicle-to-grid
VOC	volatile organic compound
WBCSD	World Business Council for Sustainable Development
WHO	World Health Organization
WPB	War Production Board
WTO	World Tourism Organization
ZEV	zero emission vehicle

Notes and References

Notes to the Preface to the Second Edition

1. See tonto.eia.doe.gov/dnav/pet/pet_pri_spt_s1_m.htm.
2. US sales of cars and light trucks are at autonews.com/section/datacenter.
3. The index began 2008 at 8,891 and ended it at 774.See wikinvest.com/wiki/Baltic_Dry_Index.
4. See worldometers.info/cars/.
5. See news.xinhuanet.com/english/2009-12/08/content_12609453.htm.
6. See motherearthnews.com/Green-Transportation/Indian-supercompact-car-market-overtakes-Japan-AP-12222009.aspx.
7. The quotation is from icaopressroom.wordpress.com/2009/12/19/air-transport-records-worst-ever-performance-in-2009/.
8. See iata.org/whatwedo/economics/ptmarchives.htm.
9. See omrpublic.iea.org/omrarchive/11dec09full.pdf and theoildrum.com/files/2009_December_Oilwatch_Monthly.pdf.
10. See energybulletin.net/node/49493.
11. This is the opinion expressed in each of the two editions of *Transport Revolutions*. It is consistent with the opinion expressed in Part 1(1) of the document at peakoiltaskforce.net/.
12. See dbstars-asia.db.com/data/20091006090702X/data/macro_note4.pdf.
13. See the statement "...global oil production is not expected to peak before 2030..." on p. 87 of *World Energy Outlook*, IEA, Paris, France, November 2009.
14. The quotation is from guardian.co.uk/environment/2009/nov/09/peak-oil-international-energy-agency.
15. See, for example, the source in Note 16 and the articles at theglobeandmail.com/news/opinions/beware-the-gold-bubble/article1401546/ and theoildrum.com/node/6025.
16. See, for example, epmag.com/Magazine/2009/11/item47352.php.
17. As well as the sources concerning China detailed in Chapter 6, see the articles at online.wsj.com/article/SB125606654494097035.html and

ft.com/cms/s/0/6fa938d6-dde2-11de-b8e2-00144feabdc0.html. For Korea, see mnn.com/transportation/cars/stories/skorea-targets-world-electric-car-market. For Germany, see greenandsave.com/green_news/green-blog/germany-boosts-electric-vehicle-development-4997. For France, see transports.edf.fr/edf-fr-accueil/transport/documentation-54010.html.

18. See chinahighlights.com/news/around-china/wuhan-guangzhou-bullet-trains-to-run-from-december-26.htm and ft.com/cms/s/0/bbdb6d5a-f304-11de-a888-00144feab49a.html.
19. The quotation is from the article at newyorker.com/reporting/2009/12/21/091221fa_fact_osnos.
20. The data and quotation are from reuters.com/article/idUSTRE5BD0KR20091214.

CHAPTER 1: Learning from Past Transport Revolutions

1. Table 1.1 is based on Table 10.3 on p. 307 of Christian (2004).
2. The economist is William Baumol of New York University, quoted in Hensel, B., "Globalization, a sea change in shipping; when containers came to Houston, it marked an industry milestone," *Houston Chronicle,* April 23, 2006, chron.com/CDA/archives/archive.mpl?id=2006_4103177.
3. The quotations in Box 1.1 are respectively from pp. 1, 1–2, 2, 5, 7, 10–11 and 268 of Levinson (2006) and are used here with permission. The first quotation describes the first related undertaking of Malcolm McLean, an entrepreneur whose foresight and perseverance were critical to widespread adoption of containerization. There were several antecedents, some of which Levinson noted, essentially dismissing them with the following: "All over the world, the main methods for handling containers in the years after World War II offered few advantages over loose freight" (p. 32). An antecedent not covered to which this dismissal would not have applied was the one that first integrated sea-rail-road container service, inaugurated in November 1955 between Vancouver and points in Yukon, via Skagway, Alaska. Containers (sized 2.1 × 2.4 × 2.4 m) were moved from Vancouver to Skagway by a ship designed to carry containers, then by train from Skagway to Whitehorse, Yukon, and then by truck to points in the Yukon (McCague 2007). White Pass & Yukon President Marvin Taylor told an interviewer, "Our system became an international showpiece and was visited by transportation executives from all over the world. Many of them later paid us the ultimate compliment by imitating elements of the design," (Elliot 1987). The rail part of the system went out of service in 1982 during a depression in Yukon's mining industry.
4. For ships arriving at Liverpool, see p. 21 of Carlson (1969).
5. For information on Liverpool and Manchester's populations, see pp. 5–6 of Booth (1969).
6. For details of Liverpool-Manchester coach traffic, see p. 23 of Carlson (1969).

7. For the return on the Duke of Bridgewater's investment, see p. 100 of Priestley (1831).
8. For the returns on shares in the Mersey & Irwell Navigation Company, see p. 27 of Carlson (1969).
9. For the price of coal at Stockton, see p. 95 of Garfield (2002).
10. The quotation from the prospectus is as relayed on p. 11 of Booth (1969).
11. The comments of James Loch are as reported on p. 134 of Garfield (2002).
12. The pre-emptive tariff reduction is noted on p. 60 of Carlson (1969).
13. The enticement of the Marquess of Stafford is noted on pp. 148–149 of Carlson (1969).
14. The quotation is from pp. 52–53 of Ransom (1990).
15. The anticipated performance is from pp. 181–182 of Carlson (1969); the actual performance is from p. 199 of Garfield (2002).
16. The quotation about canal traffic is from p. 58 of Ransom (1990).
17. For data on coach services before the railway and rail passengers after, see pp. 56–57 of Ransom (1990). The total of 460,000 passengers carried in 1831 on the Liverpool-Manchester Railway, (about 1,260 per day) may be compared with the average of about 850 rail passengers who used this route during weekday morning rush periods. (See Figures 4.40 and 4.41 of Appendix A of *North West Route Utilisation Strategy: Demand Forecasting & Appraisal, Stage 1 Report*, March 2006, at networkrail.co.uk.) The original Liverpool–Manchester route is now proposed for electrification, which will bring the minimum journey time from one end to the other down from about 45 to about 30 minutes. (See UK Department for Transport, *Britain's Transport Infrastructure: Rail Electrification, July 2009,* at dft.gov.uk/pgr/rail/pi/rail-electrification.pdf.) As noted in the text, the journey took about two hours in 1831.
18. For what happened to London coach operators, see p. 43 of Turvey (2005).
19. For gasoline rationing in Canada, see Derber (1943). For rationing and other restrictions in the UK, see Dunnage (1943).
20. The data on car ownership and use in this paragraph are from AMA (1948).
21. For the impact of the European war on car ownership in the US, see p. 132 of Cardozier (1995).
22. This quotation is from p. 34 of Gropman (1997).
23. These quotations are from p. 26 of Koistinen (2004).
24. The preceding information in this paragraph is from pp. 134–135 of Koistinen (2004).
25. This quotation is from p. 131 of Weiner (1942).
26. For the actions of the War Production Board, see p. 34 of Gropman (1997).
27. Quoted in Raskin, A. H., "Mass Magic in Detroit," in *New York Times,* March 1, 1942, p. SM4.
28. The data on vehicle production are from p. 4 of AMA (1948); those for vehicles in use are from p. 17.

29. For the information about rubber, see p. 148 of Koistinen (2004).
30. The quotation on the need to drive is from Coan, P. B., "Car Outlook is Dark: Frugal, Smart and Slow Driving is Urged on Motorists to Make Tires Last," *New York Times,* March 1, 1942, p. 1.
31. The excerpts in Box 1.2 are from "Text of Appeal of War Chiefs to Motorists on Gas," *New York Times,* April 24, 1942, p. 11 and are used here with permission.
32. See "Violators of pleasure driving ban may face neighbors at hearings," *New York Times,* January 24, 1943, p. 38. This article describes hearings of local rationing boards at which motorists accused of recreational driving were given the opportunity to say why they should not suffer confiscation of gasoline coupons. The variety of defenses offered made for colorful reporting. The hearings had just been made public, perhaps in the hope of achieving greater adherence to the regulations of the Office of Price Administration.
33. This quotation is from Coan, P. B., "Nation enters 'car sharing' era to make its tires last," *New York Times,* July 12, 1942, p. 5.
34. The poster is *When You Ride Alone You Ride With Hitler!* by Weimer Pursell, 1943. Originally in color, it was produced by the Government Printing Office for the Office of Price Administration. This copy came from the National Archives and Records Administration, Still Picture Branch (NWDNS-188-PP-42) archives.gov/exhibits/powers_of_persuasion/use_it_up/images_html/ride_with_hitler.html.
35. Nevertheless, there were almost as many cars in the US as households: just under 30 million cars in 1940 (see text associated with note 20) and some 132 million residents living in about 35 million households. (See *16th Census of the United States—1940: Housing, Volume II, Part 1, United States Summary,* www2.census.gov/prod2/decennial/documents/36911485v2p1ch1.pdf.)
36. See Gallup, G., "54% of Owners Don't Think Autos Vital, Gallup Finds in a Survey on Tire Rationing," *New York Times,* January 23, 1942, p. 14.
37. The quotation is from p. 138 of Derber (1943).
38. The data on vehicle-kilometers traveled are from p. 64 of AMA (1948).
39. These shares, and the data in Figure 2.2, are based on p. 20 of Wilson (1997). Note that intercity air travel per capita expanded by a factor of eight during this period, but still amounted to less than 2 percent of intercity travel in 1949. The actual increases in rail and bus travel per person in the US were from 592 and 261 kilometers in 1941 to 1,881 and 529 km in 1944. Meanwhile, car travel per person declined from 6,306 km in 1941 to 3,537 km in 1944 (3,397 km in 1943). Note that a majority of intercity trips were still made by car, even at the peak of the pause in the growth in motorization. However, a much lower but unknown share of trips in urban areas were made by car, and overall it is likely that trips by car were a minority of all trips, before and during the war.
40. The freight data are from p. 251 of Koistinen (2004).

41. The data on public transport ridership are from Table 6 on p. 12 of APTA (2007).
42. The transport data in Figure 2.2 are from p. 20 of Wilson (1997). Population data used to estimate per-capita values are from the US Bureau of the Census, *Historical National Population Estimates: July 1, 1900 to July 1, 1999*, census.gov/popest/archives/1990s/popclockest.txt.
43. The quotation is from p. 232 of Cardozier (1995).
44. This quotation is from p. 2 of American Transit Association (1945).
45. This quotation is from p. 157 of Cardozier (1995).
46. These data are from p. 4 of AMA (1948).
47. These data are from p. 20 of Wilson (1997).
48. The data on public transport ridership are from Table 6 on p. 12 of APTA (2007).
49. See Fox (2003), particularly Chapter 9, for transatlantic travel between 1820 and 1910.
50. Figure 1.3 is based on data in Civil Aeronautics Board (1975).
51. A transatlantic Clipper ticket cost $675 in 1939 (see *Boeing 314—Pan Am Clippers—USA,* The Aviation History On-Line Museum, updated 2006, aviation-history.com/boeing/314.html). This equaled $4,636 in 1983 dollars according to *The Inflation Calculator,* westegg.com/inflation/. In 1983, the New York–London Concorde fare was $4,417 (see "British Airways," *New York Times,* December 29, 1983, p. 1). By contrast, a ticket on the inaugural voyage of the Cunard liner *Queen Mary* cost $100 in 1936 (see "RMS Queen Mary 65th Anniversary Invitation: Pay 1936 fare," *Canada Newswire,* February 27, 2001, p. 1).
52. See Paragraph 11 of Siddiqi, A., "The Beginnings of Commercial Transatlantic Services," *History of Flight,* US Centennial of Flight Commission, Washington DC, 2003, centennialofflight.gov/essay/Commercial_Aviation/atlantic_route/Tran4.htm.
53. For Atlantic airline traffic in 1950 see p. 470 of Bender and Altschul (1982).
54. Pressurized passenger aircraft could fly at twice the altitude of flying boats and thus avoid much bad weather. Pressurization was a spinoff of military development, with the Boeing 307 Stratoliner adapting this technology from the B-17 bomber just before the start of World War Two. For details on the B-307, see boeing.com/history/boeing/stratoliner.html. For information about the Lockheed Constellation, see lockheedmartin.com/aboutus/history/Model049Constellation.html, about the Douglas DC-6, see ruudleeuw.com/dc6.htm, and about the Boeing 377 Stratocruiser, see boeing.com/history/boeing/m377.html.
55. For all-sleeper flights, see p. 22 of Speas (1955).
56. On immigrant fares, see p. 248 of Kendall (1979). In 1904, £1.50 had the purchasing power of about £110 ($180) in mid 2009; it was also the equivalent of seven or eight days of average earnings in the UK at that time.

57. For attracting the US middle class on to transatlantic planes, see p. 470 of Bender and Altschul (1982).
58. The jet aircraft of the 1950s appear to have used much more fuel than the piston-engined aircraft they replaced. Peeters et al. (2005) concluded, "The last piston-powered airliners were at least twice as fuel-efficient as the first jet-powered aircraft" and "...the last piston-powered aircraft were as fuel-efficient as the current average jet" (p. 3). However, jet fuel is less refined than fuel for piston engines and costs less.
59. See p. 15 of ICAO (1958).
60. For comparison of the Comet jet with propeller planes, see p. 19 of Serling (1982).
61. The quotation about low cost electricity for the wind tunnel is from p. 94 of Rodgers (1996).
62. For Trippe's order of the new engines, see p. 34 of Gandt (1995).
63. Trippe's words are on p. 37 of Gandt (1995).
64. For jet aircraft performance in the late 1950s, see p. 193 of Rodgers (1996).
65. For Pam Am's passenger volumes, see p. 40 of Gandt (1995).
66. The quotation about Boeing's investment is from p. 197 of Rodgers (1996).
67. On Cunard's fare discounts, see p. 256 of Kendall (1979).
68. On cruises in the 1840s, see p. 361 of Kendall (1983).
69. On ship registrations, see p. 21 of Alderton (1973).
70. Until her announced retirement in June 2007 (to become a hotel and tourist destination in Dubai in 2009), the *Queen Elizabeth 2* was still plying the Atlantic on occasion, although she served mainly as a cruise ship. Cunard's flag ship is now the *Queen Mary 2*, the world's second-largest passenger ship (the cruise ship *Freedom of the Sea* is larger), and the only passenger ship with a regular transatlantic schedule. See cunard.com.
71. For Penn Central, see Daughen and Binzen (1971).
72. The quotation on Britain's rail future is from p. 1 of British Railways Board (1963).
73. The quotation from the Commission's report is in Hosmer (1958).
74. The quotation on high-speed rail evolution is from p. 157 of Banister and Hall (1993).
75. Givoni (2006) points out that increasing railway capacity in congested corridors was a primary impetus behind the launch of high-speed rail in Japan, France and Italy. Countries with rail capacity that appeared overbuilt in relation to transport demand, including the US and the UK, did not make the commitment to advancing this mobility niche very far. Germany's efforts to add rail transport capacity were slowed by a parallel focus on even faster "trains" propelled by magnetic levitation. For a contrast between the German and French approach to developing high-speed rail, see Dunn and Perl (1994).
76. For early plans for a new line between Tokyo and Osaka, see Kakumoto (1999).

77. For reasons for the emergence of JNR, see Yoshitake (1973).
78. Smith (2003) writes that "Scientific and engineering initiatives which in the West were devoted to military projects in aviation and the development of atomic power were adapted in Japan to peaceful industrial uses. Teams of highly trained and capable engineers were recruited into the railway industry" (p. 226).
79. On the RTRI conference presentation, see p. 81 of Strobel and Straszak (1981).
80. On JNR's move beyond incrementalism, see Nishida (1977).
81. This quotation is from p. 46 of Shima (1994).
82. For JNR's traffic projections, see Knop and Straszak (1981).
83. On the JNR president's request, see p. 174 of Hosokawa (1997).
84. For the Cabinet's approval, see p. 178 of Hosokawa (1997).
85. For critics of the Shinkansen concept, see p. 81 of Strobel and Straszak (1981).
86. The quotation on timing of transport is reproduced from pp. 27–28 of Kakumoto (1999).
87. See p. 178 of Hosokawa (1997); and "Japan is Building Speedy Rail Line," *New York Times,* April 26, 1959, on p. 20. The dollar equivalencies of yen of that period assume an exchange rate of 360 yen to the dollar, and subsequent dollar inflation based on futureboy.homeip.net/fsp/dollar.fsp.
88. JNR's vice president of engineering was dispatched to Washington to persuade World Bank officials that the Shinkansen was a low-risk investment and did not rely upon experimental technology, which was explicitly prohibited from support in the Bank's lending criteria. Hideo Shima convinced the officials that the Shinkansen was a novel integration of proven rail technologies that had been developed in response to Japan's railway safety research program. See p. 48 of Shima (1994).
89. For the World Bank loan for Shinkansen, see p. 200 of Hosokawa (1987).
90. The quotation is from "Japanese Build a Super-Railroad," *Business Week,* December 1, 1962, p. 89.
91. The quotation is from "Japan's Fast Train—How it's working out," *U.S. News & World Report,* January 25, 1965, vol LVIII, no 4, pp. 69–70.
92. As with other transformative transport technologies, cities and regions that had not been touched by the initial revolution sought the new train technology so as not to be left behind in economic and social development. Shinkansen expansionary zeal reached its apex in 1970 when Japan's parliament passed the Nationwide Shinkansen Development Law which authorized a 7,000 kilometer network of high speed trains to be built by 1985. When Japan's economy was battered by the 1973 oil shock, parliament revised this plan to prioritize certain lines and put others on hold. The Shinkansen network thus grew more slowly than initially desired, but has expanded to serve most major cities on Honshu and also been extended onto Kyushu. See p. 48 of Hood (2006).
93. The quotation is from p. 2,626 of Endo (2005).
94. For the value of the Shinkansen network, see p. 13 of Okada (1994).

95. See the Japan section of *High speed lines in the world: Updated 04 June 2008*, Union internationale des chemins de fer (UIC), Paris, France, uic.org/plugins/UIC_SPIP_kit/doc_download.php?id=1359.
96. For the impacts of the introduction of Shinkansen on air travel, see Chapin, E., "Japan's Airlines Step Up Service," *New York Times*, March 14, 1965, p. S21, and "New Japan Travel Cuts Air Travel," *New York Times*, February 28, 1965, p. S18.
97. For the historical arrangements concerning SNCF, see Jones (1984).
98. For the Nora Report, see Groupe de travail du comité interministériel des entreprises publiques (1967).
99. This quotation is from p. 229 of Hayward (1986).
100. For SNCF management and the Nora Report, see Fourniau and Ribeill (1991).
101. For information on Paris and Lyons, see p. 4 of Institut National de la Statistique et des Études Économiques—Ile-De-France (1999); and p. 1 of Institut National de la Statistique et des Études Économiques—Rhône Alpes (1999).
102. For the comparison between French and German strategies, see Dunn and Perl (1994).
103. Figure 1.4 is based on Figure 1.3 of Perl (2002).
104. Figure 1.5 is based on data in Tables 3.3.7 and 3.3.8 of European Commission (2006).
105. See Jeppu, Y. V., "Story of the first airmail," Dakshina Kannada Philatelic Association (dkpna), geocities.com/dakshina_kan_pa/art4/airmail.htm; and U.S. Centennial of Flight Commission centennialofflight.gov/essay/Government_Role/1918/POL2.htm.
106. For the impact of this transport revolution, see p. 8 of UK DfT (2002).
107. This quote is from p. 157 of Rodrigue et al. (2006).
108. For how air freight was moved before the 1980s, see p. 150 of Upham et al. (2003). According to p. 78 of Airbus (2006), by 2005, 58% of air freight was moving in dedicated freighters, with the remainder moved in the belly holds of passenger aircraft. The share moving in freighters is expected to increase to 65 percent by 2025.
109. For what shippers had to do, see pp. 5–6 of Birla (2005).
110. For data on cargo revenue, see pp. 32–34 of Sigafoos (1988) and pp. 104–105 of Trimble (1993).
111. For the early history of FedEx, see pp. 2–3 of Birla (2005); pp. 40–42 of Sigafoos (1988); and pp. 126–127 of Trimble (1993).
112. This part of the history of FedEx is on p. 5 of Birla (2005); pp. 37–38 of Sigafoos (1988); and pp. 106, 108–109, 116 and 131 of Trimble (1993).
113. For the information about the launch of FedEx and its subsequent progress, see p. 1–2 of Birla (2005) and the "Corporate history" and "Our companies" sections of *About FedEx*, fedex.com. The quotation concerning COSMOS is from the latter source.
114. See "Scheduled freight tonne-kilometres," extracted from *World Air Trans-*

port Statistics, International Air Transport Association, Montreal, Quebec, 2007, iata.org/ps/publications/wats-freight-km.htm. The same source indicates that Korean Air Lines performed the most *international* schedule tonne-kilometers, with FedEx being fifth in this respect.

115. For details on technology innovation at FedEx, see various parts of Birla (2005) and *About FedEx,* fedex.com.
116. The information on these companies was from p. 50 of Direct Communication Group (2003); pp. 43–46 of Kingsley-Jones (2000); and DHL (2006) DHL International GmbH., dhl.com/publish/go/en/about/history.high .html.
117. For a discussion of high-speed rail and air freight, see pp. 157–181 of Wardman et al. (2002), who conclude that the integration may not have much potential.
118. For this aspect of FedEx's history, see pp. 212–215 and 219–220 of Trimble (1993); and *About FedEx,* fedex.com.
119. For the impact of trade liberalization of air freight, see p. 14 of *The economic and social benefits of air transport,* Air Transport Action Group, Geneva, Switzerland, 2005, iata.org/nr/rdonlyres/5c57fe77-67ff-499c-a071-4e5e2216 d728/0/atag_economic_social_benefits_2008.pdf.
120. For a discussion of integrator fulfilling orders, see p. 149 of Upham et al. (2003).
121. For the "Harry Potter" logistics, see p. 36 of Birla (2005) and *About FedEx,* fedex.com.
122. See 3plwire.com/2007/07/22/harry-potter-logistics/.
123. For a brief discussion of how rising oil prices sealed the fate of the Concorde, see p. 268 of Perl and Patterson (2004).
124. For a discussion of barriers to change posed by "lock-in," also known as "path dependence," see Geels (2004).

References

Airbus (2006) *Global Market Forecast: The Future of Flying,* 2006–2025, Blagnac, France, 90 pp., airbus.com/store/mm_repository/pdf/att00008552/media_object_file_AirbusGMF2006-2025.pdf.

Alderton, P. M. (1973) *Sea Transport: Operation and Economics,* Thomas Reed Publications Ltd., London, 268 pp.

AMA (1948) *Automobile Facts and Figures,* 28th Edition, Automobile Manufacturers Association, Detroit, MI, 80 pp.

American Transit Association (1945) *Transit Fact Book 1945,* American Transit Association, New York, NY.

APTA (2007) *Public Transportation Fact Book,* 58th edition, American Public Transportation Association, Washington DC, 107 pp., apta.com/research/stats/factbook/documents/factbook07.pdf.

Banister, D. and Hall, P. (1993) "The second railway age," *Built Environment,* vol 19, no 2/3, pp. 157–162.

Bender, M. and Altschul, S. (1982) *The Chosen Instrument,* Simon and Schuster, New York, NY, 605 pp.
Birla, M. (2005) *FedEx Delivers: How the World's Leading Shipping Company Keeps Innovating and Outperforming the Competition,* Wiley, Hoboken, NJ, 215 pp.
Booth, H. (1969) *An Account of the Liverpool and Manchester Railway,* Frank Cass and Company Limited, London.
British Railways Board (1963) *The Reshaping of British Railways Part I: Report,* Her Majesty's Stationery Office, London.
Cardozier, V. R. (1995) *The Mobilization of the United States in World War II: How the Government, Military and Industry Prepared for War,* McFarland & Company, Inc, Jefferson, NC, 277 pp.
Carlson, R. E. (1969) *The Liverpool & Manchester Railway Project 1821–1831,* Augustus M. Kelley Publishers, New York, NY, 292 pp.
Christian, D. (2004) *Maps of Time: An Introduction to Big History,* University of California Press, Berkeley, CA, 642 pp.
Civil Aeronautics Board (1975) *Handbook of Airline Statistics, United States Certificated Air Carriers,* 1975 Supplement, US Civil Aeronautics Board, Bureau of Accounts and Statistics, Washington DC.
Daughen, J. R. and Binzen, P. (1971) *The Wreck of the Penn Central,* Little, Brown, Boston, MA, 380 pp.
Derber, M. (1943) "Gasoline rationing and practice in Canada and the United States," *Journal of Marketing,* vol 8, no 2, pp. 137–144.
Direct Communications Group (2003) *Competition within the United States Parcel Delivery Market,* Silver Spring, MD, distributed by Association for Postal Commerce (PostCom), Arlington, VA, postcom.org/public/articles/2003articles/parcel_competition.pdf.
Dunn, Jr, J. A. and Perl, A. (1994) "Policy networks and industrial revitalization: High speed rail initiatives in France and Germany," *The Journal of Public Policy,* July, vol 14, no 3, pp. 311–343.
Dunnage, J. A. (1943) "Motor transport in Great Britain," *Annals of the American Academy of Political and Social Science,* vol 230, Transportation: War and Postwar, pp. 141–149.
Elliot, J. (1987) "Starting the container revolution in the Yukon," *Canadian Transportation & Distribution Management,* vol 90, no 10, pp. 36, 76.
Endo, T. (2005) "The future of high-speed train," *IEICE Transportation, Information and Systems,* vol E88-D, no 12, pp. 2,625–2,629.
European Commission (2006) *EU Energy and Transport in Figures, European Commission and Eurostat, Brussels, Belgium,* online version only at ec.europa.eu/dgs/energy_transport/figures/pocketbook/2006_en.htm.
Fourniau, M. and Ribeill, G. (1991) "La grande vitesse sur rail en France et en R.F.A.," in Brenac, E., Finon, D. and Muller, P. (eds) La grande technologie entre l'Etat et le marché, CERAT, France.

Finon, D. and Muller, P. (eds), *La grande technologie entre l'Ètat et le marchè,* CERAT, Grenoble, France.

Fox, S. (2003) *Transatlantic: Samuel Cunard, Isambard Brunel, and the Great Atlantic Steamships,* HarperCollins, New York, NY, 493 pp.

Gandt, R. (1995) *Skygods: The Fall of Pan Am,* William Morrow and Company, New York, NY, 172 pp.

Garfield, S. (2002) *The Last Journey of William Huskisson,* Faber and Faber, London, 244 pp.

Geels, F. (2004) "From sectoral systems of innovation to socio-technical systems: Insights about dynamics and change from sociology and institutional theory," *Research Policy,* vol 33, pp. 897–920.

Givoni, M. (2006) "Development and impact of modern high-speed train: a review," *Transport Reviews,* vol 26, no 5, pp. 593–627.

Gropman, A. (1997) "Industrial mobilization," in Gropman, A. (ed) *The Big "L" American Logistics in World War II,* National Defense University Press, Washington DC, 447 pp.

Groupe de travail du comité interministériel des enterprises publiques (1967) *Rapport sur les Entreprises Publiques,* La Documentation Française, Paris, France.

Hayward, J. (1986) *The State and the Market Economy: Industrial Patriotism and Economic Intervention in France,* New York University Press, New York, NY, 256 pp.

Hood, C. P. (2006) "From polling station to political station? Politics and the Shinkansen," *Japan Forum,* vol 18, pp. 45–63.

Hosokawa, B. (1997) *Old Man Thunder: Father of the Bullet Train,* Sogo Way, Denver, CO, 224 pp.

Hosmer, H. (1958) *Railroad Passenger Train Deficit,* Docket no. 31954, Interstate Commerce Commission, Washington DC.

Institut National de la Statistique et des Études Économiques—Île-de-France (1999) *Île-de-France à la page,* Mensuel, July, no 171.

Institut National de la Statistique et des Études Économiques—Rhône Alpes (1999) *La Lettre,* July, no 63.

ICAO (1958) *The Economic Implications of the Introduction Into Service of Long-Range Jet Aircraft,* Doc 7894-C/907, International Civil Aviation Organization, Montreal, Canada.

Jones, J. (1984) *The Politics of Transport in Twentieth Century France,* McGill-Queen's University Press, Kingston, Canada, 302 pp.

Kakumoto, R. (1999) "Sensible politics and transport theories?—Japan's national railways in the 20th Century," *Japan Railway & Transport Review,* vol 22, pp. 22–23, jrtr.net/jrtr22/pdf/F23_Kakumoto.pdf.

Kendall, L. C. (1979) *The Business of Shipping, Third Edition,* Cornell Maritime Press, Inc, Centreville, MD.

Kendall, L. C. (1983) *The Business of Shipping, Fourth Edition,* Cornell Maritime Press, Inc, Centreville, MD.

Kingsley-Jones, M. (2000) "Express: Europe's express package carriers have undergone tremendous change in recent years as the cargo business has boomed," *Flight International,* 19 September 19, vol 158, no 4746, pp. 43–46.

Knop, H. and Straszak, A. (1981) "The Shinkansen and national development issues," in Straszak, A. (ed) *The Shinkansen Program: Transportation, Railway, Environmental, Regional, and National Development Issues,* International Institute for Applied Systems Analysis, Laxenburg, Austria, 425 pp.

Koistinen, P. A. C. (2004) *Arsenal of World War II: The Political Economy of American Warfare 1940 –1945,* University Press of Kansas, Lawrence, KS, 657 pp.

Levinson, M. (2006) *The Box: How the Shipping Container Made the World Smaller and World Economy Bigger,* Princeton University Press, Princeton, NJ, 376 pp.

McCague, F. (2007) "Containers—50 years and counting," *CILTNA Newsletter,* Winter, Chartered Institute of Logistics and Transport in North America, Ottawa, Ontario, pp. 8–11, ciltna.com/NL%20Edition%202%20-%20Winter%202007%20Version%202d.pdf.

Nishida, M. (1977) "History of the Shinkansen," in Straszak, A. and Tuch, R. (eds) *The Shinkansen High-Speed Rail Network of Japan: Proceedings of an IIASA Conference,* June 27–30, pp. 11–20, Pergamon Press, Toronto, Canada, 464 pp.

Okada, H. (1994) "Features and economic and social effects of the Shinkansen," *Japan Railway & Transport Review,* October, no 3, pp. 9–16, jrtr.net/jrtr03/pdf/f09_oka.pdf.

Peeters, P. M., Middel, J. and Hoolhorst, A. (2005) *Fuel Efficiency of Commercial Aircraft: An Overview of Historical and Future Trends,* National Aerospace Laboratory, Amsterdam, the Netherlands, 37 pp.

Perl, A. (2002) *New Departures: Rethinking Rail Passenger Policy in the Twenty-First Century,* University of Kentucky Press, Lexington, KY, 334 pp.

Perl, A. and Patterson, J. (2004) "Will oil depletion determine aviation's response to environmental challenges?" *Annals of Air and Space Law,* vol 29, pp. 259–273.

Priestley, J. (1831) *Historical Account of the Navigable Rivers, Canals, and Railways, Throughout Great Britain, As a Reference to Nichols, Priestley & Walker's New Map of Inland Navigation,* Longman, Rees, Orme, Brown & Green, London.

Ransom, P. J. G. (1990) *The Victorian Railway and How It Evolved,* Heineman, London, 352 pp.

Rodgers, E. (1996) *Flying High: The Story of Boeing and the Rise of the Jetliner Industry,* Atlantic Monthly Press, New York, NY, 502 pp.

Rodrigue, J.-P., Comtois, C. and Slack, B. (2006) *The Geography of Transport Systems,* Routledge, London and New York, 284 pp.

Serling, R. (1982) *The Jet Age,* Time-Life Books, Alexandria, VA, 175 pp.

Shima, H. (1994) "Birth of the Shinkansen: A memoir," *Japan Railway & Transport Review,* October, pp. 45–48, jrtr.net.

Sigafoos, R. A. and Easson, R. R. (1988) *Absolutely Positively Overnight! The Unofficial Corporate History of Federal Express,* St Lukes Press, Memphis, TN, 190 pp.

Smith, R. A. (2003) "The Japanese Shinkansen: Catalyst for the renaissance of rail," *The Journal of Transport History,* vol 24, no 2, pp. 222–237.

Speas, R. D. (1955) *Technical Aspects of Air Transport Management,* McGraw-Hill, New York, NY.

Strobel, H. and Straszak, A. (1981) "Subsystems analysis," in Straszak, A. (ed) *The Shinkansen Program: Transportation, Railway, Environmental, Regional, and National Development Issues,* International Institute for Applied Systems Analysis, Laxenburg, Austria.

Trimble, V. H. (1993) *Overnight Success: Federal Express & Frederick Smith Its Renegade Creator,* Crown Publishers Inc, New York, NY, 342 pp.

Turvey, R. (2005) "Horse traction in Victorian London," *Journal of Transport History,* vol 26, no 2, pp. 38–59.

UK DfT (2002) *UK Air Freight Study Report,* Department for Transport, UK, dft.gov.uk/pgr/aviation/airports/ukairstudyreport.

Upham, P., Maughan, J., Raper, D. and Thomas, C. (2003) *Towards Sustainable Aviation,* Earthscan, London, 284 pp.

Wardman, M., Bristow, A., Toner, J. and Tweddle, G. (2002) *Review of Research Relevant to Rail Competition for Short Haul Air Routes,* EEC/ENV/2002/003, Eurocontrol Experimental Centre, Institute of Transport Studies, University of Leeds, UK, 197 pp., eurocontrol.int/eec/gallery/content/public/documents/EEC_SEE_reports/EEC_SEE_2002_003.pdf.

Weiner, J. L. (1942) "Legal and economic problems of civilian supply," *Law and Contemporary Problems,* vol 9, no 1, pp. 122–149.

Wilson, R. A. (1997) *Transportation in America: Historical Compendium 1939–1995,* Eno Transportation Foundation, Inc., Lansdowne, VA, 78 pp.

Yoshitake, K. (1973) *An Introduction to Public Enterprise in Japan,* Sage Publications, London.

CHAPTER 2: Transport Today

1. Figure 2.1 is based on Figure 1 on p. 9 of OECD (2000).The main source for the data in Figure 2.1 is Mitchell (1992–1995).
2. For a more extended discussion of the link between car ownership and democracy, see Jain and Guiver (2001), particularly pp. 577–581.
3. See Pucher and Buehler (2005) on transport in the former USSR and the countries it influenced.
4. The quotation is from p. 38 of Cullinane (2002).
5. For the factors that may determine low car ownership and use in Hong Kong, see Cullinane (2003).

6. The estimates in Box 2.1 are based on World Bank (2006), Tables 2.1, 3.12, 5.8 and 6.15; on Kenworthy and Laube (2001); and on the energy section of *OECD Statistics,* Organization for Economic Cooperation and Development, Paris, France, as available until December 2006. The last data are now available only for a fee through the International Energy Agency at data.iea.org/ieastore/statslisting.asp.
7. The Paris-based OECD was formed in 1961. The 20 initial members were: Austria, Belgium, Canada, Denmark, France, Germany, Greece, Iceland, Ireland, Italy, Luxembourg, the Netherlands, Norway, Portugal, Spain, Sweden, Switzerland, Turkey, the UK and the US. Another 10 countries have joined since 1961: Australia, the Czech Republic, Finland, Hungary, Japan, the Republic of Korea, Mexico, New Zealand, Poland and Slovakia. See oecd.org for further information.
8. For example, in Toronto, Canada, between 1986 and 2001, home–work journeys as a share of all *weekday* trips fell from 41 percent to 36 percent of the total, and journeys starting between 7–9 a.m. and 4–6 p.m. fell from 41 percent to 38 percent of the total (persons 11 years and older only). Data are from the Transportation Tomorrow Survey, jpint.utoronto.ca.
9. For more local travel in the US on Saturdays than on Mondays and Tuesdays, see Table A-13 on p. 22 of US DOT (2003).
10. Fifty kilometers (about 31 miles) is taken here to be the very approximate boundary between short-term and long-term travel, fully recognizing that, particularly in North America, many people travel farther to work. Indeed, the US Bureau of Transportation Statistics defines a long-distance trip as being 50 miles (80 km) or more in length. It notes that 13 percent of long-distance trips are to or from work (US DOT, 2006a, pp. 1–2), comprising about one in 200 of all commuting trips; see *3.3 Million Americans are "Stretch Commuters" Traveling at Least 50 Miles One-Way to Work,* 2004, bts.gov/press_releases/2004/bts010_04/html/bts010_04.html. The *average* one-way trip length of short- and long-distance trips in the US, for all purposes, was just over 20 km. In the UK, average one-way trip length is just over 10 km (UK DfT, 2005b, Table 1.3, p. 15).
11. Some statistical agencies treat a return journey as one trip, which can be a source of considerable confusion. Another source of confusion is a public transport trip requiring, say, travel by bus and then by metro (subway). Some agencies count this as one trip. Others count it as two "boardings." Yet another source of confusion concerns trips involving more than one mode, for example, a bicycle ride and then a ride by suburban train. In such cases, sometimes the whole trip is considered to be by the main mode—public transport in this example—and the other mode is neglected. Thus, much care is required when comparing data from different cities and countries.
12. Table 2.1 is based on UITP (2006). This database contains information on 52 cities, but data for four of the cities were not complete enough to be used for Table 2.1.

13. Figure 2.2 is based on data in Kenworthy and Laube (2001). This database contains information on 100 cities—60 richer cities and 40 poorer cities—but data for one richer and six poorer cities could not be used to develop Figure 2.2.
14. "Other Asian Cities" are Bangkok, Chennai, Delhi, Ho Chi Minh City, Jakarta, Kuala Lumpur, Manila, Mumbai, Seoul and Taipei. All had GDP per capita of less than $10,000 in 1995, but some would now qualify as high-income urban regions, particularly the last two.
15. "Affluent Asian Cities" are Hong Kong, Osaka, Sapporo, Singapore and Tokyo.
16. See Kenworthy and Laube (2001) for data on shares by rail, and Kenworthy (2006) for a further discussion of this point.
17. Figure 2.3 is based on the source in Note 8. Toronto's inner core includes the main business district (downtown) and the residential area around it. The "outer core" is a band about 5 km wide surrounding the core. The inner and outer cores were built up in the 19th century and the first half of the 20th century, with extensive redevelopment after 1960. The "inner suburbs" were mostly developed during the 1950s and 1960s. The "outer suburbs," now the location of most of the region's population growth, have been mostly developed since 1970. The City of Toronto comprises the inner and outer cores and the inner suburbs. Boundaries within the City of Toronto delineate planning districts. Boundaries in the outer suburbs show municipalities. The City of Toronto is about 96 percent urbanized. The outer suburbs are about 20 percent urbanized, with much of the development within 20 km of the City of Toronto boundary.
18. See Box 2.1 for the meaning of "person-kilometer," and also "tonne-kilometer."
19. Figure 2.4—showing car ownership in 32 countries—is based on data in Pucher and Buehler (2005); in Table 3.6.1 of European Commission (2006); and in Table 8.2 of Davis and Diegel (2007). Note that median values are used to represent car ownership in Eastern and Western European countries. Note too that "car" refers to all light-duty four-wheeled vehicles used primarily for personal transport, including passenger vans, sport utility vehicles and many small pick-ups, as well as regular cars.
20. The information on car ownership in this and the next few paragraphs is mostly from Tables 4.1, 16.26 and 16.27 of NBSC (2006).
21. China's car production until 2005 is in Table 14.24 of NBSC (2006). Comparable data for 2006 do not appear to be available, although data on "sedan sales" is in "China's auto market in 2006," *People's Daily Online,* February 7, 2007, english.people.com.cn/200702/07/eng20070207_348279.html, and sales have generally been close to production. On this basis, car production in each year of 2002–2006 was 1.1, 2.0, 2.3, 2.8 and 3.8 million units. Estimating production—and sales—is bedeviled by confusing terminology: the terms "passenger vehicle," "car," "passenger car" and "sedan" are not used

consistently. This may have led the *Financial Times (London)* to announce on March 8, 2007 that "China leads the US in car production." In the article by Bernard Simon with this title, it was reported that "China produced about 5.2 million cars in 2006," more than the 4.4 million produced in the US. However, the comparable figure is likely 3.8 million, not 5.2 million. Moreover, the 3.8 million produced in China includes what in the US are known as "light trucks" (SUVs, etc.), and are not included in the reported 4.4 million of passenger car production in the US. Their production totals another 4.4 million, approximately. Notwithstanding these statistical hiccups, China is clearly on a trajectory to pass the US in car production during the next decade and become the world's major producer of passenger cars.

22. The Hong Kong rate of car ownership is from Table 3.12 of World Bank (2006).
23. See p. 66 of Liu and Guan (2005) on removal of Beijing's bicycle lanes.
24. For data on cycling to work in Beijing, see Mygatt (2005).
25. This quote was from "Vice-minister speaks up for bike lanes," *Shanghai Daily News,* June 15, 2006, available as the third article in the attachment at lists.umn.edu/cgi-bin/wa?A2=indo607&L=con-pric&P=932.
26. The quotation is from Elegant, S., "Where coal is stained with blood," *Time,* March 12, 2007, time.com/time/magazine/article/0,9171,1595235,00.html.
27. Information about Singapore's Vehicle Quota System and other policies and schemes regulating car ownership and use is at the website of the Land Transport Authority, lta.gov.sg/motoring_matters/index_motoring_vo.htm.
28. For a discussion that touches on these trends, see Holder, K., "China Road," *UC Davis Magazine Online,* vol 24, no 1, Fall 2006, ucdavismagazine.ucdavis.edu/issues/fall06/feature_2.html. Data on vehicle ownership in Shanghai and Beijing are from the sources detailed in Note 20.
29. See Note 21 on car production in China.
30. The table and the chart in Box 2.2 are based on data in what was the United Nations Statistics Division's Common Database until February 2008 and is now UNdata at data.un.org/. The sources on two- and three-wheeled motorized vehicles are: p. 3 of *Emissions control of two- and three-wheel vehicles,* Manufacturers of Emission Controls Association, Washington DC, May 7, 1999, meca.org/galleries/default-file/motorcycle.pdf; and p. 26 of *Background Report: Vehicle Fuel Economy In China,* Development Research Center of the State Council, Tsinghua University Department of Environmental Science & Engineering, China Automotive Technology and Research Center, and the Chinese Research Academy of Environmental Science, November 2001, efchina.org/FReports.do?act=detail&id=84 . The bicycle information in Box 2.2 is from Mygatt (2005).
31. See Table 16.26 of NBSC (2006).

32. See "Passenger Transportation Explanatory Variables" table on pp. 114–115 of Natural Resources Canada (2006).
33. Data on the weight of light-duty vehicles sold in the US are in Appendix G ("Data stratified by vehicle type and weight class") of Heavenrich (2006).
34. See Table 3 on p.5 of Gabler and Fildes (1999).
35. The two charts in Box 2.3 are based on Singh (2006).
36. The data on travel by urban and rural US residents are from online analysis of the US Department of Transportation's 2001 *National Household Travel Survey,* bts.gov/programs/national_household_travel_survey/.
37. Figure 2.5 is based on the data for Chart 6.10 on p. 80 of UK DfT (2005a).
38. See Bayliss, D., *Buses in Great Britain: Privatisation, Deregulation and Competition,* March 1999, worldbank.org/html/fpd/transport/expopres/bayliss1 .doc, particularly Table 5.
39. In 2000–2001, London buses performed 4.7 billion passenger-kilometers; the Underground performed 7.5 billion pkm. See p. 3 of *Transport Statistics for London 2001,* Transport for London, 2001, statistics.gov.uk/STATBASE/ Product.asp?vlnk=1109.
40. Figure 2.6 is based on data in Kenworthy and Laube (2001).
41. Figure 2.7 is based on data in Kenworthy and Laube (2001).
42. The actual data for Rome, Munich and Berlin are respectively as follows: car ownership rates per 1,000 residents, 655, 469 and 354; public transport trips per person per year, 250, 266 and 263; trips by walking and cycling per person per year, 197, 314 and 358.
43. "All road motor vehicles per capita" is used as the indicator in Figure 2.8 because the data in the source for cars are inconsistent. For some jurisdiction they include vans, pick-ups and sport utility vehicles (SUVs), and for others they do not. Use of all road motor vehicles produces relatively higher numbers for poorer than richer countries than if cars or personal vehicles were used. In poorer countries, cars tend to comprise less than 80 percent of road motor vehicles. In richer countries they tend to comprise more than 80 percent. Note that in Figure 2.8 GDP is expressed in US dollars at currency conversion rates.
44. The New Zealand vehicle licensing data are from p. 50 of *New Zealand Motor Vehicle Registration Statistics 2005,* Land Transport New Zealand, landtransport.govt.nz/statistics/motor-vehicle-registration/docs/2005.pdf. Population data are from Table 13 of National Population Estimates Tables, Statistics New Zealand, stats.govt.nz/NR/rdonlyres/266BA4A7-375B-4EBD -B93E-396E8D19EF89/0/NatPopEst1a.xls.
45. Figure 2.8 is based on data in Tables 2.1, 3.12 and 4.2 of World Bank (2006). The 85 represented countries and jurisdictions include all those for which relevant data are available at the indicated source.
46. The UK data in Table 2.2 are from *Cost of Motoring Index 2005 Q3 findings,* Royal Automobile Club, rac.co.uk/know-how/owning-a-car/running-costs/

motoring-index-the-true-cost-of-motoring.htm. The US data in Table 2.2 are from Table 3-14 of US DOT (2007).

47. See p. 49 of Litman (2005) for the estimate that "variabilization" of insurance costs reduces distance driven by 10–12 percent.
48. The sources in Note 46 suggest that insurance costs comprise 10–20 percent of fixed costs. They can be thought of as an increase of this amount in the cost of car ownership. The long-run elasticity of car ownership with respect to cost is –0.9—see Table 8 of Graham and Glaister (2004)—meaning that for every 1 percent increase in price there is a 0.9 percent reduction in ownership. To the extent that kilometers travelled per vehicle are constant (see below), a 10–20 percent increase in price thus results in a 9–18 percent reduction in the number of vehicles on the road and thus the number of kilometers driven.
49. Figure 2.9 is from p. 175 of Schafer and Victor (2000). It is © 2000 Elsevier Limited, and is reproduced with permission. To the extent that the notion of travel-time constancy has validity, daily travel times may be rising to uncomfortable levels in the US. According to Figure 20 on p. 17 of Polzin (2006), the average rose from 45.7 minutes per person in 1983 to 78.5 minutes in 2001. The notion of a constant travel-time budget has been criticized on empirical and conceptual grounds. As well as the just-noted increase in daily travel times, which may not be consistent with the notion of a constant budget, there are indications that daily travel times vary with residential location and personal and household characteristics (Mokhtarian and Chen, 2004). Banerjee et al. (2007), partly in response to the work of Mokhtarian and Chen, have argued that a more useful concept than constant travel-time budget could be travel-time frontier, that is, the maximum amount of time that people are willing to allocate to travel. We feel the notion of a constant travel-time budget is at least heuristically useful, and that the constancy may apply more to aggregate populations than to particular groups of people (as may the constancy in annual distance traveled per vehicle, discussed later in the text).
50. Walking, transit and automobile cities are discussed by Newman and Kenworthy (2006).
51. The US provides perhaps the major departure from constancy in Figure 2.10, with about a 16-per-cent increase in kilometers driven annually per vehicle between 1970 and 1995, mostly in the second half of this period. This could have been because of the relatively sprawling nature of urban development in the US. For example, an analysis of development in 25 mid-sized cities in several countries across the period 1990–2000 (Schneider and Woodcock, 2008) found that, "Although all 25 cities are expanding, the results suggest that cities outside the US do not exhibit the dispersed spatial forms characteristic of American urban sprawl." Also see the discussion of Figure 2.19, particularly Note 112.

52. Figure 2.10 is from a presentation by Lee Schipper, then with the International Energy Agency, to a workshop on fuel taxation held by Transport Canada in Ottawa in March 1999. It is reproduced with the author's permission. A more up-to-date but less reproducible version of this chart is Figure 7-9 on p. 133 of IEA (2004).
53. Figure 2.5 does not directly represent distance driven per car, but a reasonable assumption is that cars owned in rural areas are driven on average more than cars owned in urban areas.
54. Table 2.3 is based on Table 22 of Hu and Reuscher (2004).
55. The data on the age of the vehicle fleet are from Table 21 of Hu and Reuscher (2004). Note that these data are not available for 1969.
56. The data on household size and vehicle ownership are from Table 2 of Hu and Reuscher (2004).
57. See Adams (2000) on the politics of car ownership.
58. For Singapore, see the source in Note 27. Other places include Shanghai, which has a scheme similar to Singapore's, and Tokyo, where a car may be licensed only if it has exclusive use of a registered parking place (which can include the upper location of a contraption that allows two cars to be parked in one spot). Governments of countries of the former Soviet Bloc limited car ownership until the 1970s or 1980s, but then succumbed to popular demand for personal motorized mobility. For a discussion of transport trends in these countries, see Suchorzewski (2005).
59. See Graham and Glaister (2004), particularly Table 8 on p. 271; and Goodwin et al. (2004), particularly Tables 6 and 7 on pp. 285 and 286.
60. Central city residents in the authors' home cities of Toronto and Vancouver own relatively few cars and make many of their local trips on foot or by bicycle (See Figure 2.3 for data on Toronto), but they appear to make more longer-distance trips than average, usually by air. Data on this point are available for different parts of Norway's Oslo region. Holden and Norland (2005) reported that in lower-density areas, energy use for local travel exceeded energy use for long-distance leisure time travel by plane, but the converse was true for higher-density areas (see especially their Figure 4 on p. 2,159). At the highest densities examined (120 housing units per hectare), per-capita annual energy use was the highest among the areas examined for leisure-time air travel and the lowest for local travel, with the former more than three times the latter. A partial explanation could be that central city residents have more disposable income because they do not own cars.
61. The EANO principle is discussed in Gilbert (2004).
62. For an elaboration of the importance of making parking as (in)accessible as public transport, see Knoflacher (2006).
63. The quotation is from p. 165 of Stern and Richardson (2005). We ask, if the evidence for a choice is particular travel behavior, why not just say that, for example, "John took a bus." Rather than "John chose to take a bus?" If the

evidence for choice is some other behavior, perhaps a response to a survey question, that behavior may be determined by different factors from those that determine travel behavior, and may have little consistency with travel behavior. This point is elaborated in Gilbert (2004).

64. Figure 2.11 is based on analysis of the online results of the 2001 *US National Household Travel Survey,* US Department of Transportation, Bureau of Transportation Statistics, Washington DC, bts.gov/programs/national_household_travel_survey/. Please see the source for definitions of the purposes ("work-related," "personal business," etc.).
65. The US data in Table 2.4 are for 2001 and are based on Table A-18b on p. 24 of US DOT (2003). The EU15 data are for 2001–2002 and are based on Table 3-XIV on p.22 of "Deliverable 7: Data Analysis and Macro Results" of *DATELINE (Design and Application of a Travel Survey for European Long-distance Trips Based on an International Network of Expertise),* July 2003, cgi.fg.uni-mb.si/elmis/docs/D7%20-%20Data%20Analysis%20and%20Macro%20Results%20110703.pdf.
66. The data in Table 2.5 are based on Table A-23 on p26 of US DOT (2003).
67. Information about the purposes of longer-distance trips is in the sources detailed in Note 65.
68. The chart referred to at the website of the World Tourism Organization is at world-tourism.org/facts/eng/vision.htm.
69. Figure 2.12 is based on Table C1 of US DOT (2006b).
70. Figure 2.13 is based on data in what was the Common Database of the United Nations Statistics Division and is now UNdata at data.un.org/, and on two 2006 news releases by the International Civil Aviation Organization (ICAO): "World Airlines Improve Operating Profits in 2005 Despite Fuel Cost Increases" (PIO 07/06, May 30) "Strong Air Traffic Growth Projected Through to 2008" (PIO 08/06, June 29). See Note 77 concerning the estimate of the impact of low-cost carriers.
71. The estimate of $12.45 assumes that Ryanair paid the going rate for jet fuel in November 2004 of $0.39 per liter and that its planes used 3.5 liters per 100 person-kilometers. With hedging, Ryanair may have paid less for the fuel used. The rate per 100pkm is from IATA's (2005) estimate of the fuel use by new planes, "IATA Critical of EC Plans to Bring Aviation into European Emissions Trading Scheme," iata.org/pressroom/pr/2005-09-28-01.htm. Ryanair's planes were not new, but they may have been a little more fully occupied than average. Note that the average fuel consumption by US passenger aircraft in 2005 and the first half of 2006 was 5.1 and 4.9 L/100pkm, down from 6.2 L/100pkm in 2004. See Heimlich, J., *US Airlines Operating in an Era of High Jet Fuel Prices,* Air Transport Association of America, Washington DC, 2006, airlines.org/NR/rdonlyres/73AADEC2-D5A2-4169-B590-1EE83A747CDA/0/Airlines_Fuel.pdf. If Ryanair's planes were consuming 5.0 L/100pkm at $0.39/L, the total fuel cost for the two flights would have been $35.56.

72. This was the finding of researchers who tracked 650,000 fares charged by low-cost and traditional airlines for comparable flights booked 1–70 days before the flight (Piga and Bachis, 2006).
73. See *Air Travel Price Index (ATPI), National-Level ATIP Series, 1995 Q1 to 2006 Q4,* US Department of Transportation, Bureau of Transportation Statistics, Washington DC, bts.gov/xml/atpi/src/datadisp.xml?t=1. Prices are about 15 percent higher in current dollars, but inflation reduced the value of the dollar by 36 percent, so prices dropped in real terms. For information on the extent of inflation in the US, see inflationdata.com/Inflation/Inflation_Rate/HistoricalInflation.aspx.
74. For airlines' fuel and labour costs, see the item by Heimlich detailed in Note 71.
75. Current and historical data are available at *Quarterly Cost Index: US Passenger Airlines,* Air Transport Association, airlines.org/economics/finance/Cost+Index.htm. Swan and Adler (2006) using the same source of data (US Department of Transport) may report fuel costs as a lower share of total costs—as opposed to operating costs—than does the Air Transport Association (ATA), specifically 7 percent in 1996–2001 vs. 12 percent. This could in part be because Swan and Adler appear to report much higher aircraft ownership costs: 32 vs. 10 percent of total costs.
76. Projected aviation activity is from the June 2009 *Industry Outlook* of the International Air Transport Association (IATA), at iata.org/NR/rdonlyres/DA8ACB38-676F-4DB1-A2AC-F5BCEF74CB2C/0/Industry_Outlook_Jun09.pdf.
77. See "Liberalisation of European Air Transport: The Benefits of Low Fares Airlines to Consumers, Airports, Regions and the Environment," European Low Fares Airline Association, 2004, elfaa.com/documents/ELFAABenefitsofLFAs2004.pdf. The estimated travel without low-cost carriers in Figure 2.13 assumes that their world share increased linearly from 1 percent in 1990 to 18 percent in 2008, with 60 percent of their share in each year being added trips.
78. High-speed rail was discussed extensively in Chapter 1.
79. For a discussion of how low-cost airlines are influencing rail services, see Sauter-Servaes and Nash (2007).
80. Tourism travel is about four times as sensitive to price as business travel. A 10-percent increase in the total price of air travel will reduce tourism travel by about 10 percent, but business travel by only about 2.5 percent. See Gillen et al. (2003).
81. For students' reasons for travel, see Kim et al. (2006).
82. The estimates of shares of energy use are based on data from the energy section of *OECD Statistics,* Organization for Economic Cooperation and Development, Paris, France, as available until December 2006. The estimates assume that the movement of people uses all road gasoline, half of aviation fuel, and a quarter of rail fuel, and that the movement of freight uses all road diesel fuel, half of aviation fuel, all marine fuel, and three quarters of rail fuel.

83. The textbook on transport is Hanson and Giuliano (2004).
84. This imbalance concerning GHG emissions from freight transport—which contributed only a quarter of transport's total in 1990—could have reflected a government position that GHGs can be reduced more cost-effectively or less disruptively from the movement of people than the movement of freight, but such reasons were not given.
85. Box 2.4 is quoted from p. 163 of McPhee (2006). The extract is © 2006 Farrar, Straus and Giroux, LLC, and is reproduced with permission.
86. Table 2.6 is based on the authors' estimates of the extent of international trade activity and values per tonne. The estimates of international trade activity assumed in Table 2.6 are, in billions of tonne-kilometers: air, 145; marine, 45,000; road, 700; rail, 470; pipeline, 230. The following values per tonne are assumed: air, $100,000; sea, $750; road, $1,500; rail, $750; pipeline, $500. These values are different from those in Table 2.7, and take into account possible differences between trade among NAFTA members and other international trade, for example, the greater amount of movement of finished products over water in the latter case. The values for freight movement are different from those underlying the right-hand panels of Figures 2.1 and 2.14, which concern all freight movement, not just international trade.
87. The freight value data in Table 2.7 are for 2004 and are derived from Tables 6-1c and 6-2c of *North American Transportation Statistics Database,* Bureau of Transportation Statistics, US Department of Transportation, nats.sct .gob.mx/nats/sys/index.jsp?i=3. The shipping cost data are for 2001 and are from Table 3-17 of US DOT (2007).
88. The cost of moving iron ore is from *The low costs of maritime transport,* Maritime International Secretariat Services, undated, marisec.org/world tradeflyer.pdf. The barge costs are from Table 3-17 of US DOT (2007).
89. The very preliminary estimates in Figure 2.14 are derived from on data on transport energy use from the energy section of *OECD Statistics,* Organization for Economic Cooperation and Development, Paris, France, as available until December 2006.
90. For oil's share of trade movements, see Table 5 of UNCTAD (2006).
91. The mode data in Figure 2.15 are from the following sources: Canada, p. 119 of Natural Resources Canada (2006); China, Table 16.2 of National Bureau of Statistics of China (2006); European Union, European Commission (2006); Japan, Ministry of Land, Infrastructure and Transport, *Summary of Transportation Statistics,* 2006, toukei.mlit.go.jp/transportation_statistics .html (this link was not available in April 2009); US, Table 1-46bM of US DOT (2007).
92. The Weyburn CO_2 Monitoring and Storage Project is an international venture coordinated by the IEA. The transported CO_2 is a by-product of the production of synthetic natural gas from coal at the Great Plains Synfuels

Plant. By March 2003, 3.5 million tonnes of CO_2 has been injected into the oil field, enhancing its yield by about a third (from 15,000 to 20,000 barrels per day, approximately). Details of the project are in *IEA GHG Weyburn: CO_2 Monitoring and Storage Project,* International Energy Agency, Cheltenham, UK, undated, ieagreen.org.uk/glossies/weyburn.pdf.

93. Information about the Black Mesa Pipeline is at blackmesapipeline.com/index.htm.
94. This statement about the state of knowledge about freight transport should perhaps be qualified by the possibility that there is much useful information in the private domain, such as the database maintained by Global Insight, globalinsight.com/ProductsServices/ProductDetail700.htm. From an analysis of what is offered, we suspect that such databases are rich in information about commodity flows and less useful as a tool for transport analysis.
95. For commercial trips in Edmonton, Alberta, see p. 385 of Hunt et al. (2004).
96. For the differences between intra- and intercity trips, see Exhibit 3.1 of *Profile of Private Trucking in Canada,* Private Truck Motor Council of Canada, 1998, strategis.ic.gc.ca/epic/internet/ints-sdc.nsf/en/fd01101e.html.
97. Figure 2.16 is reproduced with permission of the Mærsk Line. The Danish-flagged *Emma Mærsk* began service between Europe and the Far East late in 2006, the first of a fleet of seven such ships. Further information about this ship and its sister ships is at robse.dk/pages/Emma/EmmaFami.asp. According to this source, the *Emma Mærsk's* fuel consumption with a full load is equivalent to 5.5 megajoules per 100 tonne-kilometers (stated as 1 kWh per 66tkm). A heavy-duty truck typically uses 45 times as much, as discussed in Chapter 3. A fleet of larger ships appears to be under construction for Marseille, France-based CMA CGM; see Machalaba, D. and Stanley, B., "Tight squeeze for giants of the sea," *Globe and Mail,* October 10, 2006.
98. According to the Hamburg Shipbrokers Association—whose members control about 75 percent of container movements—a loaded TEU carries an average of 14 tonnes of freight; see *Hamburg Index,* vhss.de.
99. See Table 42 of UNCTAD (2006) and the corresponding tables in previous issues of this annual report.
100. This estimate, for 2005, is based on data in Tables 5 and 6 of UNCTAD (2006).
101. Box 2.5 is quoted from p. 70 of McPhee (2006). The extract is © 2006 Farrar, Straus and Giroux, LLC, and is reproduced with permission.
102. This cost is based on Corbett et al. (2006) and also Machalaba and Stanley detailed in Note 97.
103. For example, one of four papers presented at a round table organized by the European Conference of Ministers of Transport in February 2001 concluded that more than half of the economic growth of what was then West Germany during the period 1950–1990 could be attributed to growth in transport activity (Baum and Kurte, 2002). Of this contribution, only a

small part lay in the direct contribution of transport activity; most came from transport's facilitation of other activities. Another paper presented at this workshop suggested that the primary causal relationship is that economic growth causes transport growth (Vickerman, 2002).

104. The decline in distance traveled by air freight has been inferred from data in the sources detailed in Note 70.
105. Figure 2.17 is based on data in UNCTAD (2006) and earlier editions of this annual report. The finding of near constancy in trading distance over time is consistent with at least two earlier analyses. Berthleon and Freund (2004) found that the average distance that trade travels "declined slightly" between 1980 and 2000. Carrere and Schiff (2004) reported a small increase in this distance (4.0 percent) for OECD countries between 1962 and 2000, and a larger decrease (8.4 percent) across the same years for non-OECD countries.
106. The analysis of factors in the growth in road freight activity is based on data in Tables 1-32 and 1-45 of US DOT (2007).
107. In Box 2.1 on p. 27 of the SMP report (see Note 108) there is an elaboration of what is meant by "business-as-usual," under the heading "What do we mean by the phrase, 'If present trends continue'?" The thrust of this long explanation is that present conditions of human behavior, technology, economic growth and government policy will continue, with the qualification that anticipated future actions of government are taken into account if they are considered to be already having an effect.
108. The final SMP report is *Mobility 2030:Meeting the Challenges to Sustainability,* World Business Council for Sustainable Development, Geneva, Switzerland, 2004, wbcsd.org/web/publications/mobility/mobility-full .pdf. The Excel spreadsheet model that provides the projections of transport activity is at wbcsd.org/web/publications/mobility/smp-model-spread sheet.xls. Documentation for the model and the reference case is at wbcsd .org/web/publications/mobility/smp-model-document.pdf. The SMP was conducted by and for the following companies: BP, DaimlerChrysler, Ford, General Motors, Honda, Michelin, Nissan, Renault, Norsk Hydro, Shell, Toyota and Volkswagen.
109. The charts in Figure 2.18 are based on Figure 2.2 and Figure 2.5 of the first source detailed in Note 108. They are reproduced with permission of the World Business Council for Sustainable Development.
110. The actual rate of growth—for 2000–2004—is from Table 16-2 of National Bureau of Statistics of China (2006). It is compared with the model projection for 2005 in the Excel file detailed in Note 108.
111. The actual rate of growth in pkm for the US—for 2000–2004—is from Table 1-37 of US DOT (2007). This is for domestic travel only.
112. Figure 2.19 is based information at *Traffic Volume Trends,* US Federal Highway Administration, Washington DC, April 2009, at fhwa.dot.gov/ohim/

tvtw/tvtpage.cfm. Between 1976 and 2006, the number of vehicle-kilometers increased by 56 percent. This had two components: the number of light-duty vehicles per person increased by 29 percent and the number of kilometers driven per vehicle increased by 21 percent. The increase in amount driven per vehicle was noted earlier in connection with Figure 2.10 (see Note 51).

113. Figure 2.20 is based on data at the source detailed in Note 112 (vehicle movement) and on gasoline prices at *Monthly US All Grades All Formulations Retail Gasoline Prices,* Energy Information Administration, Washington DC, April 2009, at tonto.eia.doe.gov/dnav/pet/hist/mg_tt_usm.htm. Note that, unlike in Figure 2.19 total vehicle movement is shown rather than movement per capita. Totals for individual quarters are seasonally adjusted. Nominal rather than real average gasoline pump prices are shown.

References

Adams, J. (2000) "Hypermobility," *Prospect,* March, worldcarfree.net/resources/freesources/Hyperm.htm.

Banerjee, A., Ye, X. and Pendyala, R. M. (2007) "Understanding travel time expenditures around the world: Exploring the notion of a travel time frontier," *Transportation,* vol 34, pp. 51–65.

Baum, H., and Kurte, J. (2002) Untitled paper presented at Round Table 119, *Transport and Economic Development,* European Conference of Ministers of Transport, Paris, France, pp. 5–49.

Berthleon, M. and Freund, C. (2004) "On the conservation of distance in international trade," World Bank Policy Research Department Working Paper 3293, ideas.repec.org/p/wbk/wbrwps/3293.html.

Carrere, C. and Schiff, M. (2004) "On the geography of trade: Distance is alive and well," World Bank Policy Research Working Paper 3206, ideas.repec.org/p/wbk/wbrwps/3206.html.

Corbett, J. J., Winebrake, J. J. and Green, E. (2006) *Cargo on the Move Through California: Evaluating Container Fee Impacts on Port Choice,* Energy and Environmental Research Associates, Pittsford, NY, docs.nrdc.org/air/air_06081401A.pdf.

Cullinane, S. (2002) "The relation between car ownership and public transport provision: A case study of Hong Kong," *Transport Policy,* vol 9, pp. 29–39.

Cullinane, S. (2003) "Hong Kong's low car dependence: Lessons and prospects," *Journal of Transport Geography,* vol 11, pp. 25–35.

Davis, S. C. and Diegel, S. W. (2007) *Transportation Energy Data Book,* 26th Edition, Center for Transportation Analysis, Oak Ridge National Laboratory, Oak Ridge, TN, cta.ornl.gov/data/index.shtml.

European Commission (2006) *EU Energy and Transport in Figures,* European Commission and Eurostat, Brussels, Belgium, online version only at ec.europa.eu/dgs/energy_transport/figures/pocketbook/2006_en.htm.

Gabler, H. C. and Fildes, B. N. (1999) "Car crash compatibility: The prospects for international harmonization," *Society of Automotive Engineers Inc.*, Technical Paper No. 1999-01-0069, me.vt.edu/gabler/publications/990069.pdf.

Gilbert, R. (2004) "Soft measures and transport behaviour," *Communicating Environmentally Sustainable Transport: The Role of Soft Measures*, OECD, Paris, France, pp. 123–179.

Gillen, D. W., Morrison, W. G. and Stewart, C., *Air Travel Demand Elasticities: Concepts, Issues and Measurement*, Department of Finance, Government of Canada, January 2003, fin.gc.ca/consultresp/Airtravel/airtravStdy_e.html.

Goodwin, P., Dargay, J. and Hanley, M. (2004) "Elasticities of road traffic and fuel consumption with respect to price and income," *Transport Reviews*, vol 24, no 3, pp. 275–292.

Graham, D. J. and Glaister, S. (2004) "Road traffic demand elasticity estimates: A review," *Transport Reviews*, vol 24, no 3, pp. 261–274.

Hanson, S. and Giuliano, G. (eds) (2004) *The Geography of Urban Transportation*, The Guildford Press, New York, NY, 419 pp .

Heavenrich, R. M. (2006) *Light-Duty Automotive Technology and Fuel Economy Trends: 1975 Through 2006*, United States Environmental Protection Agency, Washington DC, 101 pp. plus 17 appendices, July, epa.gov/otaq/fetrends.htm.

Heimlich, J. (2006) *US Airlines Operating in an Era of High Jet Fuel Prices*, Air Transport Association of America, Washington DC.

Holden, E. and Norland, I. T. (2005) "Three challenges for the compact city as a sustainable urban form: Household consumption of energy and transport in eight residential areas in the Greater Oslo region," *Urban Studies*, vol 42, no 12, pp. 2,145–2,166.

Hu, P. S. and Reuscher, T. R. (2004) *Summary of Travel Trends: 2001 National Household Travel Survey*, US Department of Transportation, Federal Highway Administration, Washington DC, 135 pp., nhts.ornl.gov/2001/pub/STT.pdf.

Hunt, J. D., Brownlee, A. T. and Ishani, M. (2004) "Edmonton commercial movements study," *Transportation Revolutions*, Canada Transportation Research Forum, Calgary, Canada, 9–12 May, pp. 376–390.

IEA (2004) *30 Years of Energy Use in IEA Countries*, International Energy Agency, Paris, France, 218 pp.

Jain, J. and Guiver, J. (2001) "Turning the car inside out: Transport, equity and environment," *Social Policy & Administration*, vol 35, no 5, pp. 569–586.

Kenworthy, J. R. (2006) "The eco-city: Ten key transport and planning dimensions for sustainable city development," *Environment and Urbanization*, vol 18, no 1, pp. 67–85.

Kenworthy, J. R. and Laube, F. B. (2001) *The Millennium Cities Database for Sustainable Transport*, Union Internationale des Transports Publics, Brussels, Belgium (CD ROM).

Kim, K., Jogaratnam, G. and Noh, J. (2006) "Travel decisions of students at a

US university: Segmenting the international market," *Journal of Vacation Marketing*, vol 12, no 4, pp. 345–357.

Knoflacher, H. (2006) "A new way to organize parking: The key to a successful sustainable transport system for the future," *Environment & Urbanization*, vol 18, no 2, pp. 387–400.

Litman, T. (2005) "Pay as you drive pricing and insurance regulatory objectives," *Journal of Insurance Regulation*, vol 23, no 3, pp. 35–53.

Liu, R. and Guan, C.-Q. (2005) "Mode biases of urban transportation policies in China and their implications," *Journal of Urban Planning and Development*, June, vol 131, no 2, pp. 58–70 .

McPhee, J. (2006) *Uncommon Carriers*, Farrar, Straus and Giroux, New York, NY, 256 pp.

Mitchell, B. R. (1992–1995) *International Historical Statistics*, Macmillan, London, 3 volumes.

Mokhtarian, P. L. and Chen, C. (2004) "TTB or not TTB, that is the question: A review and analysis of the empirical literature on travel time (and money) budgets," *Transportation Research Part A*, vol 38, nos 9–10, pp. 643–675.

Mygatt, E. (2005) "Bicycle production remains strong worldwide," *Eco-Economy Indicators*, 13 December, Earth Policy Institute, Washington DC, earth-pol icy.org/Indicators/Bike/2005.htm.

NBSC (2006) *China Statistical Yearbook 2006*, National Bureau of Statistics of China, China Statistics Press, Beijing Info Press, Beijing, China, stats.gov .cn/tjsj/ndsj/2006/indexeh.htm .

Natural Resources Canada (2006) *Energy Use Data Handbook*, Natural Resources Canada, Office of Energy Efficiency, Ottawa, Canada, oee.nrcan.gc .ca/publications/statistics/handbook06/pdf/handbook06.pdf.

Newman, P. and Kenworthy, J. (2006) "Urban design to reduce automobile dependence," *Opolis: An International Journal of Suburban and Metropolitan Studies*, vol 2, no 1, article 3, pp. 35–52, eScholarship Repository, University of California, repositories.cdlib.org/cssd/opolis/vol2/iss1/art3.

OECD (2000) *Synthesis Report on Environmentally Sustainable Transportation (EST) Futures, Strategies and Best Practices*, OECD, Paris, France, oecd.org/ dataoecd/15/29/2388785.pdf.

Piga, C. and Bachis, E. (2006) "Pricing strategies by European low cost airlines: Or, when is it the best time to book on-line?," *Discussion Paper Series*, WP 2006-14, Department of Economics, Loughborough University, UK, lboro .ac.uk/departments/ec/RePEc/lbo/lbowps/Book_chapter.pdf.

Polzin, S. E. (2006) *The Case for Moderate Growth in Vehicle Miles of Travel: A Critical Juncture in US Travel Behavior Trends*, University of South Florida, Tampa, prepared for the US Department of Transportation, April, cutr.usf .edu/pdf/The%20Case%20for%20Moderate%20Growth%20in%20VMT -%202006%20Final.pdf.

Pucher, J. and Buehler, R. (2005) "Transport policies in central and eastern

Europe," In Button, K. J. and Hensher, D. A. (eds), *Transport Strategy, Policy, and Institutions*, Elsevier Press, Oxford, UK, 860 pp.

Sauter-Servaes, T. and Nash, A. (2007) "Applying low-cost airline pricing strategies on European railroads," *Transportation Research Record: Journal of the Transportation Research Board*, no 1995 (in press).

Schafer, A. and Victor, D. G. (2000) "The future mobility of the world population," *Transportation Research Part A*, vol 34, pp. 71–205.

Schneider, A., Woodcock, C. E. (2008), Compact, Dispersed, Fragmented, Extensive? A Comparison of Urban Growth in Twenty-five Global Cities using Remotely Sensed Data, Pattern Metrics and Census Information. *Urban Studies*, vol 45(3), pp. 659–692.

Singh, S. K. (2006) "The demand for road-based mobility in India: 1950–2030 and relevance for developing and developed countries," *European Journal of Transport and Infrastructure Research*, vol 6, no 3, pp. 247–274, ejtir.tbm.tudelft.nl/issues/2006_03/pdf/2006_03_03.pdf.

Stern, E. and Richardson, H. W. (2005) "Behavioural modelling of road users: Current research and future needs," *Transport Reviews*, March, vol 25, no 2, pp. 159–180.

Suchorzewski, W. (2005) "Society, behaviour, and private/public transport: Trends and prospects in transition economies of central and eastern Europe," in Donaghy, K., Poppelreuter, S. and Rudinger, G. (eds) *Social Dimensions of Sustainable Transport: Transatlantic Perspectives*, Ashgate, London, pp. 14–28.

Swan, W. N. and Adler, N. (2006) "Aircraft trip cost parameters: A function of stage length and seat capacity," *Transportation Research Part E*, vol 42, pp. 105–115.

UITP (2006) *Mobility in Cities Database*, Union Internationale des Transports Publics, Brussels, Belgium, (CD ROM).

UK DfT (2005a) *Focus on Personal Travel, 2005 edition*, United Kingdom Department for Transport, London, dft.gov.uk/pgr/statistics/datatablespublications/personal/focuspt/2005/.

UK DfT (2005b) *Transport Statistics Great Britain 2005*, Department for Transport, London, dft.gov.uk/pgr/statistics/datatablespublications/tsgb/edition2005.pdf.

UNCTAD (2006) *Review of Maritime Transport, 2006*, United Nations Conference on Trade and Development, New York and Geneva, Switzerland, unctad.org/en/docs/rmt2006_en.pdf.

US DOT (2003) *NHTS 2001 Highlights Report, BTS03-05*, US Department of Transportation, Bureau of Transportation Statistics, Washington DC, bts.gov/publications/highlights_of_the_2001_national_household_travel_survey/pdf/entire.pdf.

US DOT (2006a) *America on the Go... Long distance Transportation Patterns: Mode Choice*, US Department of Transportation, Bureau of Transporta-

tion Statistics, Washington DC, bts.gov/publications/america_on_the_go/long_distance_transportation_patterns/pdf/entire.pdf.
US DOT (2006b) *Estimated Impacts of September 11th on US Travel,* US Department of Transportation, Bureau of Transportation Statistics, Washington DC, bts.gov/publications/estimated_impacts_of_9_11_on_us_travel/pdf/entire.pdf.
US DOT (2007) *National Transportation Statistics 2007,* US Department of Transportation, Bureau of Transportation Statistics, Washington DC, bts.gov/publications/national_transportation_statistics/2007/index.html.
Vickerman, R. (2002) Untitled paper presented at Round Table 119, *Transport and Economic Development,* European Conference of Ministers of Transport, Paris, France, pp. 139–177.
World Bank (2006) *World Development Indicators,* World Bank, Washington DC, devdata.worldbank.org/wdi2006/contents/index2.htm.

CHAPTER 3: Transport and Energy

1. In Figure 3.1, "oil" means all petroleum liquids as elaborated below in connection with Figure 3.7. The estimate of consumption of all petroleum liquids at the end of 2008 is that of Colin Campbell, in Campbell, C., *Newsletter No. 99,* The Association for the Study of Peak Oil and Gas (ASPO), Ireland, March 2009, available at aspo-ireland.org/index.cfm/page/newsletter. The method of estimation is explained there. Campbell's estimate of the total amount of oil that could ever be extracted (by 2100) is 2.425 trillion barrels. He estimated too that daily production of total liquids peaked in 2008 at 81 billion barrels, plus refinery gains of 2–3 percent.
2. Figure 3.1 is based on a slide in a presentation by J. David Hughes of the Geological Survey of Canada entitled *Unconventional Oil—Canada's Oil Sands and their Role in the Global Context: Panacea or Pipe Dream* made at the World Oil Conference organized by the Association for the Study of Peak Oil (USA) and held in Boston, Massachusetts, October 2006, aspo-usa.com/fall2006/presentations/pdf/Hughes_D_OilSands_Boston_2006.pdf. Consumption from 1860–1964 is from the electronic data files supplied with Grübler (1998) at ftp://ftp.iiasa.ac.at/pub/ecs/ag_book/w-energy.csv. Consumption from 1965–2005 is from the historical data series workbook supplied with BP (2009), bp.com/statisticalreview. Consumption from 2006–2008 assumes the 2008 total indicated in Note 1.
3. The conventional economic argument is well reflected in the following on p. 51 of Friedman (2007): "If we were running out of coal or oil, the market would steadily push the prices up, which would stimulate innovation in alternatives. Eventually there would be a crossover, and the alternatives would kick in, start to scale, and come down in price."
4. The prices are closing prices for a barrel of West Texas Intermediate crude oil at the New York Mercantile Exchange (NYMEX) for delivery next

month at Cushing, Oklahoma. They are provided for each day since April 1983 at eia.doe.gov/emeu/international/oilprice.html. This is the most often cited "price of oil," but there are many other prices, including spot prices (i.e., oil for immediate delivery) and prices of Brent crude oil, another standard, mostly traded through an electronic exchange.

5. For oil's share of all transport fuel, see p. 492 of IEA (2006a).
6. According to the US Department of Energy at www1.eere.energy.gov/vehiclesandfuels/facts/2006_fcvt_fotw413.html, cars with diesel engines comprised 28 percent of sales in Western Europe in 1999 and 49 percent in 2005.
7. This description of an oil refinery is from p. 88 of Fischetti (2006).
8. For the energy cost of producing transport fuels from conventional crude oil, see Section 3 of EC (2006).
9. This estimate of world consumption of transport fuels is from p. 492 of IEA (2006a). Conversion to barrels is at the rate of 7.33 barrels per tonne. Bunker fuels are included among transport fuels.
10. A joule is a standard unit of energy. It is the amount required to apply a force of one newton a distance of one meter. In everyday terms, it is approximately the amount of energy required to lift a 100-gram apple a distance of one meter. Energy is also expressed in kilowatt-hours (1 kWh = 3,600 joules) and in British Thermal Units (1 BTU = 1,055.1 joules). Combustion of a liter of gasoline releases approximately 34.7 million joules (34.7 megajoules, or MJ); combustion of a liter of diesel fuel releases approximately 38.7 MJ. A barrel of oil has approximate 6,130 MJ or 6.13 gigajoules (GJ).
11. See Note 6 concerning the share of diesel cars in Europe.
12. These shares are by weight of fuel, not volume. See pp. II.38, II.128 and II.218 of IEA (2006b).
13. Figure 3.2 is based on data from the energy section of *OECD Statistics,* Organization for Economic Cooperation and Development, Paris, France, as available until December 2006.
14. See Annex A of IEA (2006a). This estimate includes international marine bunkers as a transport use and power generation as another end use.
15. See Annex A of IEA (2006a). The transport shares for North America OECD Europe, and Japan were estimated as for the world share (see Note 14) with bunker fuel amounts being estimated as a proportion of all bunker fuels prorated according to total oil use.
16. See Annex A on p. 492 of IEA (2006a) for transport's shares of oil use (other than oil used in producing and refining oil).
17. Figure 3.3 is based on Annex A of IEA (2000) for 1971 and IEA (2006a) for other years. Bunker fuels are included as transport fuels and power generation is included among other uses.
18. The charts in Figure 3.4 are based on data in Annex A of IEA (2000) for 1971 and IEA (2006a) for other years. Here, "richer countries" means OECD members; other countries are "poorer countries." Population data and pro-

jections are from IEA (2006a) and the US Census Bureau at census.gov/ipc/www/idb/worldpop.html.

19. 30 billion barrels per year (bb/y) is about 82 million barrels per day (mb/d), which is how oil production is often stated. A barrel is 159 liters, 35 imperial gallons (UK) or 42 US gallons.
20. Figure 3.5 is based on Slide 31 of a presentation by Kjell Aleklett of the University of Uppsala, Sweden, at The Energy and Environment Conference, Shijiazhuang, China, November 2006, peakoil.net/Aleklett/Aleklett_Shijiazhuang.pdf. It differs from that chart in that consumption in 2000 and projections for consumption in 2010–2030 are reference case projections from Table 3.2 and Appendix A of IEA (2006a).
21. The classification of discovered oil is from p. 314 of Haider (2000).
22. For amplification and discussion of peaks in production, see Greene et al. (2006); p. 53 of Bardi (2005); Bentley (2002); and pp. 1,674 and 1,677 of Hallock et al. (2004).
23. Figure 3.6 is reproduced from p. 108 of Smith (2007).
24. The quotation on the process of oil extraction is from Smith (2007).
25. An accessible source regarding Hubbert's work is Deffeyes (2006).
26. The number of countries in which oil production has already peaked is from p. 11 of Campbell, C., *Newsletter No. 64*, The Association for the Study of Peak Oil and Gas (ASPO), Ireland, April 2006, aspo-ireland.org/index.cfm/page/newsletter. On pp. 110–111 of Smith (2007), production is said to have peaked in 63 of 98 countries, including many that are among the world major producers: Iran, Mexico, Norway, Russia, UK and US. The estimate that peak production of conventional oil occurred in 2005 is from Campbell, C., *Newsletter No. 99*, The Association for the Study of Peak Oil and Gas (ASPO), Ireland, March 2009, aspo-ireland.org/index.cfm/page/newsletter. A more recent assessment, based on the 54 countries represented in BP (2009), indicates that in only 14 of them, responsible for about 40% of total supply, is production of all petroleum liquids still increasing. See Ghanta, P., "Is peak oil real? A list of countries past peak," at truecostblog.com/2009/07/14/is-peak-oil-real-a-list-of-countries-past-peak/.
27. Often the term "unconventional oil" is reserved for oil—such as oil from Alberta's tar sands—that is mined rather than extracted through wells. For these senses of unconventional oil, polar and deepwater oil would count as conventional oil.
28. The quotation on the steepness of the post-peak decline is from p. 101 of IEA (2004a). Also see pp. 53, 54 and 60 of Bardi (2005), and p. 63 of Hirsch (2005).
29. There have been numerous and wide ranging estimates of the date of peak production of petroleum liquids. Some are given in Note 32. We have found the estimates represented by Figure 3.7 to be the most plausible, although we are inclined to be a little conservative as to exactly when the peak might be, suggesting 2012 rather than 2010 or 2011. In any case, differences of a few

years in the timing of the global peak are likely not of consequence as the actual peak will probably have a plateau around it. Early data at IEA (2009) suggest that world production of all petroleum liquids could have peaked in mid 2008. However, given the extraordinary turns in the world economy since then, and the lag between preliminary data and more reliable final estimates of global oil production, we maintain our 2012 estimate from the first edition of this book.

30. In one test case, the Jack 2 project, reported in 2006, oil was extracted from 6,000 meters below the ocean bed, which was over 2,000 meters below the surface, see kensavage.com/index.php/archives/jack-2-project-finds-oil-in-gulf-of-mexico/.
31. Figure 3.7, based on the work of Colin Campbell, is reproduced from Slide 59 of the presentation detailed in Note 20.
32. When examining predictions of the year in which oil production will peak, care has to be taken to note what is included in "oil production." The date should be earlier if one or both of non-conventional oil and natural gas liquids are not included. Here are several predictions of peak oil production, with an indication of what is included in oil production if that is known: (i) Kenneth Deffeyes—peak in 2005 (conventional oil), see p. 48 of Deffeyes (2006); (ii) Ali Samsam Bakhtiari—peak in 2006 (all petroleum liquids), see Samsam Bakhtiari (2004 and 2006); (iii) Colin Campbell—peak in 2010 (all petroleum liquids), see the first source detailed in Note 26; (iv) Rembrandt Koppelaar—peak in 2012–2017 (all petroleum liquids), see pp. 43–44 of Koppelaar (2005); (v) Sadad al Husseini—peak in 2015 (all petroleum liquids), see Andrews, S., *Sadad al Husseini sees peak in 2015, ASPO-USA,* September 14, 2005, energybulletin.net/9498.html; (vi) Cambridge Energy Research Associates—peak after 2030 (all petroleum liquids), see Jackson, P. M., *Why the Peak Oil Theory Falls Down: Myths, Legends, and the Future of Oil Resources,* Cambridge Energy Research Associates (CERA), November 2006, summarized in a CERA press release at cera.com/aspx/cda/public1/news/pressReleases/pressReleaseDetails.aspx?CID=8444, (see also a December 2006 "open letter" rebuttal to this document by the editor of *Petroleum Review,* Christopher Skrebowski, peakoil.ie/experts/skrebowski/skrebowskiresponds.html); (vii) International Energy Agency—peak after 2030 (all petroleum liquids), see pp. 93 and 95 of IEA (2006a); (viii) Aguilera et al. (2009)—peak far in the future, if at all. These authors believe about twice as much conventional oil can be made available as is indicated in the text, and that if this should peak there is enough unconventional oil to keep production rising for the foreseeable future.
33. See Maugeri (2009), who described future dwindling and decline in oil supplies as a myth, and stated that "we are not running out of oil." Also see Lynch, M., "Peak oil is a waste of energy," *New York Times,* August 24, 2009,

at nytimes.com/2009/08/25/opinion/25lynch.html?bl&ex=1251345600&en=f0a941c577e3a21f&ei=5087%0A. Lynch argued that "Oil remains abundant, and the price will likely come down closer to the historical level of $30 a barrel as new supplies come forward in the deep waters off West Africa and Latin America, in East Africa, and perhaps in the Bakken oil shale fields of Montana and North Dakota."

34. The quotation is from p. 324 of Bardi (2009).
35. The quotation is from p. 30 of "Can coal be clean?" *The Economist, Technology Quarterly Supplement,* 2 December 2006.
36. Estimates of actual and projected world oil production represented by the solid line in the left-hand panel of Figure 3.8 are a simplified version of those given in Figure 3.7. Estimates of actual and projected world oil consumption represented by dotted line in this panel are mostly from Table 3.1 on p. 86 of IEA (2006a). Consumption data for 1990 and 2000 are based on Slide 6 of a presentation by Fatih Birol entitled "World Energy Outlook: Key Trends—Strategic Challenges," presented at the World Hydrogen Energy Conference, Lyon, France, June 2006. Also shown are the corresponding demand projections in IEA (2008), published late in 2008 well after the 2008 fall in oil consumption had begun. The crude oil price represented in the right-hand panel of Figure 3.8 is that for a barrel of West Texas Intermediate Oil as bought and sold on the New York Mercantile Exchange, as available at tonto.eia.doe.gov/dnav/pet/hist/rwtcd.htm. The solid line shows actual average spot prices for each year; the dashed line provides estimates as explained in the text.
37. The quotation and the IEA's characterization of present trends in energy use as "unsustainable" are on p. 3 of IEA (2006a).Note too that in another report published at about the same time, IEA projected continually rising demand, beyond 2030, without qualification (IEA 2006c, pp. 3–4). IEA also proposed an "alternative policy scenario" in which demand would be about 11 percent lower in 2030—but still far above present levels—the result of implementation of "all the policies [governments] are considering to curb energy imports and emissions" (IEA, 2006a, p. 6). In the 2008 *World Energy Outlook,* published in November 2008 as the economic recession that is just about to be discussed was in full flood and oil prices were plunging, IEA abandoned any such alternative policy scenario in favor of "climate policy scenarios," but at the same time refashioned the reference scenario to be similar to the previous alternative policy scenario (IEA, 2008).
38. This quotation is from p. 2 of NCEP (2005).
39. The second analysis of the extent of a shortfall-induced oil price increase is by Perry (2001).
40. The actual oil prices in Figure 3.9 are the daily spot prices for West Texas Intermediate crude oil as provided by the US Energy Information Administration at tonto.eia.doe.gov/dnav/pet/hist/rwtcd.htm. The expected prices are those in the right-hand panel of Figure 3.8.

41. Historical currency exchange rates are from oanda.com.
42. For oil shocks and currency volatility, see Amano and van Norden (1998).
43. See United States Senate, Permanent Subcommittee on Investigations, *The Role of Market Speculation in Rising Oil and Gas Prices: A Need to Put a Cop Back on the Beat,* p. 2. Available at: levin.senate.gov/newsroom/supporting/2006/PSI.gasandoilspec.062606.pdf.
44. See Maugeri (2009) for the estimate that in 2007, 1.4 billion "paper barrels" were traded on the New York Mercantile Exchange (NYMEX) while daily global oil consumption was below 85 million barrels. A paper barrel is a futures contract or other instrument that is sold before physical delivery of the oil is required. It could also be a purely financial instrument with returns linked to the price of oil. Maugeri (and others) believes that the ability to avoid delivering or receiving oil, or to bet on the price oil, affects the price of oil.
45. For the report on the experts' meeting and the comment by the chief economist of the US Commodity Futures Trading Commission, see Snow (2009).
46. For the number of speculators in oil futures markets, see Figures 10 and 11 of Interagency Task Force on Commodity Markets, *Interim Report on Crude Oil,* Washington DC: Commodity Futures Trading Commission, July 2008. Available at cftc.gov/stellent/groups/public/@newsroom/documents/file/itfinterimreportoncrudeoil0708.pdf.
47. The source cited in Note 46 includes this on p. 18: "A robust over-the-counter (OTC) market exists to trade oil contracts as well."
48. As well as Snow (2009), already cited in Note 45, see for example the editorial by Steven Kopits, "Peak Oil, Not Speculation" in *Offshore Oil & Gas News,* May 11, 2009, at energycurrent.com/index.php?id=2&storyid=17976. Snow quoted Adam Sieminski, Deutsche Bank's chief energy economist, as noting that "...stupidity can drive decisions. That's the best explanation for somebody buying a crude oil contract at $147/bbl and expecting the price to go up. Governments can't regulate against this."
49. In mid 2009 the US Commodity Futures Trading Commission has launched another inquiry into how speculation in commodities might be curbed. See Blinch R., "CFTC seeks tighter controls on speculation in commodity trading," *USA Today,* July 7, 2009, at usatoday.com/money/markets/2009–7–7-cftc-commodity-speculation-controls_N.htm.
50. The quotation is from p. 17 of Hamilton, J., *Causes and consequences of the oil shock of 2007–2008* (draft). Brookings Institution, Washington DC, 2009. Available at brookings.edu/economics/bpea/~/media/Files/Programs/ES/BPEA/2009_spring_bpea_papers/2009_spring_bpea_hamilton.pdf.
51. The changes in GDP are from the US Bureau of Economic Analysis, *Gross Domestic Product (GDP),* at bea.gov/national/xls/gdpchg.xls.
52. Figure 3.10 is based on data in Table 5.1 and Table D1 of US EIA (2008).

53. The quotation is from Reynolds, A., "It didn't start here," *New York Post*, April 9, 2009. Available at cato.org/pub_display.php?pub_id=10109.
54. The quotation is from a blog by James Hamilton on the paper detailed in Note 50, rather than from the paper itself. (See Hamilton, J., "Consequences of the Oil Shock of 2007–8," at econbrowser.com/archives/2009/04/con sequences_of.html.) He added, "...[this] is a conclusion that I don't fully believe myself. Unquestionably there were other very important shocks hitting the economy in 2007–8, first among which would be the problems in the housing sector. But housing had already been subtracting 0.94% from the average annual GDP growth rate over 2006:Q4–2007:Q3, when the economy did not appear to be in a recession. And housing subtracted only 0.89% over 2007:Q4–2008:Q3, when we now say that the economy was in recession. Something in addition to housing began to drag the economy down over the later period, and all the calculations in the paper support the conclusion that oil prices were an important factor in turning that slowdown into a recession."
55. According to the US Energy Information Administration, at tonto.eia.doe .gov/dnav/pet/pet_pri_gnd_dcus_nus_w.htm, the average retail gasoline price per gallon was $4.17 during the week of July 7, 2008. In the week of January 28, 2002, it has been $1.14. The pump price of diesel fuel had risen even more across the same period: from $1.14 to $4.73 (all current prices).
56. See Gokhale, J. (2009) *Financial Crisis and Public Policy*, Cato Institute, Washington DC, March 2009. Available at cato.org/pub_display.php?pub_ id=10066.
57. See Cortright, J., in *Driven to the Brink: How the Gas Price Spike Popped the Housing Bubble and Devalued the Suburbs*, CEOs for Cities, Chicago IL, ceosforcities.org/newsroom/pr/files/Driven%20to%20the%20Brink%20 FINAL.pdf. Cortright wrote, "The collapse of America's housing bubble—and its reverberations in financial markets—has obscured a tectonic shift in housing demand. Although housing prices are in decline almost everywhere, price declines are generally far more severe in far-flung suburbs and in metropolitan areas with weak close-in neighborhoods. The reason for this shift is rooted in the dramatic increase in gas prices over the past five years. Housing in cities and neighborhoods that require lengthy commutes and provide few transportation alternatives to the private vehicle are falling in value more precipitously than in more central, compact and accessible places."
58. The quotation is from p. 185 of Rubin (2009).
59. Rubin's argument that oil-price-spurred high interest rates contributed to problems in the US arising from questionable mortgage arrangements is on pp. 180–191 of Rubin (2009). On p. 183, he noted, "It is no coincidence that large oil-importing economies without subprime mortgage markets witnessed the same things happening in their economies that were happening

in the United States. In fact, many of those economies are doing even worse than the American economy."

60. In testimony before the Joint Economic Committee of the U.S. Congress, Hamilton stated: "One indication that [the automotive] sales decline was caused by oil prices and not other economic developments is the observation that sales of imported cars were up by 9% over this same period. Since the domestic manufacturers were more heavily reliant on sales of the less fuel-efficient vehicles, these changes represented a significant hit to the domestic auto sector. Declining production of motor vehicles and parts alone subtracted half a percent from total U.S. real GDP between 2007:Q3 and 2008:Q3. In the absence of those declines, real GDP would have clearly grown over this period and it is unlikely that we would have characterized 2007:Q4–2008:Q3 as a true economic recession. 125,000 jobs were lost in U.S. auto manufacturing between July 2007 and August 2008. If not for those losses, year-over-year total job gains for the U.S. economy would have been positive through the first year of what we now characterize as an economic recession." Available at: house.gov/jec/news/2009/Hamilton_testimony.pdf. Rubin (2009) situated the plummeting vehicle sales at the "epicentre" of the US recession (p. 184).
61. The data in Figure 3.11 are from Ward's Key Automotive Data (various months) at wardsauto.com/keydata/. Note that each data point represented average sales of that and the preceding month.
62. The quotation is from p. 33 of the source detailed in Note 50.
63. The quotations in this paragraph are from pp. 42–43 of the source detailed in Note 50.
64. The phrase "conventional wisdom" is most often associated with economist John Kenneth Galbraith, who used it to describe the often faulty expectations and assumptions of experts. See Galbraith (1958) and also Leibovich, M., "Scorecard on conventional wisdom," *New York Times,* March 9, 2008, at nytimes.com/2008/03/09/weekinreview/09leibovich.html.
65. The quotation is from Moran, E. K., "Wall Street meets Main Street: Understanding the financial crisis," *North Carolina Banking Institute Journal,* 13, pp. 5–101, 2009, at studentorgs.law.unc.edu/documents/ncbank/volume13/morann.pdf. For a similar argument, see Bosworth, B. and Flaaen, A., *America's Financial Crisis: The End of an Era* The Brookings Institution, Washington DC, April 2009, at brookings.edu/papers/2009/0414_financial_crisis_bosworth.aspx.
66. The quotation is from Kilian (2008).
67. Hall and Day (2009) suggested that the expansion of GDP of the US without concomitant energy use may have resulted from "outsourcing of much of its heavy industry" (p. 236).
68. The data in Figure 3.12 are from the sources detailed in Note 50.

69. Grant (2007) discussed addressing peak oil production as a balancing of a present potentially aversive situation with a future one. In the present are the political risks associated with advocating actions appropriate to a future decline in availability of oil. In the future are the economic and social risks associated with a decline in production for which there has been insufficient preparation.
70. For information about natural gas-fueled diesel engines, see the website of Westport Innovations Inc. at westport.com.
71. For information about the Fischer-Tropsch process, see fischer-tropsch.org/. For information about the high costs of producing such gasoline substitutes see Williams and Larson (2003); and Table 1 on p. 16 of Greene (1999).
72. For information about the Pearl GTL plant under development by Qatar Petroleum and Shell, see shell.com/home/content/qatar/news_and_library/press_releases/2009/pearl_qatar_update_05022009.html.
73. See, for example, pp. 240–242 of West and Kreith (2006) for proposals for large-scale conversion of coal and natural gas to liquid transport fuels.
74. For estimates that world production of conventional and all natural gas will peak in about 2025 and 2045, respectively, see Slide 11 of J. David Hughes, *Natural Gas in North America: Should We Be Worried?* Presentation to the World Oil Conference organized by the Association for the Study of Peak Oil (USA), Boston, October 2006, at aspo-usa.com/fall2006/presentations/pdf/Hughes_D_NatGas_Boston_2006.pdf. These estimates are based on a September 2006 personal communication from Colin Campbell. Unconventional natural gas is described in Slide 47 of the same presentation as including coal-bed methane, tight gas and shale gas.
75. According to the 2007 version of BP (2009), North American production of natural gas had peaked in 2001 at 788 billion cubic meters; by 2006, production was down to 754 billion cubic meters. See also Slide 13 of the source detailed in Note 74. However, see the next note.
76. According to the US Energy Information Administration, at tonto.eia.doe.gov/dnav/ng/hist/n9010us2a.htm, US natural gas production in 2008 was 10.6% above production in 2006 and 6.2% above 2001's peak production. Similar data are in BP (2009).
77. See, for example, Casselman, B., "US gas fields go from bust to boom," *Wall Street Journal*, April 30, 2009, at online.wsj.com/article/SB124104549891270585.html.
78. For a comment on the natural gas production data, see Slide 17 of the presentation by Matthew Simmons to the Offshore Technology Conference, May 6, 2009, at simmonsco-intl.com/files/OTC%20Topical%20Luncheon.pdf.
79. See Stern (2006) for estimation as to when European natural gas production could peak and analysis of its potential effects on competition for LNG.
80. See p. 8 of Powers (2002): "The US Coast Guard requires a two-mile moving

safety zone around each LNG tanker that enters Boston Harbor, and shuts down Boston's Logan Airport as the LNG tanker passes by.... These extraordinary precautions are taken out of concern for spectacular destructive potential of the fire and/or explosion that might result from a LNG tank rupture." Also of concern is terrorist action. A report done for the U.S Department of Energy set out some of the risks and concerns (Hightower et al., 2004). The report has been criticized as being selective; see Raines, B. and Finch, B., "Scientists say LNG review is missing critical studies," *Mobile Register,* December 23, 2003, wildcalifornia.org/pages/page-114.

81. For information on the energy density of these fuels, see envocare.co.uk/lpg_lng_cng.htm.
82. The description of methane hydrates and the quotation about their occurrence are from *Gas (Methane) Hydrates—A New Frontier,* US Geological Survey, Washington DC, September 1992, marine.usgs.gov/fact-sheets/gas-hydrates/title.html.
83. The quotation about the future of methane hydrates is from p. 359 of Sloan (2003).
84. For the most recent and what may be the most authoritative estimate of the extent of methane hydrates, see Milkov (2004).
85. The presentation, "Energy and Global Warming: Flip Sides of the Same Crisis," was by Richard Allmendinger at Cornell University on March 28, 2007, see news.cornell.edu/stories/April07/Allmendinger.cover.KA.html.
86. See pp. 1–4 of "Alaska North Slope Well Successfully Cores, Logs, and Tests Gas-Hydrate-Bearing Reservoirs," *Fire in the Ice,* Winter 2007, netl.doe.gov/technologies/oil-gas/publications/Hydrates/Newsletter/HMNewsWinter07.pdf.
87. See "Assessment of Technically Recoverable Gas-Hydrate Resources on the North Slope of Alaska," *Fire in the Ice,* Winter 2009, netl.doe.gov/technologies/oil-gas/publications/Hydrates/Newsletter/MHNewswinter09.pdf.
88. For more information about LPG, including uses for transport, see the website of the UK LP Gas Association at lpga.co.uk/.
89. Sasol, the leading producer of synthetic fuels from coal and natural gas, provides about 7 percent of the gasoline and diesel fuel used in South Africa (see sasol.investoreports.com/sasol_ar_2006/review/html/sasol_ar_2006_42.php). According to Ott (2006), it also supplies a blend of 50 percent synthetic fuel and 50 percent jet fuel for refueling jet aircraft at Johannesburg International Airport.
90. For data on fuel for generative electricity, see p. 493 of IEA (2006a).
91. According to Appendix A of IEA (2006a), burning a tonne of oil produces 2.51 tonnes of CO_2. Generating the same amount of energy from coal produces 4.39 tonnes of CO_2, i.e., 75 percent more. Generating the same energy from natural gas produces 2.26 tonnes of CO_2, i.e., 10 percent less. According to the table on p. vii of Marano and Ciferno (2001), production and use of diesel fuel from Wyoming coal results in more than twice the emissions

of greenhouse gases as production and use of the fuel from Wyoming Sweet Crude Oil.

92. There is a brief discussion in Chapter 4 of CO_2 sequestering, also known as CO_2 capture and storage or carbon capture and storage (CCS).
93. This is the "business-as-usual" projection in Table 1 of MIT (2007).
94. See IEA (2006a) for these estimates: p. 125 for coal, p. 114 for natural gas, and p. 88 for oil.
95. This recent report on coal's availability is by Zittel and Schindler (2007).
96. Reserve and consumption estimates in this and the next paragraph are based on pp. 32–35 of BP (2009), on previous editions of this source, and on Zittel and Schindler (2007).
97. Note that coal's energy balance—also known as Energy Return on Energy Invested, or EROEI—can be quite high. According to Table 1 of Roberts (2006), it is 30 at the mine gate (30 joules of usable energy for every one joule used to extract it), compared with comparable estimates of 11–23 for conventional oil and 7–11 for natural gas.
98. In June 2007, the US National Research Council produced a report saying that the coal reserves are less than popularly believed, but that "there is sufficient coal at current rates of production to meet anticipated needs through 2030," (US NRC, 2007, p. 3). Also see the presentation by David Rutledge, *Hubbert's Peak, The Question of Coal, and Climate Change,* rutledge.caltech.edu/, particularly Slide 32, which has the heading "Why are coal reserves too high?" and illustrates "the many social, environmental, and technical hindrances that are not fully taken into account in the reserve estimates."
99. US production of coal-bed methane in 2003 was about 43 billion cubic meters and rising, according to Limerick, S., *Coal bed methane in the United States: A GIS study,* US Energy Information Administration, Washington DC, June 2004, searchanddiscovery.net/documents/2004/limerick/images/limerick.pdf. US production of all natural gas in 2003 was 541 billion cubic meters and falling, according to p. 24 of BP (2009). However, US natural gas production has been increasing since 2005.
100. The estimates of reserves of coal-bed methane are from the sources detailed in Note 99.
101. For up-to-date material on coal-bed methane, see the US Environmental Protection Agency's *Coalbed Methane Extra,* epa.gov/cmop/newsroom/newsletter.html.
102. For example, according to p. 32 of BP (2009) the UK now has only about 220 million tonnes of mineable coal, of which about 10 percent is mined annually. According to a UK government report, about 17 billion tonnes of coal could be amenable to underground coal gasification; see p. 9 of DTI (2004).
103. Box 3.1 is from p. 1,031 of Orr (1979).
104. For ethanol's impacts on air pollution, see Niven (2005), von Blottnitz and Curran (2007) and Jacobson (2007).
105. See p. 385 of IEA (2006a).

106. See, for example, p. 373 of Kim and Dale (2004), who suggested that world bioethanol production could increase by a factor of 16 and displace 32 percent of global gasoline consumption (i.e., over 20 percent of today's road transport fuel and over 15 percent in 2030 if the other features of IEA's "Alternative Policy" scenario apply).
107. Ethanol has almost exactly two thirds of the energy content per liter of gasoline. For details of production and use of ethanol in Brazil, see Kojima and Johnson (2005).
108. See Urbanchuk (2004) and Lubowski et al. (2006).
109. For example, the then US president's 2007 State of the Union speech proposed production of 132 billion liters of "renewable or alternative fuels" by 2017. A supporting document indicates that this phrase is to include "corn ethanol, cellulosic ethanol, biodiesel, methanol, butanol, hydrogen, and alternative fuels [i.e., other substantially non-petroleum fuels]" although focus is likely to be of production of ethanol from corn. The current administration's target is production of 136 billion liters of "renewable fuels"—mostly biofuels other than corn ethanol—by 2022. See rsc.org/chemistry world/News/2009/May/08050902.asp.
110. See Romm (2006) as discussed by Gilbert and Perl (2007), where a more extensive analysis is available that underlies this and the next paragraph.
111. In April 2009, the Goldfield ethanol plant's operator was fined $167,600 by the US Environmental Protection Administration for violating the maximum emission of particulate matter and hazardous air pollutants. See US EPA, *Consent Agreement and Final Order,* at epa.gov/region07/businesses/consent_agree_final_order/2009/central_iowa_renewable_energy_llc_goldfield_ia_041609.pdf.
112. See Farrell et al. (2006) for a review of ethanol production and a conclusion that "current corn ethanol technologies are much less petroleum-intensive than gasoline, but have greenhouse gas emissions similar to those of gasoline."
113. However, see Note 104 on emissions resulting from ethanol use.
114. In the US, CAFE is a fuel consumption standard stated in miles per gallon. Manufacturers are required to ensure that the vehicles they produce meet or exceed the standard, on average. CAFE is discussed more in Chapter 4.
115. A useful source of information about ethanol production and use in the US, with links to other sources, is at "Ethanol: The facts, the questions," *Des Moines Register,* August 27, 2006, desmoinesregister.com/apps/pbcs.dll/article?AID=/20060827/OPINION03/608250397/1035/OPINION.
116. The ethanol production prices are from pp. 8–10 of Gallagher (2006).
117. See, for example, the website of the Iogen Corporation, iogen.ca/ and Yang and Lu (2007).
118. See Patzek (2005) and Patzek and Pimentel (2005) for in-depth analysis of challenges associated with enzymatic processing of plant material to produce ethanol.

119. For a brief indication of some aspects of the controversy, see Hanson, S., "Corn Fuels Controversy," *Council on Foreign Relations,* Washington DC, February 1, 2007, cfr.org/publication/12526/corn_fuels_controversy.html.
120. For a comparison of energy use in biofuels' production, see Table 3.7 on p. 65 of IEA (2004b).
121. For discussion of the characteristics of biodiesel, see p. 391 of IEA (2006a) and p. 110 of IEA (2004b).
122. IEA maintains a separate website on biogas initiatives at iea-biogas.net.
123. The source of the data in the two figures in Box 3.2 was the energy section of *OECD Statistics,* Organization for Economic Cooperation and Development, Paris, France, as available until December 2006. These data are now available only for a fee through the International Energy Agency at data.iea.org/ieastore/statslisting.asp. The statement for the current Swedish prime minister is in an e-mail from Linus Adolphson of the Swedish Prime Minister's office to Richard Gilbert, 27 November 2006.
124. For overviews of prospects for the "hydrogen economy," see Agrawal et al. (2005), IEA (2005) and Lovins (2005). For a view of what is happening in Europe, see European Commission, *The European Hydrogen and Fuel Cell Technology Platform,* European Commission, Brussels, Belgium, 2006, ec.europa.eu/research/energy/nn/nn_rt/nn_rt_hlg/article_1261_en.htm.
125. For discussions of fuel cells, see IEA (2005), and also Carle et al. (2005), Feitelberg et al. (2005), Marshall and Kazerani (2005) and De Bruijn (2006).
126. See IEA (2005), and also Weinert et al. (2006).
127. Figure 3.13 has been slightly adapted from Figure 9 of Bossel (2005a), and is reproduced with permission.
128. Figure 3.14 was kindly provided by Professor Judith Patterson of Concordia University, Montreal, and is reproduced with permission. In Calgary, the light-rail system is fueled entirely by electricity from a dozen 660-KW wind turbines, transmitted across the Alberta grid. The system's slogan is "Ride the Wind," displayed on the side of this two-car train.
129. See Bossel (2005a) and Mazza and Hammerschlag (2004). Bossel's analysis differs from that of Mazza and Hammerschlag in five ways: (i) Bossel includes a 5-percent loss for "conditioning" the electricity prior to electrolysis; (ii) Bossel assumes a 20-percent loss during compression rather than 8 percent (more, says Bossel, if hydrogen is liquefied); (iii) Bossel assumes a 10-percent rather than a 3-percent loss during transmission; (iv) Bossel adds a 20-percent loss for storage, etc. at the fuel cell site; (v) Bossel assumes fuel cells are 50 percent efficient rather than 60 percent; and (vi) Bossel adds a 5-percent loss for conversion of the fuel cell output. Note that Bossel provided two estimates of overall loss: 75 percent if the hydrogen is not liquefied, and 80 percent if it is. Hedström et al. (2006), in their Table 2, estimated electricity-to-electricity losses in fuel-cell system to be 45–70 percent, assuming electrolyzer efficiency to be 70–90 percent and fuel cell efficiency to be 45–60 percent, and making no allowance for hydrogen storage losses.

130. For the complexity of fuel cell vehicles and associated energy costs, see Schäfer et al. (2006).
131. The new US federal administration appears to have acted in a manner consistent with this advice. Funding for electric traction is to be given a substantial boost, while funding for mobile applications of fuel cells is to be sharply reduced. (See cfo.doe.gov/budget/10budget/Content/Highlights/FY2010Highlights.pdf.) Energy Secretary Steven Chu has been reported as saying that mobile fuel-cell applications are for the "distant future" (see technologyreview.com/business/22651/page2/).
132. Romo et al. (2005) compared the three types of storage, not only for on-vehicle use but for the related use of wayside storage for grid-connected vehicles, i.e., rail systems, notably to facilitate regenerative braking and moderate peak power demand on the grid. Batteries and supercapacitors each received the highest rating on 4 of 10 criteria; flywheels received the highest rating on six of the criteria. See also Table 2 of Hedström et al. (2006) for comparative data on the three types of storage. One emerging development concerning supercapacitors could be a "game changer" for electric traction. It concerns the prospects for what are known as "composition-modified barium titanate powders" developed by EEStor Inc, a US company. Recently, according to Zenn Motor Co. Inc, a Canadian manufacturer of electric vehicles that has invested in EEStor, an independent assessment has verified the exceptional permittivity of EEStor's particular fabrication of barium titanate. (Permittivity refers to the ability of a material to hold and transmit an electric charge.) If further development milestones are attained, the result could be a usable device with an energy density several times higher than goals for conventional batteries, at costs that are significantly lower per kilowatt-hour of electrical energy stored. (See marketwire.com/press-release/Zenn-Motor-Company-TSX-VENTURE-ZNN-993395.html.) Too little information is available so far to assess whether barium-titanate-based ultracapacitors will have a significant impact on the progress of electric traction.
133. For more on the early history of BEVs, see Anderson and Anderson (2005) and Kirsch (2000).
134. These energy densities are from Van Mierlo et al. (2006). See also Table 2 of Hedström et al. (2006), who give 0.07–0.13 MJ/kg for lead-acid batteries and 0.25–0.54 MJ/kg for lithium batteries.
135. For vehicle efficiencies, see p. 981 of Åhman (2001).
136. According to Anderson and Anderson (2005) up to 10 hours were required to fully charge a vehicle battery 100 years ago. Now, using fast-charging stations, batteries can be charged in as little as 10 minutes. Even this could be too long for some drivers. Toshiba has announced a prototype lithium ion battery that requires only one minute for recharging, see toshiba.co.jp/about/press/2005_03/pr2901.htm. Fast charging requires too high a current for regular domestic use. At-home overnight charging of electric vehicles

still requires 3–6 hours, and will continue to do so unless residences are equipped with higher-power electricity supply.

137. The information in Table 3.1 is as follows: Concerning the ICE vehicle—Honda Civic 2.2 i-CTDi—see *Honda announces all-new Civic for Europe,* Honda Motor Company, Tokyo, August 1, 2005, world.honda.com/news/2005/4050801.html. Concerning the BEV—Mitsubishi Lancer Evolution MIEV—see *Mitsubishi Motors to enter Lancer Evolution MIEV in Shikoku EV Rally 2005,* Mitsubishi Motors Corporation Tokyo, Japan, August 24, 2005, media.mitsubishi-motors.com/pressrelease/e/corporate/detail1321.html. Concerning the fuel cell vehicle—Honda ZC2—see *Specifications: The Honda FCX,* Honda Motor Company Tokyo, Japan, 2007, world.honda.com/FuelCell/FCX/specifications. The range assumes full fuel tanks and batteries run to exhaustion. The rate of energy use for the ICE is based on the stated 5.1 L/100km at 38.7 MJ/L for diesel fuel. That for the BEV is as estimated by Bossel (2005b). That for the fuel cell vehicle is based on the stated storage capacity of 3.75 kg hydrogen (at 142 MJ/kg) and the indicated range. The Mitsubishi prototype was not carried forward. A smaller all-electric vehicle, the i-MIEV, is to go on sale in Japan in 2010 (see washingtontimes.com/news/2009/may/29/mitsubishi-gives-green-light-to-electric-car/).
138. For information about the Tesla roadster, visit teslamotors.com/index.php.
139. The country/vehicle data in Table 3.2 are from Table 13 on p. 27 of An and Sauer (2004); the data on vehicle types are those in Table 3.1; and the data on the very efficient ICE vehicles are from the Loremo website at loremo.com.
140. This paragraph briefly describes the general design of almost all hybrids now being sold. Earlier hybrids had a "series" design, in which only the EM(s) drove the wheels and the sole function of the ICE was to drive the generator that powered the EM(s) and charged the battery. The plug-in hybrid proposed by General Motors, discussed later in the text, also has a series design.
141. According to Heavenrich (2005) the "urban" share of car travel in the US—as opposed to the "highway" share—peaked in 1994 at 63 percent and has been slowly declining.
142. Note that this 15-percent "shortfall" between tested and actual fuel consumption is lower that reported in a review of the literature at ECMT (2005).
143. The comparison in Table 3.3 is between two versions of the Toyota Camry sold in the US. One is a hybrid electric-ICE car; the other is the ICE version that is the most similar to the hybrid version. In Table 3.3, the EPA fuel-use ratings are from fueleconomy.gov. The *Consumer Reports* fuel ratings, and the acceleration and braking data, are from consumerreports.org/cro/cars/index.htm. The remaining information in Table 3.3 is from Toyota Motor Sales, U.S.A. Inc. at toyota.com/camry/specs.html.
144. Based on estimates in the February 2007 issue of *HybridCars* at hybridcars.com/market-dashboard/feb07-overview.html. Sales continued to rise until mid 2008 and then fell dramatically as gasoline prices fell and people's

financial circumstances tightened (see canadiandriver.com/2009/01/06/us-hybrid-sales-drop-427-per-cent-in-december.htm).

145. Porsche's "beetle" design was based on a 1933 sketch by Adolf Hitler, who ordered construction of a "volkswagen," an inexpensive "people's car" costing no more than 1,500 times the average worker's hourly wage. This would have been more affordable than the Chery QQ is today to the average worker in Beijing, as noted in Chapter 2. Although first produced in the 1930s, the Volkwagen beetle did not become available to ordinary purchasers until the 1950s.
146. For information about the Volt see chevrolet.com/electriccar/. General Motors Inc., the manufacturer, says the Volt should be called a range-extended electric vehicle rather than a hybrid because it has only one mode of traction: the electric motor. The ICE is used only to charge the battery that powers the motor. It's a matter of definition. We prefer to describe a hybrid as a vehicle that has an ICE and an electric motor, and characterize the Volt as a plug-in series hybrid. Plug-in vehicles are described later in the text.
147. The largest now is Royal Caribbean International's cruise ship, *Freedom of the Seas*. Information about the *Queen Mary 2* is at cunard.com/images/Content/QM2Technical.pdf.
148. For information about the hybrid ICE-electric tugboat, see foss.com/environment_hybridtug.html.
149. Information about the F3DM is at byd.com/showroom.php?car=f3dm. BYD Auto describes its plug-in series hybrid as a "dual-mode" vehicle.
150. See "President Bush Tries to Tackle High Fuel Prices." *Green Car Congress*, April 25, 2006, at greencarcongress.com/2006/04/president_bush_.html.
151. See US Department of Energy, 2009. *President Obama Announces $2.4 Billion in Funding to Support Next Generation Electric Vehicles.* Available at: energy.gov/news2009/7066.htm.
152. The quotation is from Romm (2006).
153. This argument is elaborated in Gilbert and Perl (2007).
154. The special demands on PHEV batteries may be this concept's Achilles' heel. For a sophisticated appraisal of several of these challenges see Smith, K., et al., "PHEV Battery Trade-Off Study and Standby Thermal Control," presentation at the 26th International Battery Seminar & Exhibit, Fort Lauderdale, Florida, March 2009, at nrel.gov/vehiclesandfuels/energystorage/pdfs/45048.pdf. The main producers of hybrid vehicles, Toyota and Honda, are not rushing to produce plug-in hybrids. A Toyota representative noted at a meeting in New York City in June 2009 that "after the batteries are depleted on a plug-in hybrid they become a heavy 'boat anchor' until the car can be recharged from a wall outlet." (Motavalli J., "Toyota: Plug-in Hybrids Will Have Limited Appeal," *New York Times*, June 7, 2009, at wheels.blogs.nytimes.com/2009/06/07/toyota-plug-in-hybrids-will-have-limited-appeal/?hp). Conversely, PHEVs can be viewed as a BEV that has to carry around a heavy ICE that may be little used.

155. The GCV is subject to a distribution loss of about 10 percent, see Figure 3.13 and source. The BEV is subject to an additional charge-discharge loss of about 30 percent, assuming an NiMH battery: see p. 99 of Matheys et al. (2006). The charge-discharge loss could be considerably lower with advanced lithium-ion batteries. Table 2 of Hedström et al. (2006) suggests a range of 2–10 percent.
156. Other things being equal, the energy required to accelerate or climb a slope is proportional to the weight of the vehicle. For example, an 800-kg vehicle carrying its own weight of batteries would require twice as much energy to move a given distance up a five-percent slope as a vehicle weighing a total of 800 kg. See pp. 224–225 of Larminie and Lowry (2003).
157. See Taplin, M., *A World of Trams and Urban Transit,* Walsall, UK, January 2006, lrta.org/world/worldind.html.
158. See the website of Trolley Motion, trolleymotion.com/en/.
159. See "Trans-Siberian railway electrification completed," *International Railway Journal,* vol 43, no 2, February 2003.
160. The rapid electrification of rail systems in China, Russia and Europe has stimulated interest in moving goods from China to eastern North America via Europe. For the 2004 report of the International Railway Union (UIC) on the project, see transportutvikling.no/NEW_report_2004.pdf. A Norwegian company, New Corridor AS, was formed at the end of 2005 to commercialize the concept.
161. See "1970–1977—Quebec Cartier Mine, Canada," *Trolley History,* Hutnyak Consulting, Elko, NV, 2001, hutnyak.com/Trolley/trolleyhistory.html#QCM.
162. These are averages for 2004 estimated from data in FTA (2006).
163. For an advocate's view of the history of PRT, see Anderson (2005). For what may be a more balanced view, see Cottrell (2005).
164. For information about the Heathrow system see "Pod power for airport," *The Engineer Online,* December 21, 2006, theengineer.co.uk/Articles/297584/Pod+power+for+airport.htm, and also the Heathrow Airport website at heathrowairport.com/. The "pods" in the Heathrow system are battery-electric vehicles that are not grid-connected while in motion, but they have the other attributes of grid-connected systems. For a reference to a possible system in Dubai, see difc.ae/district/facts_and_figures/. Concerning PRT in Abu Dhabi, see Fox, J., "Abu Dhabi to Debut Personal Rapid Transit 'Podcars' Later This Year," at treehugger.com/files/2009/02/masdar-prt-interview.php. For information about PRT plans for Daventry, UK, see daventrydc.gov.uk/environment-and-planning/regeneration/prt-personal-rapid-transport/plans-on-prt-in-daventry/. For an indication of the controversial nature of this proposal, see daventrytowncouncil.gov.uk/Article/Detail.aspx?ArticleUid=8ceba1d0-0f61-4266-8e74-972317760a51.
165. The quotation is from p. 35 of NETMOBIL (2005).
166. The analysis for Corby is reported in Bly and Teychenne (2005).

167. The quotations are from p. 474 of Vuchic (2007). At the heart of Vuchic's argument is the suggestion that for a PRT system to provide the same level of service as a more conventional system "whose vehicles stop at all stations with a certain frequency" the PRT would have to provide up to $(n-1)^2$ more vehicle trips, where n is the number of stations in each system, and that the PRT systems operating costs, speed and reliability would be "closely correlated" with the $(n-1)^2$ factor. Thus, Vuchic seems to be saying that for 21-station systems PRT would be up to 400 times as expensive, slow and unreliable as a conventional system.
168. The quotation is from p. 18 of the transcript of the question-and-answer session after a presentation by Peter Hall, *The Sustainable City: A Mythical Beast?* L'Enfant Lecture on City Planning and Design, American Planning Association and the National Building Museum, Washington DC, December 15, 2005, at planning.org/lenfant/2005/transcript.htm.
169. The Calthorpe quotation is of a remark made at the 2005 conference of the Congress for the New Urbanism, Pasadena, CA, cities21.org/conspirators .htm.
170. The German firm Siemens makes a line of "trolley assist" heavy-duty trucks chiefly for use in mines. See www2.sea.siemens.com/Industry+Solutions/ Mining/mining-solutions/Trolley-Assist-Haul-Trucks.htm.
171. For a description of this aspect of the history of motoring in Canada, see Nicol (1999).
172. The comparative trolley bus cost data are from APTA (2007) and from Schuchmann, A., *Management of Costs and Financing.* Presentation at a workshop of the UITP Trolleybus Group, Salzburg, Austria, April 2006, trolleymotion.com/common/files/uitp/Schuchmann_S2RConsulting.pdf.
173. The data on capital and operating costs, particularly for Landkrona, Sweden, are from Andersson, P. G., *Trolleybus Landskrona: the world's smallest trolleybus system,* Presentation at a workshop of the UITP Trolleybus Group, Salzberg, Austria, April 2006, trolleymotion.com/common/ files/uitp/Anderson_Landskrona.pdf.
174. See the second source detailed in Note 172.
175. According to APTA (2007), the 81,033 in-service diesel buses in the US in 2004 moved an average of 133 km/day; the 597 trolley buses moved an average of 98 km/day. In Singapore, another jurisdiction for which such data are readily available, the 3,131 in-service diesel buses traveled an average of 263 km/day in 2004; see *Land Transport Statistics in Brief 2006,* Land Transport Authority, Singapore lta.gov.sg/corp_info/doc/Stats%20In%20 Brief%20(2006).pdf.
176. For information about Kathmandu's trolley bus system, see Pradhan et al. (2006) and Shrestha (2004).
177. For Belgium, see Van Mierlo et al. (2006). For the UK, see the Department for Transport's website at dft.gov.uk/pgr/scienceresearch/technology/lctis/ lowcarbontis?page=12.

178. See p. 4 of E.ON Netz (2005) for the following quote: "Wind energy is only able to replace traditional power stations to a limited extent. [Its] dependence on the prevailing wind conditions means that wind power has a limited load factor even when technically available. It is not possible to guarantee its use for the continual cover of electricity consumption. Consequently, traditional power stations with capacities equal to 90% of the installed wind power capacity must be permanently online in order to guarantee power supply at all times." E.ON Netz is a distribution company based in northern Germany that handles almost half of the wind power generated in Germany, which is the world's largest producer.
179. See Slide 6 of Kempton, W., "Plug-in Hybrid & Battery Electric Vehicles for Grid Integration of Renewables," University of Delaware, March, presented at the meeting of the Utility Wind Integration Group, Arlington, VA, April 5–7, 2006, uwig.org/Arlington/Kempton.pdf.
180. For information about the large NiCd battery, see Steven Eckroad, *Golden Valley Cooperative Project in Alaska—40 MW Nickel-Cadmium Battery,* Presentation to a California Energy Commission Staff workshop, "Meeting California's Electricity System Challenges through Electricity Energy Storage," February 24, 2005, energy.ca.gov/research/notices/2005–2–4_workshop/05%20Eckroad-EPRI%20on%20BESS.pdf.
181. For the sodium-sulfur batteries, see Mears, D., *Overview of NAS Battery for Load Management,* at the workshop detailed in Note 180, energy.ca.gov/research/notices/2005-2-4_workshop/11%20Mears-NAS%20Battery%20Feb05.pdf.
182. For information about Halton Hills Hydro's system, see oee.nrcan.gc.ca/industrial/technical-info/library/newsletter/archives-2006/Vol-X-no-20-oct15.cfm?attr=24.
183. For information about the King Island system and vanadium flow batteries generally, see Thwaites (2007) and also *Remote Area Power Systems: King Island,* vrbpower.com/docs/casestudies/RAPS%20Case%20Study%20(King%20Isl)%20March%202006%20(HR).pdf.
184. For information about the Donegal system, see the document "VRB ESS™ Energy Storage & the development of dispatchable wind turbine output: Feasibility study for the implementation of an energy storage facility at Sorne Hill, Buncrana, Co. Donegal," at vrbpower.com/docs/news/2007/Ireland%20Feasibility%20study%20for%20VRB-ESS%20March%202007.pdf.
185. Table 3.4 is based on data in IEA (2006e).
186. These data on coal are from Table 5.1 (p. 127) of IEA (2006a).
187. The report on coal's availability is Zittel and Schindler (2007). See also US NRC (2007).
188. See MIT (2007), and also the US Department of Energy's FutureGen project, at fossil.energy.gov/programs/powersystems/futuregen/.
189. These data of nuclear generation of electricity are from IEA (2006a).
190. The report on uranium is by Zittel and Schindler (2006).

191. This paragraph reflects the report by Zittel and Schindler (2006).
192. For information about thorium, see IAEA (2005).
193. The most recent elaboration of the Gaia hypothesis is in Lovelock (2006).
194. The quotation is from James Lovelock, "Nuclear power is the only green solution," *The Independent,* (UK), May 24, 2004. This article is available at Lovelock's website, ecolo.org/lovelock/.
195. The quotation is from p. 2 of Czisch (2006a).
196. The conclusions are based on Czisch (2006b).
197. For this estimate, see p. 6 of Czisch (2006a). For the current proposal to supply Europe with electricity, see *Clean Power from Deserts,* Desertec Foundation, Hamburg, Germany, February 2009, at desertec.org/filead min/downloads/DESERTEC-WhiteBook_en_small.pdf. Also see Richter, C., Teske S., Short R., *Global Concentrating Solar Power Outlook 2009,* Greenpeace International, Amsterdam, the Netherlands, May 2009, at greenpeace.org/raw/content/international/press/reports/concentrating -solar-power-2009.pdf.
198. These are the authors' estimates based on the following: The present annual electricity consumption of each of the US and China is approximately 4.5 and 3.0 petawatt-hours, respectively (IEA, 2006a). For average generation of 1.0 megawatt-hour per square meter per year, land areas of respectively 4,500 and 3,000 square kilometers would be required. The land areas of each of the southwest US and western China that receive sufficient insolation are many times the size of these areas. In the US in particular, all parts of the states of Arizona and New Mexico meet this insolation criterion, even in December (see rredc.nrel.gov/solar/old_data/nsrdb/redbook/atlas/). The combined land area of these states is over 600,000 square kilometers. For further discussion of this matter, see the presentation by David Rutledge detailed in Note 98. His Slide 66 is a world map suggesting among other things that a relatively small area of Californian desert could be used to supply "sufficient energy to replace the whole US grid" from solar thermal installations.
199. For a readily available discussion of the potential of concentrating solar power in relation to other renewable source of electricity, see pp. 177–85 of Mackay (2009). See also Zweibel et al. (2008) and Mehos et al. (2009).
200. For the review of marine energy, see Kerr (2007).
201. For information about the La Rance tidal barrage facility, see esru.strath.ac .uk/EandE/Web_sites/01–2/RE_info/tidal1.htm.
202. Testing is near completion of the world's largest marine current (tidal stream) generator (in Northern Ireland's Strangford Lough), with a maximum output of 1.2 megawatts (producing perhaps five gigawatts of power annually); see theengineer.co.uk/Articles/309448/Tidal+power.htm. A very much larger (150MW) installation is planned for the extremely powerful deep tidal currents in Scotland's Pentland Firth. See Webb T., "Boost for

huge Scots tide-power plan," *The Observer,* March 29, 2009. The power would be provided to a large data center to be located in this remote, cool place. At the upper end of estimates, generators submerged in Pentland Firth could provide more than 80 TWh annually.

203. For total electricity generation in the UK, see p. II.574 of IEA (2006e).

References

Agrawal, R., Offutt, M. and Ramage, M. P. (2005) "Hydrogen economy—an opportunity for chemical engineers?," *AIChE Journal,* vol 51, no 6, pp. 582–1,589.

Aguilera, R. F., Eggert, R. G., Lagos, G., and Tilton, J. E. (2009), *The Energy Journal,* vol 30, no 1, pp. 141–174.

Åhman, M. (2001) "Primary energy efficiency of alternative powertrains in vehicles," *Energy,* November, vol 26, no 11, pp. 973–989.

Aleklett, K. (2006) "Oil: A bumpy road ahead," *World Watch,* January/February 2006, vol 19, no 1, pp. 10–12, World Watch Institute, worldwatch.org/ww/peakoil/.

Amano, R. A. and van Norden, S. (1998) "Exchange rates and oil prices," *Review of International Economics,* vol 6(4), pp. 683–694.

An, F. and Sauer, A. (2004) *Comparison of Passenger Vehicle Fuel Economy and Greenhouse Gas Emission Standards Around the World,* Pew Center on Climate Change, Washington DC, pewclimate.org/docUploads/Fuel%20Economy%20and%20GHG%20Standards_010605_110719.pdf.

Anderson, J. E. (2005) *The future of high-capacity personal rapid transit,* Presentation at the Advanced Automated Transit Systems Conference, Bologna, Italy, November 7–8, 2005, gettherefast.org/documents/FutureofHCPRT-Jan606.doc.

Anderson, J. and Anderson, C. D. (2005) *Electric and Hybrid Cars: A History,* McFarland & Company, Jefferson, NC, 189 pp.

APTA (2007) *Public Transportation Fact Book,* 58th edition, American Public Transportation Association, Washington DC, 107 pp. apta.com/research/stats/factbook/documents/factbook07.pdf.

Bardi, U. (2005), "The mineral economy: A model for the shape of oil production curves," *Energy Policy,* January, vol 33, no 1, pp. 53–61.

Bardi, U. (2009) "Peak oil: The four stages of a new idea" *Energy,* vol 34, pp. 323–26.

Bentley, R. W. (2002) "Global oil & gas depletion: An overview," *Energy Policy,* February, vol 30, no 3, pp. 189–205.

Bly, P. and Teychenne, R. (2005) "Three financial and socio-economic assessments of a personal rapid transit system," *Proceedings of the 10th International Conference on Automated People Movers,* Orlando, FL, doi:10.1061/40766(174)39, at atsltd.co.uk/uploads/Documents/economic_assessments.doc.

Bossel, U. (2005a) *Does a Hydrogen Economy Make Sense?* European Fuel Cell Forum, Oberrohrdorf, Switzerland, efcf.com/reports/E13.pdf.

Bossel, U. (2005b) *On the Way to a Sustainable Energy Future,* European Fuel Cell Forum, Oberrohrdorf, Switzerland, efcf.com/reports/E17.pdf.

BP (2009) *Statistical Review of World Energy,* BP plc, London, bp.com/statistical review.

Carle, G., Axhausen, K., Wokaun, A. and Keller, P. (2005) "Opportunities and risks during the introduction of fuel cell cars," *Transport Reviews,* November, vol 25, no 6, pp. 739–760.

Cottrell, W. D. (2005) "Critical review of the personal rapid transit literature," *Proceedings of the 10th International Conference on Automated People Movers,* Orlando, FL, 1–4 May 2005, American Society of Civil Engineers, Washington DC.

Czisch, G. (2006a) *Low Cost but Totally Renewable Electricity Supply for a Huge Supply Area: A European/Trans-European Example,* Institute for Electrical Engineering, University of Kassel, Germany, iset.uni-kassel.de/abt/w3-w/projekte/LowCostEuropElSup_revised_for_AKE_2006.pdf.

Czisch, G. (2006b) *Joint Renewable Electricity Supply for Europe and its Neighbours—Transfer to Other World Regions and China,* Institute for Electrical Engineering, University of Kassel, Germany, iset.uni-kassel.de/abt/w3-w/projekte/RenElSupEU_Trans_ot_WR+Ch.pdf.

De Bruijn, F. (2006) "The current status of fuel cell technology for mobile and stationary applications," *Green Chemistry,* vol 7, pp. 135–150.

Deffeyes, K. S. (2006) *Beyond Oil: The View from Hubbert's Peak,* Hill and Wang, New York, NY.

DTI (2004) *Review of the Feasibility of Underground Coal Gasification in the UK,* Department of Trade and Industry, London, UK, 47 pp., dti.gov.uk/files/file19145.pdf.

E.ON Netz (2005) *Wind Report 2005,* E.ON Netz GmbH, Germany, windaction.org/?module=uploads&func=download&fileId=232.

EC (2006) *Well-to-Wheels Analysis of Future Automotive Fuels and Powertrains in the European Context: Well-to-Tank Report,* European Commission, Joint Research Centre, ies.jrc.ec.europa.eu/WTW.

ECMT (2005) *Making Cars More Fuel Efficient: Technology for Real Improvements on the Road,* European Conference of Ministers of Transport, Paris, France, 82 pp.

Farrell, A. E., Plevin, R. J., Turner, B. T., Jones, A. D., O'Hare, M. and Kammen, D. M. (2006) "Ethanol can contribute to energy and environmental goals," *Science,* 27 January, vol 311, pp. 506–508 (and erratum 23 June 2006).

Feitelberg, A. S., Stathopoulos, J., Qi, Z., Smith, C. and Elter, J. F. (2005) "Reliability of plug power GenSys™ fuel cell systems," *Journal of Power Sources,* vol 147, pp. 203–207.

Fischetti, M. (2006) "Carbon hooch," *Scientific American,* June, vol 296, no 6, pp. 88–89.

Friedman, T. L. (2007) "The power of green," *New York Times Magazine,* 15 April, nytimes.com/2007/04/15/magazine/15green.t.html.

FTA (2006) *National Transit Database*, US Federal Transit Administration, Washington DC, ntdprogram.gov/ntdprogram/data.htm.

Gallagher, P. (2006) *Effects of International Markets and Potential Policy Changes on the US Ethanol Industry: Will Sector Policies Liberalize Soon?*, Federal Reserve Bank Conference on Globally Competitive Agriculture & the Midwest, 29 September, Chicago, IL, chicagofed.org/news_and_conferences/conferences_and_events/files/2006_global_ag_Gallagher.pdf.

Gilbert, R. and Perl, A. (2007) "Grid-connected vehicles as the core of future land-based transport systems," *Energy Policy*, vol 35, pp. 3,053–3,060.

Grant, L. K. (2007) "Peak oil as a behavioural problem." *Behavior and Social Issues*, vol 16, pp. 65–88.

Greene, D. (1999) *An Assessment Of Energy And Environmental Issues Related To The Use Of Gas-To-Liquid Fuels In Transportation*, ORNL/TM-1999/258, prepared by the Oak Ridge National Laboratory for the US Department of Energy, www-cta.ornl.gov/cta/Publications/Reports/ORNL_TM_1999_258.pdf.

Greene, D. L., Hopson, J. and Li, J. (2006) "Have we run out of oil yet? Oil peaking analysis from an optimist's perspective," *Energy Policy*, March, vol 34, no 5, pp. 515–531.

Grübler, A. (1998) *Technology and Global Change*, Cambridge University Press, Cambridge, UK, ftp://ftp.iiasa.ac.at/pub/ecs/ag_book/w-energy.csv.

Haider, G. M. (2000) "World oil reserves: Problems in definition and estimation," *OPEC Review*, 1 June, vol 24, no 4, pp. 305–327.

Hall, C. A. S. and Day, J. W. (2009) "Revisiting the limits to growth after peak oil," *American Scientist*, vol 97, pp. 230–37.

Hallock, J. L. Jr., Tharakan, P. J., Hall, C. A. S., Jefferson, M. and Wua, W. (2004) "Forecasting the limits to the availability and diversity of global conventional oil supply," *Energy*, September, vol 29, no 11, pp. 1,673–1,696.

Heavenrich, R. M. (2005) "Light-duty automotive technology and fuel economy trends: 1975 through 2005," Appendix C, *City and Highway Driving Data*, US Environmental Protection Agency, Washington DC.

Hedström, L., Saxe, M., Folkesson, A., Wallmark, C., Haraldsson, K., Bryngelsson, M. and Alvfors, P. (2006) "Key factors in planning a sustainable energy future including hydrogen and fuel cells," *Bulletin of Science Technology Society*, vol 26, no 4, pp. 264–277.

Hightower, M., Gritzo, L., Luketa-Hanlin, A., Covan, J., Tieszen, S., Wellman, G., Irwin, M., Kaneshige, M., Melof, B., Morrow, C. and Ragland, D. (2004) *Guidance on Risk Analysis and Safety Implications of a Large Liquefied Natural Gas (LNG) Spill Over Water*, SAND2004-258, Sandia National Laboratories, Albuquerque, NM, fe.doe.gov/programs/oilgas/storage/lng/sandia_lng_1204.pdf.

Hirsch, R. L. (2005) "Shaping the peak of world oil production," *World Oil*, vol 226, no 10, pp. 61–65.

IAEA (2005) *Thorium Fuel Cycle—Potential Benefits and Challenges*, International

Atomic Energy Agency, Vienna, Austria, 104 pp., www-pub.iaea.org/MTCD/publications/PDF/TE_1450_web.pdf.

IEA (2000) *World Energy Outlook 2000,* IEA, Paris, France, 457 pp.

IEA (2004a) *World Energy Outlook 2004,* IEA, Paris, France, 570 pp.

IEA (2004b) *Biofuels for Transport: An International Perspective,* IEA, Paris, France, 210 pp.

IEA (2005), *Prospects for Hydrogen and Fuel Cells,* IEA, Paris, France, 253 pp.

IEA (2006a) *World Energy Outlook 2006,* IEA, Paris, France, 596 pp.

IEA (2006b) *Energy Statistics of OECD Countries 2003–2004,* IEA, Paris, France, 402 pp.

IEA (2006c) *Energy Technology Perspectives 2006,* IEA, Paris, France, 479 pp.

IEA (2006d) *Oil Market Report,* IEA, Paris, France, 10 November, omrpublic.iea.org/omrarchive/10nov06full.pdf.

IEA (2006e) *Electricity Information,* IEA, Paris, France, 605 pp.

IEA (2008) *World Energy Outlook 2008,* IEA, Paris, France, 578 pp.

IEA (2009) *Oil Market Report,* IEA, Paris France, June 11, at omrpublic.iea.org/omrarchive/11jun09full.pdf.

Jacobson, M. Z. (2007) "Effects of ethanol (E85) versus gasoline vehicles on cancer and mortality in the United States," *Environmental Science & Technology,* vol 41, no 11, pp. 4,150–4,157.

Kerr, D. (2007) "Marine energy," *Philosophical Transactions of the Royal Society,* vol 365, pp. 971–992.

Kilian L., "The Economic Effects of Energy Price Shocks," *Journal of Economic Literature,* vol 46(4), pp. 871–09. Available as a pre-publication version at www-personal.umich.edu/~lkilian/jel052407.pdf.

Kim, S. and Dale, B. E. (2004) "Global potential bioethanol production from wasted crops and crop residues," *Biomass and Bioenergy,* vol 26, no 4, pp. 361–375.

Kirsch, D. A. (2000) *The Electric Vehicle and the Burden of History,* Rutgers University Press, New Brunswick, NJ, 291 pp.

Kojima, M. and Johnson, T. (2005) *Potential for Biofuels for Transport in Developing Countries,* Joint UNDP / World Bank, Energy Sector Management Assistance Programme (ESMAP), October, Washington DC, esmap.org/filez/pubs/31205BiofuelsforWeb.pdf.

Koppelaar, R. H. E. M. (2005) *World Oil Production and Peaking Outlook,* Peak Oil Netherlands Foundation, peakoil.nl/wp-content/uploads/2006/09/asponl_2005_report.pdf.

Larminie, J. and Lowry, J. (2003) *Electric Vehicle Technology Explained,* Wiley, New York, NY.

Lovelock, J. (2006) *The Revenge of Gaia: Why the Earth Is Fighting Back—and How We Can Still Save Humanity,* Allen Lane, Santa Barbara, CA, 177 pp.

Lovins, A. B. (2005) *Twenty Hydrogen Myths,* Rocky Mountain Institute, Snowmass, Colorado, rmi.org/images/other/Energy/E03-5_20HydrogenMyths.pdf.

Lubowski, R. N., Vesterby, M., Bucholtz, S., Baez, A. and Roberts, M. J. (2006) *Major Uses of Land in the United States, 2002,* Economic Information Bulletin no 14, US Department of Agriculture, Washington DC, ers.usda.gov/publications/EIB14/eib14.pdf.

Mackay, D. J. C. (2009) *Sustainable Energy—without the hot air,* UIT Cambridge Ltd., Cambridge, UK. Available online at withouthotair.com.

Marano, J. J., Ciferno, J. P. (2001) *Life-Cycle Greenhouse-Gas Emissions Inventory For Fischer-Tropsch Fuels,* Energy and Environmental Solutions LLC for the US Department of Energy, National Energy Technology Laboratory, 186 pp., netl.doe.gov/technologies/coalpower/gasification/pubs/pdf/GHG finalADOBE.pdf.

Marshall, J. and Kazerani, M. (2005) "Design of an efficient fuel cell vehicle drivetrain, featuring a novel boost converter," *Industrial Electronics Society, IECON 2005, 32nd Annual Conference of IEEE,* pp. 1,299–1,234.

Matheys, J., Timmermans, J.-M., Van Autenboer, W., Van Mierlo, J., Maggetto, G., Meyer, S., De Groot, A., Hecq, W. and Van den Bossche, P. (2006) "Comparison of the environmental impact of 5 electric vehicle battery technologies using LCA," *Proceedings of the 13th CIRP International Conference On Life Cycle Engineering,* May–June 2006, Leuven, Belgium, pp. 97–102, mech.kuleuven.be/lce2006/010.pdf.

Maugeri, L. (2009) "Understanding Oil Price Behavior through an Analysis of a Crisis," in *Review of Environmental Economics and Policy,* vol 3, pp. 147–166.

Mazza, P. and Hammerschlag, R. (2004) *Carrying the Energy Future:Comparing Hydrogen and Electricity for Transmission, Storage and Transportation,* Institute for Lifecycle Environmental Assessment, Seattle, WA, iere.org/ILEA/downloads/MazzaHammerschlag.pdf.

Mehos, M., Kabel, D., and Smithers, P. (2009) "Planting the seed: Greening the grid with concentrating solar power," *IEEE Energy and Power Magazine,* May–June, pp. 55–62.

Milkov, A. V. (2004) "Global estimates of hydrate-bound gas in marine sediments: How much is really out there? *Earth-Science Reviews,* vol 66, no 3–4, pp. 183–197.

MIT (2007), *The Future of Coal,* Massachusetts Institute of Technology, Cambridge, MA, 192 pp., web.mit.edu/coal/The_Future_of_Coal.pdf.

NCEP (2005) "Oil shockwave: Oil crisis executive simulation," US National Commission on Energy Policy, Washington DC, energycommission.org/files/contentFiles/oil_shockwave_report_440cc39a643cd.pdf.

NETMOBIL (2005) *EU Potential for Innovative Personal Urban Mobility,* NETMOBIL project, University of Southampton, UK, eukn.org/binaries/eukn/dg-research/research/2005/10/netmobil-d7-eu-potential-final.pdf.

Nicol, J. (1999) *The All-Red Route,* McArthur and Co., Toronto, Canada, 358 pp.

Niven, R. K. (2005) "Ethanol in gasoline: Environmental impacts and sustainability review article," *Renewable and Sustainable Energy Reviews,* vol 9, pp. 535–555.

Orr, D. W. (1979) "US energy policy and the political economy of participation," *Journal of Politics,* November, vol 41, no 4, pp. 1,027–1,056.
Ott, J. (2006) 'Synthetics soar," *Aviation Week & Space Technology,* vol 165, no 5, pp. 54–56.
Patzek, T. W. (2005) "The United States of America meets the planet Earth," Presented 23 August, at the National Press Club Conference, Washington DC, berkeley.edu/news/media/releases/2005/08/NPC_briefing_Patzek.pdf.
Patzek, T. W. and Pimentel, D. (2005) "Thermodynamics of energy production from biomass," *Critical Reviews in Plant Sciences,* September, vol 24, no 5–6, pp. 327–364.
Perry, G. L. (2001) *The War on Terrorism, the World Oil Market and the US Economy,* Analysis Paper #7, The Brookings Institution, Washington DC, brookings.edu/papers/2001/1024terrorism_perry.aspx.
Powers, B. (2002) "Assessment of potential risk associated with location of LNG receiving terminal adjacent to Bajamar and feasible alternative locations," prepared for Bajamar Real Estate Services, Baja California, "LNG terminal white paper—2002," Border Power Plant Working Group, borderpower plants.org/pdf_docs/lng_position_paper_june2002_english.pdf.
Pradhan, S., Ale, B. B., Amatya, V. B. (2006) "Mitigation potential of greenhouse gas emission and implications on fuel consumption due to clean energy vehicles as public passenger transport in Kathmandu Valley of Nepal: A case study of trolley buses in Ring Road," *Energy,* vol 31, pp. 1,748–1,760.
Roberts, S. (2006) "Energy as a driver of change," *The Arup Journal,* vol 41, no 2, pp. 22–28, arup.com/_assets/_download/download630.pdf.
Romm, J. (2006) "The car and fuel of the future," *Energy Policy,* vol 34, no 17, pp. 2,609–2,614.
Romo, L., Turner, D. and Brian, L. S. (2005) "Cutting traction power costs with wayside energy storage systems in rail transit systems," *Proceedings of the Joint Rail Conference 2005,* held in Pueblo, Colorado, American Society of Mechanical Engineers, New York, NY, pp. 187–192.
Rubin, J. (2009) *Why your world is about to get a whole lot smaller,* Random House Canada, Toronto, Ontario.
Samsam Bakhtiari, A. M. (2004) "World oil production capacity model suggests output peak by 2006–07," *Oil & Gas Journal,* 26 April, vol 102, no 16, pp. 18–20.
Samsam Bakhtiari, A. M. (2006) Evidence at a hearing on *Australia's Future Oil Supply and Alternative Transport Fuels,* Senate, Commonwealth of Australia, 11 July, Sydney, Australia, aph.gov.au/hansard/senate/commttee/S9515.pdf.
Schäfer, A., Heywood, J. B. and Weiss, M. A. (2006) "Future fuel cell and internal combustion engine automobile technologies: A 25-year life cycle and fleet impact assessment," *Energy,* September, vol 31, no 12, pp. 1,728–1,751.
Shrestha, R. S. (2004) "Pre-feasibility report on trolley bus development in Ring Road of the Kathmandu Valley," PREGA and Winrock International, clean airnet.org/caiasia/1412/articles-58936_resource_1.pdf.

Sloan, E. D. Jr. (2003) "Fundamental principles and applications of natural gas hydrates," *Nature,* vol 426, pp. 353–359.
Smith, M. R. (2007) "Resource depletion: Modeling and forecasting oil production," pp. 107–113 of Greene, D. L. (ed) *Modeling the Oil Transition: A Summary of the Proceedings of the DOE/EPA Workshop on the Economic and Environmental Implications of Global Energy Transitions,* Oak Ridge National Laboratory, Oak Ridge, TN, 193 pp., www-cta.ornl.gov/cta/Publications/Reports/ORNL_TM_2007_014_EnergyTransitionsWorkshopSummary.pdf.
Snow, N. (2009) "CFTC: Cause of 2008's oil-price surge remains elusive," *Oil & Gas Journal,* vol 107(15), p. 25.
Stern, J. (2006) *The New Security Environment for European Gas: Worsening Geopolitics and Increasing Global Competition for LNG,* Oxford Institute for Energy Studies, NG 15, October, oxfordenergy.org/pdfs/NG15.pdf.
Thwaites, T. (2007) "A bank for the wind," *New Scientist,* vol 193, no 2586, pp. 39–41.
Urbanchuk, J. M. (2004) *The Contribution of the Ethanol Industry to the American Economy in 2004,* LECG, Wayne, PA, agmrc.org/media/cms/EthanolEconomicContributionREV_363935769CAA2.pdf.
US EIA (2008) *Annual Energy Review 2007,* DOE/EIA-0384, Energy Information Administration, Washington DC, June 2008, eia.doe.gov/aer/pdf/aer.pdf.
US NRC (2007) *Coal: Research and Development to Support National Energy Policy,* National Academies Press, Washington DC, 144 pp.
Van Mierlo, J., Maggetto, G. and Lataire, Ph (2006) "Which energy source for road transport in the future? A comparison of battery, hybrid and fuel cell vehicles," *Energy Conversion and Management,* September, vol 47, no 17, pp. 2,748–2,760.
von Blottnitz, H. and Curran, M. A. (2007) "A review of assessments conducted on bio-ethanol as a transportation fuel from a net energy, greenhouse gas, and environmental life cycle perspective," *Journal of Cleaner Production,* vol 15, pp. 607–619.
Vuchic, V. R. (2007) *Urban Transit Systems and Technology,* John Wiley & Sons, New York, NY, 624 pp.
Weinert, J. X., Ogden, J. M., Shaojun, L. and Jianxin, M. (2006) *Hydrogen Refueling Station Costs in Shanghai,* [online], Institute of Transportation Studies, University of California, Davis, CA, its.ucdavis.edu/publications/2006/UCD-ITS-RR-06-4.pdf.
West, R. E., and Kreith, F. (2006) "A vision for a secure transportation system without hydrogen or oil," *Journal of Energy Resources Technology,* Transactions of the ASME, September, vol 128, pp. 236–243.
Williams, R. H. and Larson, E. D. (2003) "A comparison of direct and indirect liquefaction technologies for making fluid fuels from coal," *Energy for Sustainable Development,* vol 7, no 4, pp. 103–124.
Yang, B. and Lu, Y. (2007) "The promise of cellulosic ethanol production in China," *Journal of Chemical Technology and Biotechnology,* vol 82, pp. 6–10.

Zittel, W. and Schindler, J. (2006) *Uranium Resources and Nuclear Energy,* Energy Watch Group, Ottobrun, Germany, lbst.de/publications/studies__e/2006/EWG-paper_1-6_Uranium-Resources-Nuclear-Energy_03DEC2006.pdf.

Zittel, W. and Schindler, J. (2007) *Coal: Resources and Future Production,* Energy Watch Group, Ottobrun, Germany, energywatchgroup.org/fileadmin/global/pdf/EWG_Report_Coal_10-7-2007ms.pdf.

Zweibel, K., Mason, J., and Fthenakis, V. (2008) "A solar grand plan," *Scientific American,* January, pp. 64–73.

CHAPTER 4: Transport's Adverse Impacts

1. The quotation on mobility benefits and costs is from p. 177 of Greene and Wegener (1997).
2. For use of the term "hypermobility," see p. 32 of Kay (1997) and p. 95 of Adams (2000).
3. Figure 4.1 is an elaboration of Box 3 on p. 17 of OECD (2002).
4. Figure 4.2 was part of a presentation by Ernst-Ulrich von Weizsäcker at the OECD International Conference held in Vienna, Austria, 4–6 October, 2000. It is Box 2 on p. 20 of the February 2002 *Report on the OECD Conference Environmentally Sustainable Transport (EST): Futures, Strategies And Best Practice,* olis.oecd.org/olis/2001doc.nsf/LinkTo/env-epoc-wpnep-t(2001)8-final. This report is © 2002 OECD. It references the original German source.
5. GHGs reflect back to earth some of the radiation emitted from the earth's surface, helping maintain the average temperature of the surface at about 30°C above what it would otherwise be, that is, an average of about +15°C rather than –15°C or colder. The principal GHGs are water vapor, including clouds, which may account for about three quarters of the baseline warming, and carbon dioxide (CO_2), which accounts for most of the remaining baseline warming (see realclimate.org/index.php?p=142). Increases in CO_2 concentrations as a result of human activity, chiefly fossil fuel burning, are believed to account for the largest part of the current increase in warming (see Note 11). The reflecting back of the radiation is known as *radiative forcing* and gases that do it are said to be *radiatively* active. Colloquially the effect is known as the greenhouse effect, although greenhouses are warmer because they trap warm air rather than because their surfaces are radiatively active.
6. Box 4.1 is based on Svensmark (2007) and Svensmark and Calder (2007). Also see Marsh and Svensmark (2000), Fichtner et al. (2006), Svensmark et al. (2007), Svensmark et al. (2009) and Bershadskii (2009). See too Kirby, J., "Cosmic rays and climate," Presentation at the June 2009 CERN colloquium, at wattsupwiththat.files.wordpress.com/2009/07/.pdf. The quotation in the last paragraph of Box 4.1 is from p. 226 of Svensmark and Calder (2007). The cooling of Antarctica noted in Box 4.1 is reported by the US National Aeronautics and Space Administration (NASA) at data.giss.nasa.gov:

compared with 1950–1981, there was on average warming at all latitudes in 2006 except 75–90°S, where there was cooling. IPCC (2007c) has also noted Antarctic cooling across the period 1970–2004; see Figure SPM-1. There is an explanation of the Antarctic cooling on pp. 19–20 of Svensmark and Calder (2007):

> Cloud tops have a high albedo and exert their cooling effect by scattering back into the cosmos much of the sunlight that could otherwise warm the surface. But the snows on the Antarctic ice sheets are dazzlingly white, with a higher albedo than the cloud tops. There, extra cloud cover warms the surface, and less cloudiness cools it. Satellite measurements show the warming effect of clouds on Antarctica, and meteorologists at far southern latitudes confirm it by observation. Greenland too has an ice sheet, but it is smaller and not so white. And while conditions in Greenland are coupled to the general climate of the northern hemisphere, Antarctica is largely isolated by vortices in the ocean and the air. The cosmic-ray and cloud-forcing hypothesis therefore predicts that temperature changes in Antarctica should be opposite in sign to changes in temperature in the rest of the world. This is exactly what is observed, in a well-known phenomenon that some geophysicists have called the polar see-saw, but for which "the Antarctic climate anomaly" seems a better name.

An alternative explanation of the ongoing Antarctic cooling—although not the longer-term "polar see-saw"—is that it is due to the ozone hole, which allows otherwise trapped solar radiation to escape (see "Nematode Herders and Climate Change," global-warming.accuweather.com/2006/11/nematode_herders_and_climate_c_2.html). For a strong criticism of cosmoclimatology, see Pearce, F., "Climate myths: It's all down to cosmic rays," *New Scientist,* May 16, 2007, environment.newscientist.com/article.ns?id=dn11651. IPCC (2007b) observed on p. 193 that the level of scientific understanding of cosmic ray influences is very low. Cosmoclimatology is not the only alternative theory of climate change. The adiabatic theory, of Russian origin (Sorokhtin et al., 2007), is based on the assumption that the main factor determining the Earth's climate is atmospheric pressure. For these authors, "a common perception of climate warming as a result of CO_2 and other 'greenhouse' gases accumulating in the atmosphere is a myth" (p. 9). See also Chilingar et al. (2009) on the adiabatic theory.

7. Although there is little doubt that the late 1980s and the 1990s contained many unusually warm years, the proposition that global temperatures are rising remains an open question. Both the main sources of global temperature data suggest that since about 2002 global temperatures have been falling. One is the US National Aeronautics and Space Administration. NASA data are at data.giss.nasa.gov/gistemp/. The other is the Climate Research

Unit at the UK's University of East Anglia. CRU's data are at cru.uea.ac.uk/cru/data/temperature/#datdow. Data from these sources show too that global temperatures fell from about 1940 until about 1975, even though atmospheric carbon dioxide levels rose throughout that period.

8. The quotation is from p. 10 of IPCC (2007a). A footnote on p. 4 of this document indicates that "very likely" means that the assessed likelihood of the relationship is more than 90 percent, and "likely" means it is more than 66 percent. "Most" is not defined, but can perhaps be taken to mean "more than 50 percent."
9. For example IPCC (2007c) noted on p. 6 that "globally, the potential for food production is projected to increase with increases in local average temperature over the range of 1–3°C" and particularly at mid to high latitudes. Also, "globally, commercial timber productivity rises modestly with climate change in the short- to medium-term, with large regional variability." Svensmark and Calder (2007, p. 30) observed, "...among the thousands of human generations, ours may be the first that was ever frightened by a warming."
10. For the adverse effects of climate change, see pp. 4–5 of EEA (2004), Chapter 1 of McCarthy et al. (2001) and pp. 5–12 of IPCC (2007c).
11. For more information about the Protocol to the United Nations Framework Convention on Climate Change, known as the Kyoto Protocol, see unfccc.int/2860.php. The six GHGs—or groups of GHGs—covered by the Protocol are carbon dioxide (CO_2), methane (CH_4), nitrous oxide (N_2O) and three industrial chemicals: sulfur hexafluoride, hydrofluorocarbons and perfluorocarbons. According to Table 2.1 on p. 141 of IPCC (2007b), in 2005, CO_2 contributed 63 percent of radiative forcing by long-lived radiatively active gases in the atmosphere (see Note 5 for an explanation of these terms). Other contributions were made by CH_4 (18 percent), N_2O (6 percent), the other gases covered by the Kyoto Protocol (0.6 percent), and long-lived radiatively active gases not covered by the Kyoto Protocol (12 percent). GHGs not covered in the Kyoto Protocol—in this case because they had already been addressed in the Montreal Protocol (see Note 51)—include chlorofluorocarbons (CFCs) and hydrochlorofluorocarbons (HCFCs). Note that this radiative forcing was said to be augmented by contributions from short-lived gases, notably tropospheric ozone, and considerably offset by the net radiative forcing of aerosols in the atmosphere; see Figure 2.22 on p. 206 of IPCC (2007b). Confusingly, another IPCC document gives the contribution of CO_2 as 77 percent of "total anthropogenic GHG emissions in 2004" (p. 3 of IPCC 2007d), without an indication as to how this estimate was reached. Note that these estimates of the components of possible human contributions to climate change do not include radiative forcing by water vapor, characterized in Note 5 as the principal component of baseline warming and a potential factor in human-induced warming.

12. Table 4.1 is based on data at the website of the Secretariat of the United Nations Framework Convention on Climate Change, unfccc.int/ghg_emis sions_data/items/3800.php. Note that the data for "all sources" in Table 4.1 do not include data on land use, land use changes and forests (LULUCF), which are required by the Kyoto Protocol to be reported but are not included in national targets. Omitted Annex 1 countries are small or have incomplete data sets.
13. Each country was assigned a GHG emissions target, but, by agreement, the 15 members of the EU at the time have an overall target of an 8 percent reduction from 1990, varying from a 28 percent reduction for Luxembourg to a 27 percent increase for Portugal. The US target in Table 4.2 was agreed to at the time of the signing of the Kyoto Protocol in 1997, but, as noted in this paragraph, does not bind the US.
14. Table 4.2 is based on the source detailed in Note 12.
15. Catalytic converters are used only in vehicles with gasoline engines. Thus the share of CO_2 in GHGs emitted from the operation of diesel engines is higher than gasoline engines, although, as noted in the text, diesel engines are responsible for less GHGs overall per unit of energy realized. See also the source detailed in Note 18.
16. The shares of GHGs from transport are in CO_2 equivalents, as given on p. xvi of US EIA (2006). HFC-134a is discussed below as an ozone-depleting substance.
17. These average CO_2 emissions from fuels are from Table A13–5 on p. 435 of Canada (2006).
18. See Monahan and Friedman (2004), who note the following on p. 2: "Diesel vehicles can help a car travel 30 to more than 40 percent farther on a gallon of diesel fuel. However, this advantage is only partly due to the higher efficiency of diesel engines, which offer a 15 to 25 percent improvement over gasoline. The remaining increase is due to the fact that diesel fuel contains 13 percent more energy than a gallon of gasoline."
19. According to the US Department of Energy, cars with diesel engines comprised 28 percent of sales in Western Europe in 1999 and 49 percent in 2005, *Fact of the Week,* February 27, 2006, FreedomCAR and Vehicle Technologies Program, www1.eere.energy.gov/vehiclesandfuels/facts/2006_fcvt_fot w413.html.
20. For an account of the "altitude effect" on aviation's GHG emissions, see pp. 558–560 of Sausen et al. (2005).
21. The exact amount could depend on the flight distance and thus the portion of a flight spent at a cruising height. Energy use is much higher while climbing but the additional effects noted in this paragraph occur while cruising at a considerable height. A complicating matter is the large weight of fuel that must be carried for long flights. The net result is a minimum fuel use per seat-kilometer at just over 2000 km (see Figure 4–II on p. 25 of RCEP,

2002), but there is no simple function that describes how GHG emissions per kilometer change with flight distance.

22. These estimates are based on the data on energy use by fuel type, as in the energy section of *OECD Statistics,* OECD, Paris, France, until December 2006. These data are now available only for a fee through the International Energy Agency at data.iea.org/ieastore/statslisting.asp. Use of fuel consumption by weight is assumed to be a reasonable surrogate for CO_2 emissions and therefore GHG emissions.
23. For the suggestion that bunker-fuel induced low-level clouds could counteract a global warming effect, see pp. 745–746 of Capaldo et al. (1999).
24. For the other analysis of ships' clouds and global warming, see p. 20 of Endresen et al. (2003). Note that the cooling by the low-level clouds that line shipping routes is the same effect as cooling that could be due to cosmic ray penetration, discussed above, albeit with a different cause.
25. Hansen (2006, p. 960) among others has argued that greater intensity and frequency of tropical storms is an ongoing feature of climate change. He noted that the US National Oceanic and Atmospheric Administration does not agree and, perhaps in violation of the US constitution, has instructed its scientists not to dispute its opinion.
26. See Kennett, J., *Katrina, Rita Cost to Oil Industry Rises to Record $17 billion,* August 22, 2006, bloomberg.com/apps/news?pid=20601109&sid=aZjgqvrTSgrs&refer=news.
27. Calgary's wind-powered light-rail system was discussed in Chapter 3. Infrastructure issues are discussed in more detail later in this chapter.
28. The EC proposal is in Commission of the European Communities, *Limiting Global Climate Change to 2 degrees Celsius: The way ahead for 2020 and beyond,* Communication from the Commission to the European Council, the European Parliament, the European Economic and Social Committee and the Committee of the Regions, January 10, 2007, ec.europa.eu/environment/climat/pdf/com_2007_2_en.pdf.
29. See Commission of the European Communities, *An Energy Policy for Europe,* Communication from the Commission to the European Council and the European Parliament, January 10, 2007, ec.europa.eu/energy/energy_policy/doc/01_energy_policy_for_europe_en.pdf.
30. The quotation is from p. 15 of the source detailed in Note 29.
31. The strategy for light-duty vehicles is in *Results of the review of the Community Strategy to reduce CO_2 emissions from passenger cars and light-commercial vehicles,* Communication from the Commission to the Council and the European Parliament, February 7, 2007, ec.europa.eu/environment/co2/pdf/com_2007_19_en.pdf.
32. The quotation is from p. 5 of the source detailed in Note 31.
33. Fontaras and Samaras (2007) provide details of the voluntary agreement between the EC and the European Automobile Manufacturers Association and show how it is being met by "dieselization" of the car fleet.

34. The increased power and weight of European vehicles is discussed by Van den Brink and Van Wee (2001).
35. The gasoline surplus arises not from poor anticipation of demand but from practical limits as to how much of each of diesel fuel, gasoline, etc. can be produced from a given batch of crude oil.
36. Canada is in a similar situation. It is heading to be, after Spain, the most egregious defaulter among signatories to the Kyoto Protocol (see Note 11). A substantial portion of Canada's GHG emissions come from its oil and natural gas production, particularly from oil sands, much of which is exported to the US (Canada, 2006).
37. Discussion of and data on the CAFE standards can be found in Perl and Dunn (2007), Heavenrich (2006) and NHTSA (2004). The description of the CAFE standards here is greatly simplified and does not cover, among many features, provisions for dual-fuel vehicles and for credits whereby one year's performance can be used to offset another's.
38. Figure 4.3 is based on data in Tables 1 and 2 of Heavenrich (2006) and Table II-6 of NHTSA (2004). Note that the rated fuel consumption data are those of the US National Highway Traffic and Safety Administration (NHTSA) and not those of the US Environmental Protection Agency (EPA), except for 1975–1977 and 2005–2006.
39. The charts in Box 4.2 are based on data in Tables 3.7, 4.1, 4.2, 4.5 and 4.6 of Davis and Diegel (2008). (Some data for the 1970s come from an earlier version of this source.) Information about the SCRAP-IT program is at scrapit.ca.
40. These changes are not cumulative, and estimating their combined effect is a complex matter. Our estimate assumes that 10-percent changes in weight and power produce respectively 6-percent and 3-percent changes in fuel consumption, other things being equal (Van den Brink and Van Wee, 2001).
41. See the corresponding road tests at consumerreports.org and the corresponding fuel economy ratings at fueleconomy.gov.
42. The legislation increasing CAFE standards from 2011 to 2020 was the *Energy Independence and Security Act of 2007*, at frwebgate.access.gpo.gov/cgi-bin/getdoc.cgi?dbname=110_cong_bills&docid=f:h6enr.txt.pdf.
43. For the background on the CAFE standards announcement, see whitehouse.gov/the_press_office/Background-Briefing-on-Auto-Emissions-and-Efficiency-Standards/.
44. For President Obama's announcement and its supporters, see Allen and Javers, "Obama announces new fuel standards," at dyn.politico.com/printstory.cfm?uuid=5481D758-18FE-70B2-A8AF6C70885AA5D5.
45. The notice of joint rulemaking is at epa.gov/fedrgstr/EPA-AIR/2009/May/Day-22/a12009.pdf.
46. CAFE standards and GHG emissions standards essentially amount to the same thing, except that the latter take account of vehicle air-conditioning and differences between gasoline and diesel fuel. For discussion of this see

Ponticel, P., "US joins greenhouse-gas game and ups CAFE ante," *Society of Automotive Engineers Vehicle Engineering Online*, June 25, 2009, at sae.org/mags/sve/6545.

47. Note that the California standard would concern emissions of all GHGs, whereas the EU voluntary agreement discussed above and the proposed US federal regulation concerns CO_2 emissions only. The California standard would not apply to light-duty vehicles that are not passenger cars and that weigh more than 1700 kg; for them a different standard is to apply. For further details, see An and Sauer (2004) and Perl and Dunn (2007). Note that the equivalent gasoline consumption has been estimated by the present authors and is not provided by the State of California.
48. For an announcement of the US EPA's granting of a waiver of provisions of the Clean Air Act and California's response, including its apparent acceptance of the federal standards, see yosemite.epa.gov/opa/admpress.nsf/bd4379a92ceceeac8525735900400c27/5e448236de5fb369852575e500568e1b!OpenDocument. Also see the source detailed in Note 46.
49. See Table 12 and Figure 9 of An and Sauer (2004).
50. See *Japan Proposes Tougher Fuel Economy Regulations; Passenger Car Fuel Economy to Increase 23.5% by 2015*, Green Car congress, December 15, 2006, greencarcongress.com/2006/12/japan_proposes_.html. Note that, to allow comparisons with the US, the fuel consumption standard is stated in the text in the manner used in connection with the US CAFE standards, not as in the indicated new report or the underlying press release. See pp. 22–23 and 25 of An and Sauer (2004) for details as to how to make such conversions.
51. Note that CFCs and the related HCFCs are also GHGs, as observed in Note 11.
52. For ODSs, see the website of the Ozone Secretariat of the United National Environment Programme, ozone.unep.org.
53. According to p. 1 of UNEP (2006), "The Montreal Protocol is working. The concentrations of ozone depleting substances in the atmosphere are now decreasing.... Global ozone is still lower than in the 1970s, and a return to that state is not expected for several decades."
54. For the previous mention of HFC-134a, see Note 16 and associated text. For alternatives to HFC-134a, see, for example, Wongwises et al. (2006).
55. For data on air conditioning in German cars, see p. 79 of Schwartz (2005).
56. For the fuel impacts of vehicle air conditioning, see a presentation by John Rugh, Valerie Hovland, and Stephen Andersen, *Significant Fuel Savings and Emission Reductions by Improving Vehicle Air Conditioning*, to the 15th Annual Earth Technologies Forum and Mobile Air Conditioning Summit, April 2004, nrel.gov/vehiclesandfuels/ancillary_loads/pdfs/fuel_savings_ac.pdf. For a more general discussion of vehicle air conditioning see Chapter 6 of IPCC (2005).
57. The quotation is from the POPs part of the website of the United Nations Environment Programme, chem.unep.ch/pops.

58. For what is in the blood of polar bears see, for example, Sonne et al. (2004).
59. For an estimate of transport's share of dioxin emissions in Australia, see p. 46 of Smit et al. (2004).
60. For a discussion of some of the issues of ships' discharges, although with few data, see *Towards a strategy to protect and conserve the marine environment,* a communication from the European Commission to the European Parliament, 2002, europa.eu/legislation_summaries/other/l28129_en.htm.
61. For a comprehensive review of ocean acidification's causes and effects see Doney et al. (2009). See also the collection of 11 papers on ocean acidification edited and introduced by Alain Vézina and Ove Hoegh-Guldberg (2008). See also Cooley and Doney (2009).
62. These conclusions concerning ocean acidification are quoted from the first source detailed in Note 61.
63. For example, a public opinion poll conducted in November 2004 found that climate change and air pollution were included equally often among the five main environmental issues of concern in what in 2005 were the 25 EU countries. In several countries, including Italy and the UK, air quality was the issue most often identified. See *The Attitudes of European Citizens towards the Environment,* European Commission, April 2005, ec.europa.eu/environment/barometer/pdf/report_ebenv_2005_04_22_en.pdf.
64. The quotation is the beginning of the chairman's summary of the "First Governmental Meeting on Urban Air Quality in Asia," held in Yogyakarta, Indonesia, in December 2006, cleanairnet.org/baq2006/1757/articles-69980_summary.doc. The report on economic costs appears to be forthcoming.
65. For example, a statement by the Alliance of Automobile Manufacturers (US) claimed, "Today's vehicles produce 99% fewer smog-forming emissions than did cars from the 1970s," autoalliance.org/index.cfm?objectid=2FD49034-1D09-317F-BB6BB9218034096F. In reviewing the history of the California Air Resources Board (ARB), another California state agency noted, "Through ARB regulations, today's new cars pollute 99 percent less than their predecessors did thirty years ago. Still, over half of the state's current smog-forming emissions come from gasoline and diesel-powered vehicles." See *The History of the California Environmental Protection Agency,* arb.ca.gov/knowzone/history.htm.
66. A claim that vehicles can be air cleaners is reported in Gertner, J., "From 0 to 60 to world domination," *The New York Times Magazine,* February 18, 2007, select.nytimes.com/preview/2007/02/18/magazine/1154665179108.html. Takahiro Fujimoto, "a management professor at the University of Tokyo and a longtime Toyota observer," was reported on p. 58 as saying that in the heavy intersections of Tokyo where air quality is poor, "the emission gas of some advanced cars is cleaner than intake air."
67. The air quality standard for nitrogen oxides (NOx) in Tokyo is 0.6 parts per million (ppm, daily average of hourly readings). In 2003–2004, this

standard was met about half the time at measuring stations set up along roads where large amounts of automobile emissions are present. See metro .tokyo.jp/ENGLISH/TOPICS/2005/ftf78300.htm. According to Heck and Farrauto (2001, p. 443), for a car the "typical exhaust gas composition at the normal engine operating conditions" includes NOx at a concentration of 900 ppm. According to Table 1 of Olsson and Anderssen (2004, p. 90), the "typical composition of car exhaust" (for a gasoline engine) includes NOx at a concentration of between 50 and 5000 ppm. Thus, even if the NOx concentration in a Toyota car exhaust were only a quarter of the lowest typical value (i.e., it was 12.5 ppm), and even if the NOx concentration of ambient air at a particular Tokyo intersection were four times the standard, i.e., 2.4 ppm, the exhaust gas would still have more than five times as much NOx per liter as ambient air. The car would *not* be acting as an air cleaner. More plausible is the limited assertion by Twigg (2007, p. 8) that "the most demanding legislation in the world today, California's SULEV HC limit is in some cases lower than ambient air." (HC refers to hydrocarbon emissions. SULEV—Super Low Emission Vehicle—is described later in the text.)

68. On p. 352 of IPCC (2001) there is the following: "While the Northern to Southern Hemisphere ratio of the solar and well-mixed greenhouse gas forcings is very nearly 1 [e.g., those of CO_2 and methane], that for the fossil fuel generated sulphate and carbonaceous aerosols and tropospheric O_3 is substantially greater than 1 (i.e., primarily in the Northern Hemisphere), and that for stratospheric O_3 and biomass burning aerosol is less than 1 (i.e., primarily in the Southern Hemisphere)."
69. According to one news report, "urban pollutants such as ozone and carbon monoxide...rise into the upper atmosphere and are blown eastwards at 150 mph [240 km/h] by the jet stream." See Leak, J. and Sadler, R., "US toxic gases push British pollution over the safety limit," *Sunday Times,* January 21, 2007, timesonline.co.uk/tol/newspapers/sunday_times/britain/article 1294982.ece. The report is based on measurements by Lewis et al. (2007), which builds on earlier modeling work, e.g., Jonson et al. (2006) and Auvray and Bey (2005). The last analysis suggested that the North American contribution to European smog levels may be higher than the European contribution. It also noted a substantial contribution from Asia.
70. See the source concerning the ARB in Note 65.
71. Nitric oxide (NO) is produced during combustion in the presence of air. NO reacts with the oxygen in the air to form nitrogen dioxide (NO_2), which is usually the most prominent among nitrogen oxides (NOx). These oxides should not be confused with nitrous oxide (N_2O), also produced during combustion in air in small quantities. N_2O, which is a potent GHG, can be generated by the action of the three-way catalytic converters, again in small quantities. It is also a general anaesthetic known as "laughing gas."
72. Current concern is with smaller particles, PM2.5 and even PM1.0, which are

more hazardous because they penetrate more deeply into respiratory tissue, are produced mainly from diesel engines, and were not measured in 1970. PM2.5 emissions are positively correlated with PM10 emissions.

73. Table 4.3 is based on data from the US *National Emissions Inventory* at epa.gov/ttn/chief/trends/. Not shown are emissions of lead compounds, which were substantial in 1970, when tetraethyl lead was added to gasoline to reduce premature explosion (knocking). Lead compounds began to be phased out in the US in 1973 and were banned as a gasoline ingredient in 1996 (to be replaced in part by the carcinogen benzene). Table 4.3 includes data from vehicles with gasoline and diesel engines. Few light-duty vehicles used diesel fuel throughout this period, but the share of heavy-duty vehicles using diesel fuel likely increased substantially, from about 55 percent of the total to about 90 percent (see Table A.6 of Davis and Diegel, 2008).
74. Table 4.4 is based on data from the US *National Emissions Inventory* at epa .gov/ttn/chief/trends/.
75. Also see Note 67 concerning the SULEV standard.
76. The US emissions standards set out in Table 4.5 are from Tables 4-29 and 4-30a of US DOT (2007). The California standards are from Table 2 of Twigg (2007). Both sets of standards apply to regular cars and to other light-duty vehicles, e.g., vans and SUVs with an unladen weight less than 1,288 kg (3,450 lb). Until the 2007 model year, "light trucks" had more relaxed standards. Note that other states are permitted to adopt the more stringent California standards rather than the federal standard. New York, New Jersey, Vermont and Massachusetts have done so.
77. An aspect of current standards is that emission control systems have to be certified not to decline in performance for up to 15 years or 240,000 km (150,000 miles).
78. The quotation concerning real-world emissions is from p. 514 of Ross et al. (1998).
79. The data on real-world emissions are from Ross and Wenzel (1997).
80. The information on the difference between standards and real-world emissions is in the chart on p. 3 of DeCicco (2000).
81. The quotation on emissions standards and real-world emissions is from p. 36 of MacLean and Lave (2003).
82. For the deployment of emissions measuring devices, see the July 2003 interview with US EPA official Christopher Grundler at challengebibendum .com/challengeBib/AfficheServlet?Rubrique=20071022160629&Langue =EN&Page=20070904183410_040920071834 34_115.
83. The US emissions standards for heavy-duty vehicles set out in Table 4.6 are from Table 4-32b of US DOT (2007). The original standards are in grams per brake horsepower-hour, a measure of an engine's energy output. There is no straightforward way to compare the heavy-duty vehicle standards in Table 4.6 with those for light-duty vehicles in Table 4.5.

84. The air quality standards and data in Table 4.7 are from information at the website of the US EPA, respectively at epa.gov/air/criteria.html and epa .gov/air/airtrends/. The annual standard for PM10 was revoked in December 2006, "due to a lack of evidence linking health problems to long-term exposure to coarse particle pollution." A 24-hour standard for PM10 is still in place, as are annual and 24-hour standards for PM2.5 (see Note 86).
85. Air quality data for the US do not appear to be readily available for years before 1980. Emissions data for the US are available for the years from 1900 (see US EPA, 2000).
86. Two criteria pollutants are not shown in Table 4.7: lead, atmospheric concentrations of which have been at or close to zero since the early 1990s, and very fine particulate matter (<2.5 μm), known as PM2.5, for which records are available only since 1990.
87. Total VOCs are reported in other countries. For Canada, for example, Environment Canada has reported that the average concentration of total VOCs as recorded at monitoring stations was 126 micrograms/cubic meter in 1990 and 77 $\mu g/m^3$ in 2001, a decline of 39 percent. See p. 72 of *Air Quality in Canada: 2001 Summary and 1990–2001 Trend Analysis, May 2004*, etc-cte .ec.gc.ca/publications/naps/NAPS_2001%20Summary_1990-2001_Trend_ Analysis.pdf.
88. Summer smog also contains particulates and other pollutants. Winter smog contains more particulates and other pollutants and less ozone. The "peas-souper" smogs of London in the 1950s, which one of the authors lived through, were winter smogs caused mostly by residential open-fire coal burning. For a good discussion of smogs, see Elsom (1996).
89. Canada (1997, p. 49) suggested that ozone formation in urban cores tends to be VOC-limited while formation in downwind suburban and rural areas tends to be NOx-limited. See also Wang et al. (2005).
90. Table 4.8 is based on a summary produced by Denmark's National Environmental Research Institute, www2.dmu.dk/AtmosphericEnvironment/Ex post/database/docs/AQ_limit_values.pdf. For each pollutant, the standard offering maximum comparability is illustrated. However, key features of some standards are not represented, and Table 4.8 must be regarded as illustrative only. For example, the US standard for SO_2 may be exceeded only once a year and the much more stringent EU standard for SO_2 may be exceeded three times a year. For some pollutants, China has different standards for residential, commercial and industrial areas. In these cases the middle standard, that is, the one for commercial areas, is shown. The WHO guidelines shown for PM10 and SO_2 are not those in the Danish summary but are from the more recent WHO (2006). WHO has only an 8-hour average guideline for ozone: 100 $\mu g/m^3$. Japan has only a 24-hour average standard for NO_2: 115$\mu g/m^3$. Japan's ozone standard is actually for "photochemical oxidants," as published at env.go.jp/en/air/aq/aq.html.
91. The data in Table 4.9 were downloaded from Eurostat, *Statistics in Focus,*

European Commission, Brussels, Belgium, 2007, epp.eurostat.ec.europa.eu/portal/page?_pageid=1090,30070682,1090_30298591&_dad=portal&_schema=PORTAL.

92. The data in Table 4.10 were downloaded from Eurostat, *Statistics in Focus*, European Commission, Brussels, Belgium, 2007, epp.eurostat.ec.europa.eu/portal/page?_pageid=1090,30070682,1090_30298591&_dad=portal&_schema=PORTAL.
93. This phrase was used on p. 42 of WHO (2006) in connection with the data represented in Figure 4.4.
94. For European trends in transport-related air quality, see pp. 18–19 of EEA (2007a).
95. For a discussion of ozone levels in Europe, see pp. 21–22 of EEA (2007b).
96. Figure 4.4 is Figure 6 on p. 44 of WHO (2006) and is reproduced with permission.
97. For PM10 levels in numerous cities, see particularly Figure 4 on pp. 38–39 of WHO (2006).
98. For contributions of transport to air quality in the Pearl River Delta, see Wang et al. (2005). Also see Shao et al. (2006).
99. The quotation on the importance of vehicle emissions for the air quality of Asian cities is from p. 5 of ADB (2006).
100. The data on air quality in poorer cities in this paragraph is from Tables 1 and 2 of WHO (2006).
101. For air quality impacts of non-road transport, see Box 2.3 on pp. 68–69 of WHO (2005).
102. The quotation on transport's health effects is from p. 125 of WHO (2005).
103. The quotation on health risk is from p. 235 of Samet and Krewski (2007).
104. Figure 1 of Samet and Krewski (2007) provides a more complex representation of factors than appears in Figure 4.5, while focusing only on emissions of particulates. That figure points to five components (sources, concentrations, exposures, doses and responses) and numerous linking factors, including chemical transformation in the air, human activity patterns and variation in human responses to chemical assault.
105. The quotation on living near a major road is from Bayer-Oglesby et al. (2006). The study found that income was not a factor in urban areas. Females and sometime smokers were a little less affected.
106. For elevation pollution levels at 1,500 m, see Jerret et al. (2007).
107. For vehicle-related pollution falling rapidly with distance from vehicle, see Gilbert et al. (2003) and particularly Chapter 3 (pp. 85–123) and Table 3.1 (pp. 96–97) of WHO (2005).
108. See *Review of Vertical Exhausts*. Austroads (Association of Australian and New Zealand road transport and traffic authorities), Sydney, Australia, January 1993, onlinepublications.austroads.com.au/script/Freedownload.asp?DocN=AR0000055_0904.
109. "On the curb side" means positioned closer to the curbside rear wheel than

to the offside rear wheel. About 30 cars with twin exhausts were counted among the curbside group.

110. For documentation on the increase in prevalence, see Bousquet et al. (2007).
111. For further discussion of these points, see WHO (1999), particularly Table 4.1 of that document where the full guidelines are set out. For more recent consideration of transport noise see Schade (2003), Entec (2006) and Miedema (2007). "dB" as used here should be taken to mean dB(A), referring to a weighting of acoustic frequencies that approximates the sensitivity of the normal human ear.
112. Figure 4.6 is a version of Figure 22 on p. 25 of OECD (2000).
113. According to Michaud et al. (2005), only 7 percent of the US population is exposed to road traffic noise of over 65 dB, compared with 8 percent in Australia and 20 percent in Europe. (Note that Figure 4.6 suggests that the EU exposure is 15 rather than 20 percent—still much higher than in the US and Australia.)
114. The quotations about noise are from p. 1 of Toronto (2000).
115. For annoyance by noise, see the source in Note 113.
116. For an indication of the improvement in aircraft noise, see Fleming et al. (1999).
117. For these US noise production limits, which do not appear to have changed since the 1990s, see US DOT (2000) particularly Figure 5-30 and associated text.
118. See Flandez, R., "Blind pedestrians say quiet hybrids pose safety threat," *Wall Street Journal,* February 13, 2007, online.wsj.com/public/article/SB117133115592406662-7BH5dNRG2MssUH28WlvpqNMnCy8_20080212.html.
119. Figure 4.7 is a redrawn version of Figure 1 of Revitt (2004), and is reproduced by permission of the Royal Society of Chemistry.
120. According to Winkler (2005), treatment of run-off is required by EU Directive 2000/60/EC.
121. For an elaboration of methods of treating highway run-off, see Stengel et al. (2006).
122. This is the conclusion of Stengel et al. (2006). These authors noted a Canadian study in which 20 percent of samples of run-off from heavily traveled roads (over 100,000 vehicles per day) were found to be severely toxic.
123. For an account of some of the impacts of sea ports, see Mann (2006).
124. For the environmental impacts of airports, see Kelly and Allan (2006). The largest airports appear to be those serving Riyadh, Saudi Arabia (over 20,000 hectares) and Denver, Colorado, US (over 13,000 ha). The busiest airports are not the largest. The busiest overall, the airport serving Atlanta, Georgia, US, occupies only about 1,500 ha. London, UK, is the busiest aviation center. Two of its five airports together carry 20 percent more passengers than the airport serving Atlanta. These two airports occupy a total of less than 2,000 ha.

125. For a history of MFA and MIPS, see Røpke (2005). For how the related notion of "ecological rucksack" compares with the parallel concept of "ecological footprint," see Røpke (2005). For an indication of the political importance of MFA, see Giljum et al. (2005).
126. See Røpke (2005) for discussion of the notions of ecological rucksack and ecological footprint. Also see the "Definitions and Concepts" page of the website of the International Institute for Sustainable Development, iisd.org/susprod/principles.htm, and the Wuppertal Institute's website, wupperinst.org/FactorFour/FactorFour_FAQ.html.
127. Figure 4.8 is based on Figures 1 and 2 of Federici and Basosi (2007). The infrastructure cost per kilometer for road does not vary with distance because it is apportioned by distance. The same applies to rail but not air modes. Aircraft construction inputs were apportioned per kilometer; aviation infrastructure inputs were apportioned per trip.
128. How aircraft fuel use varies with distance is also discussed in Note 21.
129. von Rozycki et al. (2002) conducted an MFA for the high-speed rail line between Hanover and Würzburg in Germany, which is tunneled for 120 km of its 327 km length and estimated a MIPS value of only 0.5 kilogram per person-kilometer compared with the more than 3.0 kg/pkm shown in Figure 4.8. von Rozycki et al. do not provide data for other modes, and there is insufficient information in the two reports to account for the discrepancy.
130. For more on impacts of transport on wildlife, see Forman et al. (2003), O'Brien (2006) and Seiler and Heildin (2006).
131. For more on restoring habitat connectivity, see Forman et al. (2003) and van der Grift and Pouwels (2006).
132. Figure 4.9 is from a presentation by Richard Forman entitled "Environmental impacts of road infrastructure—experience and policy implications from the USA," given to an OECD "high-level" meeting on "Tackling transport's externalities by performance-related instruments—the role of regulatory, economic and information/communication instruments," November 12, 2003, OECD Headquarters, Paris, France. It was originally published as Figure 4-2 on p. 177 of TRB, *Special Report 251: Toward a Sustainable Future: Addressing the Long-Term Effects of Motor Vehicle Transportation on Climate and Ecology,* Transportation Research Board, National Research Council, Washington DC. This latter report is © 1997 TRB, and Figure 4.9 is reproduced from it with permission.
133. According to Margaret Chan, director-general of the WHO, "Road traffic crashes are not accidents. We need to challenge the notion that road traffic crashes are unavoidable and make room for a pro-active, preventive approach to reducing death on our roads" (Toroyan and Peden, 2007, p. vi). The use of "objective language" in transport discourse perhaps began with a memorandum sent by Michael Wright, City Administrator for the City of West Palm Beach, Florida, to his managers in November 1996, part

of which read as follows: "Accidents are events during which something harmful or unlucky happens unexpectedly or by chance. Accident implies no fault. It is well known that the vast majority of accidents are preventable and that fault can be assigned. The use of accident also reduces the degree of responsibility and severity associated with the situation and invokes an inherent degree of sympathy for the person responsible. Objective language includes collision and crash." Objective language had been introduced to the City of West Palm Beach by transport planner Ian Crawford, who had been persuaded of its importance while a student of the late Professor John Braaksma at Carleton University, Ottawa.

134. For road modes' share of fatalities in the EU, see p. 13 of ETSC (2003).
135. Table 4.11 is based on Tables 1 and 2 of ETSC (2003).
136. Per-capita road fatalities in the US fell by 36 percent between 1980 and 2004. These data are from *International Road Traffic and Accident Database,* oecd.org/document/53/0,3343,en_2649_34351_2002165_1_1_1_1,00.html. In the 15 countries that were EU members before 2004, per-capita transport fatalities fell by 26 percent between 1994 and 2004, from 12.5 to 9.2 per 100,000 residents. These EU15 data on road fatalities were downloaded from Eurostat, epp.eurostat.ec.europa.eu/portal/page/portal/product_details/dataset?p_product_code=TSDTR420. Note the similarity to Canadian road fatality rates shown in Figure 4.10, even though per-capita road travel is considerably higher in Canada. Note too that the EU15 data vary considerably among countries. In 2004, the two lowest fatality rates per 100,000 residents were 5.4 (Sweden) and 5.7 (UK), and the two highest were 15.5 (Portugal) and 16.4 (Greece). Although the UK has one of the lowest rates overall, it may have one of the highest rates of one type of fatality: child pedestrian deaths (see "Suffer the little children," *The Economist,* March 3, 2007, pp. 60–61).
137. The fatality and serious injury data represented in Figure 4.10 are from *Canadian Motor Vehicle Traffic Collision Statistics 2004,* Transport Canada, Ottawa, Canada, tc.gc.ca/roadsafety/tp/tpage 3322/2004/page1.htm. A serious injury is one that required hospitalization for treatment or observation. Population data represented in Figure 4.10 are from *Estimated population of Canada, 1605 to present,* Statistics Canada, Ottawa, Canada, statcan.ca/english/freepub/98-187-XIE/pop.htm.
138. For this suggestion about injury outcome, see p. 1,535 of Ameratunga et al. (2006).
139. The quotation is from p. 1,534 of Ameratunga et al. (2006).
140. Road fatalities estimates and projections represented in Table 4.12 are from Table 2.5 of Peden et al. (2004). Population estimates and projections are from Table 2.1 of the World Bank's *2006 World Development Indicators,* site resources.worldbank.org/DATASTATISTICS/Resources/table2-1.pdf. An anomaly in Peden et al. (2004) is the claim on p. 3 that "Worldwide, the number of people killed in road traffic crashes each year is estimated at al-

most 1.2 million." As reproduced here in Table 4.12, Table 2.5 of Peden et al., provides an estimate of about 850,000 fatalities for 2004 and 1.2 million *for 2020*. The anomaly is repeated in Ameratunga et al. (2006, p. 1,533) — "In 2002, an estimated 1.2 million people were killed...in road-traffic crashes worldwide"—and in Toroyan and Peden (2007, p. 1): "Each year nearly 1.2 million people die and millions more are injured or disabled as a result of road traffic crashes." The anomaly may be explained on pp. 30–31 of Kopits and Cropper (2003), who provided the lower estimate of total fatalities reproduced in Table 2.5 of Peden et al. (2004).

141. The data on types of road user described in this paragraph are from Figure 2.7 of Peden et al. (2004).
142. See the discussion of data challenges on pp. 52–60 of Peden et al. (2004).
143. The work on non-reporting of pedestrian injuries was noted on p. 1,535 of Ameratunga et al. (2006).
144. The information about costs comes from Table 2.9 of Peden et al. (2004).
145. For causes of fatalities in children and young people, see Table 1 of Toroyan and Peden (2007); for all ages see Table 1.1 of Peden et al. (2004).
146. For example, Peden et al. (2004) mention data to the effect that the probability that a pedestrian aged over 64 years will die as a result of being hit by a car traveling at 75 km/h is more than 60 percent, whereas the corresponding probability for a person aged under 15 years is about 20 percent. These authors also note, in their Figure 3.3, data that the all-age probability of a pedestrian's death as a result of being hit by a car traveling at 30 km/h or less is near zero. However, it rises steeply with increasing speed: with probabilities of death of 35 percent at 40 km/h, 55 percent at 50 km/h and at or near 100 percent at 80 km/h.
147. The basic data for Figure 4.11 are in Figure 2 of Toroyan and Peden (2007), supplemented by data from Table A.2 on p. 172 of WHO (2004). The population estimates used to compute the rates per 100,000 persons are from the United Nation's *World Population Prospects: The 2006 Revision* at esa .un.org/unpp/page 2kodata.asp. Note that the rates per 100,000 have been reduced to make them consistent with those in Table 4.12. See Note 140 for pointers to the reasons for the reductions. See WHO (2009), who.int/ violence_injury_prevention/road_safety_status/2009/en/, for more recent data and analysis.
148. The road fatality rate (for 2002) is from Table A.2 on p. 172 of WHO (2004). The homicide rate (for 2000) is from Table A.3 on p. 274 of Krug et al. (2002).
149. The regional and country homicide data in Table 4.13, all for 2000, are respectively from Tables A.3 and A.8 of Krug et al. (2002). The regional and country road fatality data, all for 2002, are respectively from Tables A.2 and A.4 of the *Statistical Annex* detailed in Note 147.
150. This exercise was conducted by John Adams of University College London and is reported in Adams (2000) and also in Gilbert (2002).

151. The quotation is from p. 5 of EEA (2006).
152. The quotations are from the back cover and from p. 27 of Morris (2005).
153. The quotations are from pp. 144 and 225 of Bruegmann (2005).
154. The assessment of sprawl and social relationships is in Freeman (2001).
155. The quotation is from p. 1 of Huttenmoser and Meierhofer (1995).
156. The study of aircraft noise and reading is Clark et al. (2006).
157. The study of the sleep quality of Swedish children and adults is Öhrström et al. (2006).
158. For example, a paper presented at a round table organized by the European Conference of Ministers of Transport in February 2001 concluded that more than half of the economic growth of what was then West Germany during the period 1950–1990 could be attributed to growth in transport activity (Baum and Kurte, 2002). Of this contribution, only a small part lay in the direct contribution of transport activity; most came from transport's facilitation of other activities.
159. In the round table mentioned in Note 158, Vickerman (2002) suggested that economic growth causes transport growth.
160. In May 2007, the Quebec government indicated it would intervene in a day-old strike of public transport workers in Montreal. The threat led to early settlement of outstanding issues. However, in Vancouver, which has much lower per-capita public transport use than Toronto or Montreal, public transport workers struck for four months in 2001 before the British Columbia government intervened.
161. For transport's share of GDP in the US, see Table 3-2b of US DOT (2007).
162. These data for EU15 are from Figure 2-46 on p. 63 of EC (2005).
163. The data in Table 4.14 are based on Kenworthy and Laube (2001).
164. This felicitous phrase appeared in Pill, J., "Metro's future: Vienna surrounded by Phoenix?" *Toronto Star,* February 15, 1990. It refers to the Toronto region's oddity in having a core of about a million people living in some of the densest configurations in North America surrounded by what were then about three million (now 4.5 million) living in some of the least dense configurations.
165. Figure 4.12 is Figure 1-I of RCEP (2007), and is Crown copyright 2007. It is reproduced with permission of the UK Office of Public Sector Information and of the originator of the diagram, Professor Judith Petts of the University of Birmingham, UK.
166. The energy use data in Table 4.15 are based chiefly on Tables 1-37M, 1-46bM and 4-6M of US DOT (2007), which show respectively for the US for several years total pkm by mode, total tkm by mode and total energy use by mode. Values are extrapolated to 2007. Exceptions and qualifications are as follows. The value for "Local public transport (electric)" includes pkm for light rail, heavy rail, trolley bus and half of commuter rail. Other pkm for local and regional public transport are assigned to "Local public transport (ICE)." The passenger rail values in US DOT (2007) are not used because of the

limited US experience with this mode. Instead, Finnish data for 2001 are used for inter-city rail (i.e., not high speed), from *Unit emissions of vehicles in Finland,* Technical Research Centre of Finland, lipasto.vtt.fi/yksikkopaa stot/indexe.htm. The value for passenger marine is the authors' "guestimate" that has regard for the marine freight values discussed later in this note. Passenger air modes use the values in Table 4-21M of US DOT (2007) extrapolated to 2007. The freight estimate for "Rail (ICE)" is from US DOT (2007) extrapolated to 2007 and can be considered reliable; it (0.24MJ/tkm) is used rather than the Finnish estimate (0.43MJ/tkm). The freight estimate "Rail (electric)" comes from the Finnish source. The estimate for high-speed rail is based on information provided in Tables 9.1 (Germany), 9.4 (France) and 9.6 (Spain) of Jørgensen and Sorenson (1997) using respective average train occupancies of 323, 184, and 157 (50 percent in each case), and also the estimate for German ICE trains in Table 3 of von Rozycki et al. (2003), which assumes 309 passengers per train, i.e., 46 percent occupancy. The freight estimate for "Marine (domestic)" is not the one derived from US DOT (2007) because that value (1.5MJ/tkm) seems implausibly high. Other values are lower, e.g., the equivalent of 0.3MJ/tkm at donauschifffahrt.info/en/pub lic_relations/advantages_of_inland_waterway_transport/environment/ energy_consumption/, and the range of 0.12–0.25MJ/tkm for short-sea shipping at advancedmaritimetechnology.aticorp.org/short-sea-shipping/ create3s-european-commission-initiative/EUSSS_04-2004.pdf. A problem with "Marine (domestic)" is that it covers a wide variety of shipping including coastal (i.e., short-sea), lake, intraport, river and canal movement. The value shown is positioned to be closer to the just-cited values. The estimate for "Marine (international)" is based on the worldwide consumption in 2004 of about 165 million tonnes of bunker and other fuel (IEA, 2006) to perform about 40 billion tkm of ocean freight movement (UNCTAD, 2006). The air freight estimates are from the Finnish source.

167. Occupancy data are from Tables 4-24M and 4-22M of US DOT (2007). In case the bus occupancy level should be regarded as unusually low, the average occupancy of buses in London, UK, should be noted. According to *London Travel Report 2006* (Transport for London, tfl.gov.uk/corporate/ about-tfl/publications/1482.aspx) this was 14.7 passengers in 2005–2006. The bus size was not stated but if the fleet mix of single- and double-decked buses has an average capacity of 66 passengers per bus, the overall occupancy level would be the same as for the U.S (22 percent). The high value for "Local public transport (ICE)" in Table 4.15 also reflects inclusion of some inefficient commuter rail, demand-responsive and ferry boat activity.

References

Adams, J. G. U. (2000) "The social implications of hypermobility," in *The Economic and Social Implications of Sustainable Transportation,* Organization for Economic Cooperation and Development, Paris, France, pp. 95–133,

olis.oecd.org/olis/1999doc.nsf/c16431e1b3f24c0ac12569fa005d1d99/c125685b002f5004c125686b005cb510/$FILE/00071363.PDF.
ADB (2006) *Urban Air Quality Management: Summary of Country/City Synthesis Reports across Asia,* Asia Development Bank, Manila, The Philippines, 25 pp., adb.org/Documents/Reports/Urban-Air-Quality-Management/overview.pdf.
Ameratunga, S., Hijar, M. and Norton, R. (2006) "Road traffic injuries: Confronting disparities to address a global-health problem," *The Lancet,* vol 367, pp. 1,533–1,540.
An, F. and Sauer, A. (2004) *Comparison of Passenger Vehicle Fuel Economy and Greenhouse Gas Emission Standards Around the World,* Pew Center for Global Climate Change, Arlington, VA, pewclimate.org/docUploads/Fuel%20Economy%20and%20GHG%20Standards_010605_110719.pdf.
Auvray, M. and Bey, I. (2005) "Long-range transport to Europe: Seasonal variations and inplications for the European ozone budget," *Journal of Geophysical Research,* vol 110, D11303, doi:10.1029/2004JD005503.
Baum, H., and Kurte, J. (2002) Untitled paper presented at Round Table 119, *Transport and Economic Development,* European Conference of Ministers of Transport, Paris, France, pp. 5–49.
Bayer-Oglesby, L., Schindler, C., Hazenkamp-von Arx, M. E., Braun-Fahrländer, S., Keidel, D., Rapp, R., Künnzli, N., Braendli, O., Burdet, L., Liu, L.-J. S., Leuenberger, P., Ackermann-Liebrich, U. and the SAPALDIA Team (2006) "Living near main streets and respiratory symptoms in adults: The Swiss cohort study on air pollution and lung diseases in adults," *American Journal of Epidemiology,* vol 164, no 12, pp. 1,190–1,198.
Bershadskii, A. (2009) Transitional solar dynamics and global warming, *Physica A,* vol 388, pp. 3,213–3,224.
Bousquet, J., Dahl, R. and Khaltaev, N. (2007) "Global alliance against chronic respiratory diseases," *European Respiratory Journal,* vol 29, pp. 233–239.
Bruegmann, R. (2005) *Sprawl: A Compact History,* University of Chicago Press, Chicago, IL, 301 pp.
Canada (1997) *Summary for Policy Makers: A Synthesis of the Key Results of the NOx/VOC Science Program,* Environment Canada, Ottawa.
Canada (2006) *National Inventory Report: Greenhouse Gases and Sinks in Canada, 1990–2004,* Environment Canada, Ottawa, 482 pp., ec.gc.ca/pdb/ghg/inventory_report/2004_report/2004_report_e.pdf.
Capaldo, K., Corbett, J. J., Kasibhatla, P., Fishbeck, P. S. and Pandis, S. N. (1999) "Effects of ship emission on sulfur cycling and radiative climate forcing over the ocean," *Nature,* vol 400, pp. 743–746.
Chilingar, G. V., Khilyuk, L. F. and Sorokhtin, O. G. (2009) Cooling of atmosphere due to CO_2 emission, *Energy Sources Part A,* vol 30, pp. 1–9.
Clark, C., Martin, R., van Kempen, E., Alfred, T., Head, J., Davies, H. W., Haines, M. M., Lopez Barrio, I., Matheson, M. and Stansfeld, S. A. (2006)

"Exposure-effect relations between aircraft and road traffic noise exposure at school and reading comprehension: The RANCH project," *American Journal of Epidemiology*, vol 163, pp. 27–37.

Cooley, S. R. and Doney, S. C. (2009) Anticipating ocean acidification's economic consequences for commercial fisheries. *Environmental Research Letters*, vol 4 (June 1, 2009) doi:10.1088/1748-9326/4/2/024007, at iop.org/EJ/article/1748-9326/4/2/024007/erl9_2_024007.html.

Davis, S. C. and Diegel, S. W. (2008) *Transportation Energy Data Book*, 27th edition, Center for Transportation Analysis, Oak Ridge National Laboratory, Oak Ridge, TN, cta.ornl.gov/data/index.shtml.

DeCicco, J. (2000) *It's Not (just) Technology, It's the Market (stupid!)*, American Council for an Energy-Efficient Economy, Washington DC, 13 pp., aceee.org/briefs/nottech.pdf.

Doney, S. C., Fabry, V. J., Feely, R. A. and Kleypas, J. A. (2009) Ocean Acidification: The Other CO_2 Problem, *Annual Review of Marine Science*, vol 1, pp. 169–192.

EC (2005) *Annual Energy and Transport Review 2003*, Office for Official Publications of the European communities, Luxembourg, 164 pp., ec.europa.eu/dgs/energy_transport/figures_archive/energy_review_2003/doc/full_report.pdf.

EEA (2004) *Impacts of Europe's Changing Climate: An Indicator-based Assessment*, European Environment Agency, Copenhagen, Denmark, reports.eea.europa.eu/climate_report_2_2004/en.

EEA (2006) *Urban Sprawl in Europe: The Ignored Challenge*, European Environment Agency, Copenhagen, Denmark, 56 pp., reports.eea.europa.eu/eea_report_2006_10/en.

EEA (2007a) *Transport and Environment: On the Way to a New Common Transport Policy*, European Environment Agency, Copenhagen, Denmark, 38 pp., reports.eea.europa.eu/eea_report_2007_1/en/eea_report_1_2007.pdf.

EEA (2007b) *Air Pollution by Ozone in Europe in Summer 2006*, European Environment Agency, Copenhagen, Denmark, 26 pp., reports.eea.europa.eu/technical_report_2007_5/en/eea_technical_report_5_2007.pdf.

Elsom, D. (1996) *Smog Alert: Managing Urban Air Quality*, Earthscan Publications Ltd, London, 192 pp.

Endresen, Ø., Sørgård, E., Sundet, J. K., Dalsøren, S. B., Isaksen, I. S. A., Berglen, T. F. and Gravir, G. (2003) "Emission from international sea transportation and environmental impact," *Journal of Geophysical Research*, vol 108, no D17, pp. 4560–4581.

Entec (2006) "Development of a methodology to assess population exposed to high levels of noise and air pollution close to major transport infrastructure," Report prepared for the European Commission, Entec UK Ltd, London, ec.europa.eu/environment/air/pdf/hot_spots/final_report_main.pdf.

ETSC (2003) *Transport Safety Performance in the EU: A Statistical Overview,* European Transport Safety Council, Brussels, Belgium, 32 pp., etsc.be/oldsite/statoverv.pdf.

Federici, M. and Basosi, R. (2007) "Air versus terrestrial transport modalities: An environmental comparison," in Ulgiati, S., Bargigli, S., Brown, M. T., Giampietro, M., Herendeen, R. A. and Mayumi, K. (eds) *Advances in Energy Studies: Perspectives on Energy Future,* Base 2001, Milan, Italy, 720 pp.

Fichtner, H., Scherer, K. and Heber, B. (2006) "A criterion to discriminate between solar and cosmic ray forcing of the terrestrial climate," *Atmospheric Chemistry and Physics Discussions,* vol 6, pp. 10811–10836.

Fleming, G. G., Armstrong, R. E., Stusnik, E., Polcak, K. D. and Lindeman, W. (1999) "Transportation-related noise in the United States," TRB Millennium Paper, Transportation Research Board, Washington DC, onlinepubs.trb.org/onlinepubs/millennium/00134.pdf.

Fontaras, G., and Samaras, Z. (2007) "A quantitative analysis of the European Automakers' voluntary commitment to reduce CO_2 emissions from new passenger cars based on independent experimental data," *Energy Policy,* vol 35, no 4, pp. 2,239–2,248.

Forman, R. T. T., Sperling, D., Bissonette, J. A., Clevenger, A. P., Cutshall, C. D., Dale, V. H., Fahrig, L., France, R., Goldman, C. R., Heanue, K., Jones, J. A., Swanson, F. J., Turrentine, T. and Winter, T. C. (2003) *Road Ecology: Science and Solutions,* Island Press, Washington DC.

Freeman L. (2001) "The effects of sprawl on neighborhood social ties: An explanatory analysis," *Journal of the American Planning Association,* vol 67, pp. 69–77.

Gilbert, R. (2002) "Social implications of sustainable transport," in Black, W. R. and Nijkamp, P. (eds), *Social Change and Sustainable Transport,* Indiana University Press, Bloomington, IN, pp. 63–69.

Gilbert, N. L., Woodhouse, S., Stieba, D. M. and Brook, J. R. (2003) "Ambient nitrogen dioxide and distance from a major highway," *Science of the Total Environment,* vol 312, pp. 43–46.

Giljum, S., Hak, T., Hinterberger, F. and Kovanda, J. (2005) "Environmental governance in the European Union: Strategies and instruments for absolute decoupling," *International Journal of Sustainable Development,* vol 8, nos 1/2, pp. 31–46.

Greene, D. L. and Wegener, M. (1997) "Sustainable transport," *Journal of Transport Geography,* vol 5, no3, pp. 177–190.

Hansen, J. E., (2006) "Can we still avoid dangerous human-made climate change?" *Social Research,* vol 73, no 3, pp. 949–971.

Heavenrich, R. (2006) *Light-Duty Automotive Technology and Fuel Economy Trends: 1975 Through 2006,* US Environmental Protection Agency, Washington DC, 101 pp., epa.gov/otaq/cert/mpg/fetrends/420r06011.pdf.

Heck, R. M. and Farrauto, R. J. (2001) "Automobile exhaust catalysts," *Applied Catalysis A: General,* vol 221, no 1–2, pp. 443–457.

Huttenmoser, M. and Meierhofer, M. (1995) "Children and their living surroundings: Empirical investigations into the significance of living surroundings for the everyday life and development of children," *Children's Environments*, vol 12, no 4, pp. 1–17.
IPCC (2001) *Climate Change 2001: The Scientific Basis*, Cambridge University Press, Cambridge, UK and New York, 881 pp., grida.no/climate/ipcc_tar/wg1/index.htm.
IPCC (2005) *IPCC/TEAP Special Report on Safeguarding the Ozone Layer and the Global Climate System: Issues Related to Hydrofluorocarbons and Perfluorocarbons*, Clodic, D. F., Baker, J. A., Chen, J., Hirata, T. H., Hwang, R. J., Köhler, J., Petitjean, C. and Suwono, A. (eds of Chapter 6) Cambridge University Press, Cambridge, UK and New York, NY, 478 pp., arch.rivm.nl/env/int/ipcc/pages_media/SROC-final/SpecialReportSROC.html.
IPCC (2007a) *Climate Change 2007: The Physical Science Basis; Summary for Policymakers; Contribution of Working Group I to the Fourth Assessment Report*, Intergovernmental Panel on Climate Change, Geneva, Switzerland, ipcc.ch/SPM2feb07.pdf.
IPCC (2007b) "Changes in atmospheric constituents and in radiative forcing" Chapter 2 of *Climate Change 2007: The Physical Science Basis; Final Report; Contribution of Working Group I to the Fourth Assessment Report*, Intergovernmental Panel on Climate Change, ipcc.ch/.
IPCC (2007c) *Climate Change 2007: Impacts, Adaptation and Vulnerability; Summary for Policymakers; Contribution of Working Group II to the Fourth Assessment Report*, Intergovernmental Panel on Climate Change, Geneva, Switzerland, ipcc.ch/.
Jacobs, J. (2004) *Dark Age Ahead*, Random House, New York, NY, 256 pp.
Jerrett, M., Arain, M. A., Kanaroglou, P., Beckerman, B., Crouse, D., Gilbert, N. L., Brook, J. R., Finkelstein, N. and Finkelstein, M. M. (2007) "Modeling the intraurban variability of ambient traffic pollution in Toronto, Canada," *Journal of Toxicology and Environmental Health, Part A*, vol 70, pp. 200–212.
Jonson, J. E., Simpson, D., Fagerli, H. and Solberg, S. (2006) "Can we explain the trends in European ozone levels?," *Atmospheric Chemistry and Physics*, vol 6, pp. 51–66.
Jørgensen, M. W. and Sorenson, S. C. (1997) *Estimating emissions from railway traffic*, inrets.fr/infos/cost319/MEETDeliverable17.PDF.
Kay, J. H. (1997) *Asphalt Nation: How the Automobile Took Over America and How We Can Take It Back*, Crown Publishers, New York, NY, 417 pp.
Kelly, T. and Allan, J. (2006) "Ecological effects of aviation," in Davenport, J. and Davenport, J. L. (eds), *The Ecology of Transportation: Managing Mobility for the Environment*, Springer, the Netherlands, pp. 5–24.
Kenworthy, J. and Laube, F. (2001) *The Millennium Cities Database for Sustainable Transport*, Union Internationale des transports publics (UITP), Brussels, Belgium (CD-ROM).

Kopits, E. and Cropper, M. (2003) *Traffic Fatalities and Economic Growth*, World Bank, Washington DC, 48 pp., ntl.bts.gov/lib/24000/24400/24490/25935_wps3035.pdf.

Krug, E. G., Dahlberg, L. L., Mercy, J. A., Zwi, A. B. and Lozano, R. (2002) *World Report on Violence and Health*, World Health Organization, Geneva, Switzerland, 372 pp.

Lewis, A. C., Evans, M. J., Methven, J., Watson, N., Lee, J. D., Hopkins, J. R., Purvis, R. M., Arnold, S. R., McQuaid, J. B., Whalley, L. K., Pilling, M. J., Heard, D. E., Monks, P. S., Parker, A. E., Reeves, C. E., Oram, D. E., Mills, G., Bandy, B. J., Stewart, D., Coe, H., Williams, P. and Crosier, J. (2007) "Chemical composition observed over the mid-Atlantic and the detection of pollution signatures far from source regions," *Journal of Geophysical Research*, vol 112, no D10, D10S39 10.1029/2006JD007584.

MacLean, H. L. and Lave, L. B. (2003) "Evaluating automobile fuel/propulsion system technologies," *Progress in Energy and Combustion Science*, vol 29, pp. 1–69.

Mann, R. (2006) "The local costs to ecological services associated with high seas global transport," in Davenport, J. and Davenport, J. L. (eds) *The Ecology of Transportation: Managing Mobility for the Environment*, Springer, the Netherlands, pp. 25–38.

Marsh, N., and Svensmark, H. (2000) "Cosmic rays, clouds, and climate," *Space Science Reviews*, vol 94, pp. 215–230.

McCarthy, J. J., Canziani, O. F., Leary, N. A., Dokken, D. J. and White, K. S. (eds) (2001) *Climate Change 2001: Impacts, Adaptation, and Vulnerability; Contribution of Working Group II to the Third Assessment Report of the Intergovernmental Panel on Climate Change*, Cambridge University Press, Cambridge, UK, 1032 pp., grida.no/climate/ipcc_tar/wg2/index.htm.

Michaud, D. S., Keith, S. E. and McMurchy, D. (2005) "Noise annoyance in Canada," *Noise & Health 2005*, vol 7, no 27, pp. 39–47.

Miedema, H. M. E. (2007) "Annoyance caused by environmental noise: Elements for evidence-based noise policies," *Journal of Social Issues*, vol 63, no 1, pp. 41–57.

Monahan, P. and Friedman, D. (2004) *The Diesel Dilemma: Diesel's Role in the Race for Clean Cars*, Union of Concerned Scientists, Cambridge, MA, ucsusa.org/assets/documents/clean_vehicles/dieseldilemma_fullreport.pdf.

Morris, D. E. (2005) *It's A Sprawl World After All*, New Society Publishers, Gabriola Island, BC, Canada, 243 pp.

NHTSA (2004) *Automotive Fuel Economy Program: Annual Update, Calendar Year 2004*, US National Highway Traffic and Safety Administration, Washington DC, nhtsa.dot.gov/staticfiles/DOT/NHTSA/Vehicle%20Safety/CAFE/2004_Fuel_Economy_Program.pdf.

O'Brien, E. (2006) "Habitat fragmentation due to transport infrastructure: Practical considerations," in Davenport, J. and Davenport J. L. (eds) *The Ecology of Transportation: Managing Mobility for the Environment,* Springer, the Netherlands, pp. 191–204.

OECD (2000) *Synthesis Report: EST: Environmentally Sustainable Transport—futures, strategies and best practices,* Organization for Economic Cooperation and Development, Paris, France, oecd.org/dataoecd/15/29/2388785.pdf.

OECD (2002) *Policy Instruments for Achieving Environmentally Sustainable Transport, Organization for Economic Cooperation and Development,* Paris, France, 172 pp.

Olsson, L. and Anderssen, B. (2004) "Kinetic modelling in automotive catalysis," *Topics in Catalysis,* vol 28, nos 1–4, pp. 89–98.

Öhrström, E., Hadzibajramovic, E., Holmes, M. and Svensson, H. (2006) "Effects of road traffic noise on sleep: Studies on children and adults," *Journal of Environmental Psychology,* vol 26, pp. 116–126.

Peden, M., Scurfield, R., Sleet, D., Mohan, D., Hyder, A. A., Jarawan, E. and Mathers, C. (2004) *World Report on Road Traffic Injury Prevention,* World Health Organization, Geneva, Switzerland, 244 pp., who.int/violence_injury_prevention/publications/road_traffic/world_report/en/index.html.

Perl, A. and Dunn Jr, J. A. (2007) "Reframing automobile fuel economy policy in North America: The politics of punctuating a policy equilibrium," *Transport Reviews,* vol 27, no 1, pp. 1–35.

RCEP (2002) *The Environmental Effects of Civil Aircraft in Flight,* Royal Commission on Environmental Pollution (UK), London, rcep.org.uk/aviation/av12-txt.pdf.

RCEP (2007) *The Urban Environment,* Royal Commission on Environmental Pollution (UK), London, rcep.org.uk/urban/report/urban-environment.pdf.

Revitt, D. M. (2004) "Water pollution impacts of transport," *Issues in Environmental Science and Technology,* vol 20, pp. 81–110.

Røpke, I. (2005) "Trends in the development of ecological economics from the late 1980s to the early 2000s," *Ecological Economics,* vol 55, pp. 262–290.

Ross, M., Goodwin, R., Watkins, R., Wenzel, T. and Wang, M. Q. (1998) "Real-world emissions from conventional passenger cars," *Journal of the Air and Waste Management Association,* vol 48, pp. 502–515.

Ross, M. and Wenzel, T., "Real-world emissions from conventional passenger cars," Chapter 2 of DeCicco, J. and Delucchi, M. (eds) *Transportation, Energy, and Environment: How Far Can Technology Take Us?,* American Council for an Energy-Efficient Economy, Washington DC, 1997, pp. 21–49.

Samet, J. and Krewski, D. (2007) "Health effects associated with exposure to ambient air pollution," *Journal of Toxicology and Environmental Health, Part A,* vol 70, no 3, pp. 227–42.

Sausen R., Isaksen, I., Grewe, V., Hauglustaine, D., Lee, D. S., Myhre, G., Köhler, M. O., Pitari, G., Schumann, U., Stordal, F. and Zerefos, C. (2005) "Aviation radiative forcing in 2000: An update on IPCC (1999)," *Meteorologische Zeitschrift,* vol 114, no 4, pp. 555–561.

Schade, W. (2003) "Transport noise: A challenge for sustainable mobility," *International Social Science Journal,* vol 55, pp. 279–294.

Schwartz, W. (2005) *Emissions, Activity Data, and Emission Factors of Fluorinated Greenhouse Gases (F-Gases) in Germany 1995–2002,* Umweltbundesamt, Berlin, Germany, 287 pp., umweltdaten.de/publikationen/fpdf-l/2903.pdf.

Seiler, A. and Helldin, J. O. (2006) "Mortality in wildlife due to transportation," in Davenport, J. and Davenport, J. L. (eds) *The Ecology of Transportation: Managing Mobility for the Environment*, Springer, the Netherlands, pp. 165–189.

Shao, M., Tang, X., Zhang, Y. and Li, W. (2006) "City clusters in China: Air and surface water pollution," *Frontiers in Ecology and the Environment*, vol 4, pp. 353–361.

Smit, R., Zeise, K., Caffin, A. and Anyon, P. (2004) *Dioxins Emissions from Motor Vehicles in Australia*, National Dioxins Program Technical Report No. 2, Department of the Environment and Heritage, Canberra, Australia, environment.gov.au/settlements/publications/chemicals/dioxins/report-2/index.html.

Sonne, C., Dietz, R., Born, E. W., Riget, F. F., Kirkegaard, M., Hyldstrup, L., Letcher, R. J. and Muir, D. C. G. (2004) "Is bone mineral composition disrupted by organochlorines in east Greenland polar bears (*Ursus maritimus*)?" *Environmental Health Perspectives*, vol 112, no 17, pp. 1,711–1,716.

Sorokhtin, O. G., Chilingar, G. V., Khilyuk, L. and Gorfunkel, M. V. (2007) "Evolution of the Earth's global climate," *Energy Sources, Part A: Recovery, Utilization, and Environmental Effects,* vol 29, no 1, pp. 1–19.

Stengel, D. B., O'Reilly, S. and O'Halloran, J. (2006) "Contaminants and pollutants," in Davenport, J. and Davenport, J. L. (eds) *The Ecology of Transportation: Managing Mobility for the Environment*, Springer, the Netherlands, pp. 361–388.

Svensmark, H. (2007) "Cosmoclimatology: A new theory emerges," *Astronomy & Geophysics*, vol 48, no 1, pp. 18–24.

Svensmark, H., and Calder, N. (2007) *The Chilling Stars: A New Theory of Climate Change*, Icon Books Ltd, Cambridge, UK, 246 pp.

Svensmark, H., Pederson, J. O. P., March, N. D., Enghoff, M. B. and Uggerhøj, U. I. (2007) "Experimental evidence for the role of ions in particle nucleation under atmospheric conditions," *Proceedings of the Royal Society A*, vol 463, pp. 385–396.

Svensmark H., Bondo, T. and Svensmark, J. (2009), Cosmic ray decreases affect atmospheric aerosols and clouds, *Geophysical Research Letters*, doi:10.1029/2009GL038429, in press.

Toronto (2000), *Health Effects of Noise*, Medical Officer of Health, City of Toronto, Canada, city.toronto.on.ca/health/hphe/pdf/noiserpt_attachment march23.pdf.

Toroyan, T. and Peden, M. (2007) *Youth and Road Safety*, World Health Organization, Geneva, 40 pp., whqlibdoc.who.int/publications/2007/9241595116_eng.pdf.

Twigg, M. V. (2007) "Progress and future challenges in controlling automotive exhaust gas emissions," *Applied Catalysis B Environmental*, vol 70, pp. 2–15.

UNCTAD (2006) *Review of Maritime Transport*, 2006, United Nations Conference on Trade and Development, New York and Geneva, Switzerland, unctad.org/en/docs/rmt2006_en.pdf.

UNEP (2006) *Environmental Effects of Ozone Depletion and its Interactions with Climate Change: 2006 Assessment*, United Nations Environment Programme, Nairobi, Kenya, 233 pp., ozone.unep.org/Assessment_Panels/EEAP/eeap-report2006.pdf.

US DOT (2000) *The Changing Face of Transportation*, US Department of Transportation, Washington DC, bts.gov/publications/the_changing_face_of_transportation/.

US DOT (2007) *National Transportation Statistics 2007*, Bureau of Transportation Statistics, US Department of Transportation, Washington DC, bts.gov/publications/national_transportation_statistics/index.html.

US EIA (2006) *Emissions of Greenhouse Gases in the United States 2005*, US Energy Information Administration, Washington DC, eia.doe.gov/oiaf/1605/ggrpt/index.html.

US EPA (2000) *National Air Pollutant Emission Trends, 1900–1998*, US Environmental Protection Agency, Washington DC, 238 pp., epa.gov/ttn/chief/trends/trends98/trends98.pdf.

Van den Brink, R. M. M. and Van Wee, B. (2001) "Why has car-fleet specific fuel consumption not shown any decrease since 1990? Quantitative analysis of Dutch passenger car-fleet specific fuel consumption," *Transportation Research Part D*, vol 6, no 2, pp. 75–93.

van der Grift, E. and Pouwels, R. (2006) "Restoring habitat connectivity across transport corridors: Identifying high-priority locations for de-fragmentation with the use of an expert-based model," in Davenport, J. and Davenport, J. L. (eds) *The Ecology of Transportation: Managing Mobility for the Environment*, Springer, the Netherlands, pp. 205–231.

Vézina, A. F. and Ove Hoegh-Guldberg, O. (2008) Effects of ocean acidification on marine ecosystems, *Marine Ecology Progress Series*, vol 373, pp. 199–309, December 23, 2008, doi: 10.3354/meps07868, at int-res.com/articles/theme/m373_ThemeSection.pdf.

Vickerman, R. (2002) Untitled paper presented at Round Table 119, *Transport and Economic Development*. European Conference of Ministers of Transport, Paris, France, pp. 139–177.

von Rozycki, C., Koeser, H. and Schwarz, H. (2003) "Ecology profile of the German high-speed rail passenger transport system, ICE," *International Journal of Life Cycle Analysis*, vol 8, pp. 83–91.

Wang, X., Carmichael, G., Chen, D., Tang, Y. and Wang, T. (2005) "Impacts of different emissions sources on air quality during March 2001 in the Pearl River Delta (PRD) region," *Atmospheric Environment*, vol 39, no 239, pp. 5,227–5,241.

WHO (1999) *Guidelines for Community Noise,* World Health Organization, Geneva, Switzerland, who.int/docstore/peh/noise/guidelines2.html.

WHO (2005) *Health Effects of Transport-related Air Pollution,* Krzyzanowski, M., Kuna-Dibbert, B. and Schneider, J. (eds) World Health Organization Regional Office for Europe, Copenhagen, Denmark, 205 pp., euro.who.int/document/e86650.pdf.

WHO (2006) *WHO Air Quality Guidelines for Particulate Matter, Ozone, Nitrogen dioxide and Sulfur dioxide: Global Update 2005*, World Health Organization, Geneva, Switzerland, euro.who.int/Document/E90038.pdf.

WHO (2009) *Global Status Report on Road Safety,* World Health Organization, Geneva, Switzerland, 301 pp., who.int/violence_injury_prevention/road_safety_status/2009/en/.

Winkler, M. (2005) *The Characterization of Highway Runoff Water Quality,* Technical University, Graz, Austria, portal.tugraz.at/portal/page/portal/Files/i2150/download/Diplomarbeiten/DA_Winkler.pdf.

Wongwises, S., Kamboon, A. and Orachon, B. (2006) "Experimental investigation of hydrocarbon mixtures to replace HFC-134a in an automotive air conditioning system," *Energy Conversion and Management,* vol 47, nos 11–12, pp. 1,644–1,659.

CHAPTER 5: The Next Transport Revolutions

1. See, for example, Jacobs (2004), Diamond (2005) and Kunstler (2005). In reprising his 2006 book, *The Upside of Down: Catastrophe, Creativity, and the Renewal of Civilization,* Thomas Homer-Dixon described the crux of the current predicament as this: "Our global system is becoming steadily more complex, yet the high-quality energy we need to cope with this complexity will soon be steadily less available." (Homer-Dixon, T., "Prepare today for tomorrow's breakdown," *Globe & Mail* (Toronto), 14 May 2007.
2. For oil consumption and import data, see BP (2009) and earlier editions of this document.
3. For example, according to US CIA (2009), China's GDP in 2006 was $7.8 trillion, after only that of the European Union ($14.8 trillion) and the US ($14.3 trillion). However, China's *per capita* GDP was $6,000, ranking 132 among 229 economies, behind El Salvador, Turkmenistan and Albania, although ahead of Egypt, Jordan and Bolivia. The world average per capita GDP in 2006 was $10,400. These statements of GDP in US dollars involve application of purchasing power parities, which reflect the cost of goods within the respective economies and compensate for exchange rate anomalies.
4. Mandarin is the most widely spoken language and shares a writing system with other languages, notably Wu and Cantonese. English may be second, third or fourth among first languages spoken (Spanish and Arabic being the other contenders) but is overwhelmingly the leading second language.
5. The data in Table 5.1 are Appendix A of IEA (2008). The amounts for poorer

and richer countries are from the same sources, counting OECD members and countries in Eastern Europe and Eurasia among the richer countries.

6. Table 5.1 shows that richer countries were actually responsible for 74 percent of total oil consumption in 1990, and that the 2025 shortfall of expected supply in relation to BAU consumption will be 11.3 billion barrels. Allocating this shortfall as proposed in the text, richer countries use about 7.5 billion barrels less than the BAU projection, and poor countries would use about 3.7 billion barrels less. Hence, the targets for 2025 in Table 5.1.
7. For non-fuel uses of oil, see Tables 1.15 and 1.3 of US EIA (2007a).
8. The US population estimate and projection are from *World Population Prospects: The 2008 Revision,* United Nations Population Division, New York, NY, esa.un.org/unpp.
9. In Table 5.2, person-kilometers (pkm) for 2007 are extrapolated from data for 1995–2004 in Table 4.21M (aviation) and Table 1.37M (other modes) of US DOT (2007a). The split between ICE and electric intercity rail was estimated from the energy-use data in Table 4.6M of the same source and an assumption that electric locomotives use about one third of the energy used by diesel-electric locomotives. The rates of energy use in 2007 are those in Table 4.15 of Chapter 4 and are derived from the sources noted there. Suggested rates of energy use in 2025 are discussed in the text.
10. The electronic spreadsheets used to construct the scenarios in Table 5.2 to Table 5.5 are available from the authors, whose contact information is at the book's website: transportrevolutions.info. The spreadsheets can be used to construct alternative scenarios for 2025, and to apply different estimates of person- and tonne-kilometers and of energy use per pkm and tkm for 2007.
11. Jitneys are minibuses offering trips along fixed routes, sometimes with little regulation, often between unmarked stops, and always for payment. For anticipated energy consumption by PRT, see Gustavsson (1995).
12. The suggested unit fuel use by airships is speculative. A current operating company reports use of 1.3 MJ/pkm for a 15-passenger vehicle (see Airship Management Services, airshipman.com/faq.htm). Larger airships and use of solar power from canopy-mounted connectors could reduce liquid fuel consumption substantially.
13. For towing kite applications for ships, see SkySails GmbH & Co, skysails .info.
14. In Table 5.3, tonne-kilometers (tkm) for 2007 for movement by road, rail, pipeline and domestic marine are extrapolated from data for 1995–2004 in Table 1.46bM of US DOT (2007a). The authors estimated tkm for international marine by noting that (i) according to Table III.16 of WTO (2006), US trade outside North America was about 9 percent of world trade by value, and thus might comprise about 9 percent of world tkm; and (ii) according to Table 5 of UNCTAD (2006), total world trade involved about 44 trillion tkm in 2004, which we extrapolated to 48 trillion tkm in 2007. Aviation tkm

are extrapolated from data in UNdata, United Nations Statistics Division, unstats.un.org/unsd/databases.htm. Rates of energy use are those in Table 4.15 of Chapter 4 and are derived from the sources noted there. Suggested pkm and rates of energy use in 2025 are discussed in the text.

15. Within the US—that is, excluding Marine (international) and Air (international)—more than half of tkm would be electrified (actually, 54 percent).
16. As we noted in the Introduction, other people than US residents regard themselves as "Americans," and to this point in the book we have tried to avoid confusion by not using the terms America, American and Americans. However, from this point, use only of "US" became stilted or ungainly, and we have more often used America to mean the US, American to mean pertaining to the US, and Americans to mean residents of the US.
17. In January 1980, the Carter doctrine identified the Persian Gulf as a "vital interest" of the country and declared that "an attempt by an outside force to gain control of the Persian Gulf region would be regarded as an assault on the vital interests of the United State." (O'Rourke and Connolly, 2003, p. 588). According to Table 15.12 of Delucchi and Murphy (2006), in 2004 the US spent between $14.9 billion and $52.5 billion defending the flow of oil from the Persian Gulf, and US oil interests there, and an additional $22.1 billion to $35.2 billion on other military attention to this region. In these authors' Table 15.11, the value of US imports from the region in 2004 was given as $41.7 billion.
18. For state aid to canals and roads in the 18th century, see Hartz (1948) and Goodrich (1960).
19. For government incentives to build a continental railway network, see Fogel (1960).
20. For early involvement of the US government in highway development, see Seely (1987).
21. For the development of the US Interstate Highway System, see Rose (1990).
22. The "Futurama" book is Bel Geddes (1940). According to Gelernter (1995, p. 364), the book's vision of transcontinental superhighways was described by Robert Moses, the leading highway builder of the day, as "plain bunk." We hope our proposals share a similar fate.
23. An Army Corps of Engineers report by Fournier and Westervelt (2005) concluded, "The days of inexpensive, convenient, abundant energy sources are quickly drawing to a close" (p. 57).
24. The 2008 budget amount was obtained from dot.gov/bib2009/crosswalk .htm.
25. The quotation is from p. 131 of Bardach (1976).
26. This figure combines the Federal Highway Administration's $41.26 billion in its Enacted 2008 Budget for Federal-Aid Highways Obligation Limitation noted at dot.gov/bib2009/htm/FHA.html with the FAA's Enacted 2008 budget amounts of $2.514 billion for "Facilities and Operation" and $3.515 billion for "Air Traffic Organization" noted at dot.gov/bib2009/htm/FAA.html.

27. For state and local spending on transport, see US Census Bureau, *State and Local Government Finances by Level of Government and by State: 2005–06*, census.gov/govs/www/estimate.html.
28. The term "travelport" was coined by Hank Dittmar in "Travelports are way to go," *The Los Angeles Times*, 30 December 2002.
29. For an account of the Anderson campaign, see Bisnow (1983). For a contemporary comment on the fuel tax proposal, see Lewis, A., "A 50-cent tax on gasoline?" *New York Times*, 10 December 1979, p. A23, nytimes.com/2007/02/26/us/26seniors.html?_r=1&scp=1&sq=26th%20Feb.%202007,%20Making%20the%20return%20trip&st=cse. These pensions, funded by payroll taxes, are known in the US as Social Security pensions.
30. For an account of the fuel price escalator, see Leicester (2006).
31. See, for example, the results of the Harris Poll, "Americans Would Like to See a Larger Share of Passengers and Freight Going By Rail in Future," 8 February 2006, harrisinteractive.com/harris_poll/index.asp?PID=638.
32. On p. 64 of Francis and Wheeler (2006), there is a description of a program run since 2002 by De Lijn, Belgium's Flemish public transport agency. Families giving up their cars receive three-year passes for use on De Lijn's system. If it is a second car, the pass is for one person. If it is the only car, each family member receives the pass. During the first three years, almost 30,000 families took advantage of the program and only 800 returned to owning a car.
33. A completely different approach would involve limiting fuel consumption by rationing, which has merit in that richer and poorer citizens could be affected more equally. We favor a new tax because of the revenues it would provide, and we would urge that the alternative transport arrangements supported by these revenues be initially focused on places where people with low incomes live.
34. The ongoing reverse migration of elderly Americans from retirement communities in the South to live closer to their families has begun to yield a net outflow of population from this region for the first time since the 1930s. Between 2000 and 2005, 121,000 people aged 75 and older left the South, while 87,000 arrived. See Roberts, S., "Making the return trip: Elderly head back North," *New York Times*, 26 February 2007, select.nytimes.com/gst/abstract.html?res=F60610F73A5A0C758EDDAB0894DF404482. Nevertheless, the response of some communities in the North is to focus on consolidating their decline rather than reversing it. For an extreme case, see Streitfeld D., "An effort to save Flint, Mich., by shrinking it," *New York Times*, April 22, 2009, nytimes.com/2009/04/22/business/22flint.html.
35. Levinson et al. (1997) estimated an annual 5.65 billion pkm for 677 kilometers of two-way track between Los Angeles and San Francisco/Sacramento.
36. The data on railway track in the US are in Table 1.1 of US DOT (2007a).
37. Levinson et al. (1997) estimated these costs for California to total about $16 million per km of two-way track in 1994$, or about $22 million in 2007$.

Of this, about 13 percent would be for rolling stock. Costs elsewhere in 2007$ have ranged from $14 million to $70 million per km, and another estimate for California was $33 million per km (see *California High-Speed Rail Implementation Plan* at cahighspeedrail.ca.gov/images/chsr/20080123171537_ImplementationPlan.pdf). The unit cost assumed here is $40 million per km. See also cahighspeedrail.ca.gov/faqs/financing.htm.

38. This estimate assumes a real annual cost of borrowing of two percent that over 30 years would add about a third to the total capital cost.
39. For information about operating costs see the two sources cited in Note 35.
40. According to Table 3.14 of US DOT (2007a) the extrapolated 2007 cost of car operation in the US per kilometer is $0.36.
41. Based on an analysis of the deployments of 115 successful technologies, Nakićenović (2007) concluded that reducing costs by half requires from 2 to 7 doublings of cumulative production. A critical factor was production skill, which improved by about 20 percent with each doubling.
42. The estimates of investment in transport were extrapolated from data in Table 3.2a of US DOT (2007a). They are addressed again later in this chapter.
43. This estimate of the US GDP is extrapolated from Table 3.2a of US DOT (2007a) and produces a slightly different result from the estimate in Note 3.
44. The US rail network reached a peak in about 1930, when it had about twice the current extent in terms of track kilometers and carried more than five times as much passenger traffic as today, although less than a fifth as much freight traffic. The decade with the highest rate of track construction was the 1880s, when about 180,000 *track* kilometers were added. (See Slide 18 of the presentation by Steve Heminger entitled "National surface transportation policy and revenue study commission" to a meeting of the Associated General Contractors of America, 30 September 2006, bicycling.511.org/meetings/presentations/AGC_9-30-06_SH.ppt#256,1,Slide 1.) The above proposals could involve addition of about 50,000 km of track and electrification and upgrading of perhaps an additional 100,000 track kilometers, all in the space of about 15 years.
45. Information about the indicated efforts concerning high-speed rail in the US is available respectively at fra.dot.gov/us/content/515 (federal); cahighspeedrail.ca.gov/ (California); floridahighspeedrail.org/ (Florida); lib.utexas.edu/taro/tslac/20071/20071-P.html (Texas); and dot.state.mn.us/passengerrail/onepagers/midwest.html (Midwestern states consortium).
46. Figure 5.1 is from the website of the US Federal Railroad Administration at fra.dot.gov/us/content/203.
47. The Acela can operate as fast as 240 km/h over a 56 km stretch of the 734 km route. Over the rest of this route, top speeds range from 145 to 217 km/h.
48. Information on different uses of the NEC infrastructure assets that are currently bundled together under various forms of government ownership can be found in Voorhees Transportation Center, *Northeast Corridor Action*

Plan, New Brunswick, New Jersey: Rutgers University, Bloustein School of Planning and Public Policy, 2006, policy.rutgers.edu/vtc/documents/ProgEval.Amtrak_FINAL%20REPORT.pdf. The rail infrastructure condominium concept, described below, could be used to enhance the NEC's capacity by enabling different users (e.g., intercity, commuter and freight rail carriers) to invest in specific assets with ownership privileges.

49. This concept was initially articulated on p. 245 of Perl (2002). In North America, the most common use of the word "condominium" is to describe an apartment building in which the units are owned individually and the common spaces and services are owned collectively by the owners of the units. The infrastructure condominium would be similar in effect to what pertains in the UK and Sweden, where operation of the tracks and the trains is separate, and many train operators, public and private, can use the tracks. It would differ in that in those countries the tracks and rights-of-way are owned by companies that do not operate trains (a quasi-public company in the UK, a public company in Sweden), whereas in the US, the infrastructure assets on a rail right-of-way could be owned by numerous entities, particularly train operators.
50. Of the 575 certified airports, 316 serve regularly scheduled airlines with sufficient frequency to feature in the monthly reports of the US Department of Transportation (US DOT) on aircraft delays. (In April 2007, 76 percent of the 613,740 flights by scheduled aircraft arrived on time.) Of the 316 airports, 31 are characterized as "major airports." The US actually had 19,279 airports in 2005. Of these, 14,584 were for private use only, and 57 (the 2004 total) were military airports. Data are from Table 1.3 of US DOT (2007a) and from the website of Bureau of Transportation Statistics of the US DOT, *Airline On-Time Statistics and Delay Causes,* transtats.bts.gov/OT_Delay/ot_delaycause1.asp?display=data&pn=1.
51. The data on rail access to airports are from Table 92 of APTA (2007).
52. For an example of where airport security searches are heading, see "New airport X-rays scan bodies, not just bags," *New York Times,* 24 February 2007, select.nytimes.com/gst/abstract.html?res=FB0A13FE3A5A0C778ED DAB0894DF404482.
53. For a summary of data on China's consumption, see p. 9 of Brown (2006).
54. For China's oil consumption, see BP (2009).
55. According to the United Nations Statistics Division, at unstats.un.org/unsd/snaama/dnllist.asp, among world economies in 2005 China ranked first in agricultural production, third in manufacturing, construction, capital formation and exports (after the US and Japan or, in the case of exports, the US and Germany), fourth in imports (after the US, Germany and the UK), and sixth or seventh in other sectors including internal trade and household consumption. These comparisons are on an exchange rate basis; on a purchasing parity basis China may not have such high rankings.

56. According to Table 43 on p. 78 of UNCTAD (2006), in 2005 three of the world's four busiest and seven of the 20 busiest container ports were in China.
57. China's population data and projections are from the online version of *World Population Prospects: The 2008 Revision, Population Database,* United Nations Population Division, esa.un.org/unpp/.
58. The quotation is from French, H. W., "Big, gritty Chongqing, city of 12 million, is China's model for future," *New York Times,* 1 June 2007.
59. In 2005, Shanghai was said to have almost twice as many skyscrapers as New York City—commercial and residential—and was set to add a thousand more (25 percent of the 2005 total) by 2010. See Barboza, D., "China builds its dreams, and some fear a bubble," *New York Times,* 18 October 2005, nytimes.com/2005/10/18/business/worldbusiness/18bubble.html?_r=1&scp=1&sq=China%20builds%20it's%20dreams%20Barboza&st=cse.
60. The quotation is from p. 1,185 of Liu and Diamond (2005).
61. The quotations in this sentence are from pp. 60 and 63 of Engler and Mugyenyi (2005).
62. For a listing of the world's largest shopping centers, see the "Shopping Mall Studies" website of Eastern Connecticut State University, nutmeg.easternct.edu/~pocock/MallsWorld.htm. The South China Mall has a gross leasable area (GLA) of 660,000 m^2. Second, with a GLA of 560,000 m^2 is Beijing's Jin Juan Mall, also known as the Great Mall of China. Third, with a GLA of 386,000 m^2, is the SM Mall of Asia, Pasay City, The Philippines. Fourth, with a GLA of 350,000 m^2, is Dubai Mall. Fifth, and largest in North America, with a GLA of 350,000 m^2, is the West Edmonton Mall. Sixth, with a GLA of 348,000 m^2, is Cevahir Istanbul, which appears to be the largest shopping center in Europe. (It is on the west side of the Bosphorus Strait.) Of the 25 shopping centers listed as the world's largest, six are in China (and another six are elsewhere in Asia). Information about the South China Mall can be found at southchinamall.com.cn/english/index1.jsp, including the detail that the Mall has 8,000 parking spaces. The West Edmonton Mall boasts the "world's largest parking lot," providing for over 20,000 vehicles (see westedmall.com/plan/parking.asp).
63. The population estimate and projection for China are from the source detailed in Note 57.
64. In Table 5.4, person-kilometers (pkm) for 2007 are mostly extrapolated from data for 1996–2005 in Tables 11.11, 16.7 and 16.37 of NBSC (2006). In the absence of other information, highway pkm were evenly divided between "Personal vehicle (ICE)" and "Bus (intercity, ICE)." These totals would also include some movement within urban regions. Rail pkm were equally divided between ICE and electric propulsion, as suggested in the text. The pkm for 2007 for "Personal vehicle (electric)" represent use of

battery-powered scooters and bicycles. Weinert et al. (2007) noted that four times as many of these vehicles are being produced in China as cars. We have assumed that their annual distance traveled per vehicle—2,400 km, as reported by Cherry (2006)—represents one tenth of the pkm associated with each car. Thus, the "Personal vehicle (electric)" pkm are set at 40 percent of the car pkm. Extrapolation for public transport pkm was based on 2003–2005 only. In the absence of other information, rates of energy use in Table 5.4 are set at 75 percent of the corresponding rates for the US in Table 5.2, except for aviation, for which the rates are assumed to be the same as the US. The main factor contributing to such a difference is likely to be higher occupancies in China. The value for "Personal vehicle (electric)" is actually 0.047MJ/pkm from Slide 8 of Cherry (2006), which is for an electric bicycle but could also represent the higher-powered, higher-occupancy electric scooter. In their Table 5, Weinert et al. (2007) gave a slightly higher value (0.015 kWh/km or 0.0504 MJ/km), noting that even this value for electric bicycles is lower than that of the food energy required to bicycle one km. Suggested pkm and rates of energy use in 2025 are discussed in the text.

65. In Table 5.5, tonne-kilometers (tkm) for 2007 are mostly extrapolated from data for 1996–2005 in Tables 16.9 and 16.37 of NBSC (2006). In the absence of other information, rail tkm are assumed to be equally divided between ICE and electric propulsion. Energy use rates are those of the US in Table 4.15 (Chapter 4) and Table 5.3.
66. See "China has the world's second largest electric railway network," *People's Daily,* 29 September 2006, english.people.com.cn/200609/29/eng200609 29_307504.html
67. For information about Dafeng City, see Wang and Ye (2004).
68. For information about Dongtan, see Pearce (2006) and Normile (2008). For the contrary information gathered in 2008 and relayed in the next paragraph, see Moore, M., "China's pioneering eco-city of Dongtan stalls," *Daily Telegraph,* October 18, 2008, telegraph.co.uk/news/worldnews/asia/ china/3223969/Chinas-pioneering-eco-city-of-Dongtan-stalls.html.
69. For information about Jiading, see metrasys.de/region/region_de.html. The Toronto region has now passed the Detroit region as North America's main automotive center, so we should perhaps say "the Toronto of Asia."
70. The Chinese Grand Prix and its significance was discussed by French, H. W., "With a raceway, China motors toward the modern age," *New York Times,* 26 September 2004, p. 3, select.nytimes.com/gst/abstract.html?res=F 40B12FA345D0C758EDDA00894DC404482.
71. The quotation is from p. 15 of Zi (2001).
72. Information about Beijing's satellite cities is in Jing, L., "Beijing's new growth plan addresses resources bottleneck," *China Daily,* 8 November 2004, chinadaily.com.cn/english/doc/2004-11/08/content_389278.htm.

73. The quotation is from "Beijing plans to build 11 new cities," *SinoCast China Business Daily News,* 3 January 2006, findarticles.com/p/articles/mi_hb5562/is_200601/ai_n22726696/.
74. The walking and bicycling estimate is from Table 4 of Kenworthy and Hu (2002).
75. The debt level is reported in Zhou and Szyliowicz (2006).
76. These fiscal details are from Oliver (2006) and Feng (2006).
77. For the "Community Energy Management" approach see Sadownik and Jaccard (2002), and p. 15 of the same source for the quotation.
78. The quotation is from p. 5 of Lieberthal (1997).
79. The quotation is from p. 424 of Zhu (2004).
80. The quotation is from p. 310 of Molotch (1976).
81. The quotation is from p. 443 of Friedmann (2006).
82. The three conditions are from p. 6 of Lieberthal (1997).
83. The phrase is from the title of Kunstler (1993).
84. Table 5.6 is based on data and projections in Appendix H of US EIA (2007b). Note that the values in Table 5.6 are lower than those in IEA (2006), particularly for China in 2025. The reason for the differences is unclear. It could be because the values in US EIA (2007b) are for production by "central producers" only, whereas those in IEA (2006) may be for all electricity production, including that by industries for their own use. In making the conversion from the units in the scenario tables, note that one exajoule (EJ) is the same as 279 terawatt-hours (TWh).
85. For comparison of investments in conservation, see Lin (2007), who noted that in 2004 much of China experienced curtailment of electricity supply "causing widespread disruption to industrial production and huge economic losses" (p. 916). Investment in conservation in China is at a lower level than in the US and declining as a percentage of investment in electricity generation. It also appears to be declining in the US, although good data are available only on public-sector investment of this type.
86. For a review of the effectiveness of conservation measures see Gellings et al. (2006). These authors note that "the biggest savings in electricity can be attained in a few areas: lights, motor systems, and the refrigeration of food and rooms" (p. 60). They conclude that energy conservation is usually cost-effective but note analyses arriving at contrary conclusions.
87. According to US EIA (2007b), 66 percent of electricity is generated from fossil fuels worldwide. The breakdown by fuel type is oil 5 percent, natural gas 20 percent, coal 43 percent, nuclear 17 percent and renewables 17 percent.
88. The quotation is from p. 10 of ACORE (2007). Note that whereas current and proposed capacity, as set out in Table 5.6, provides about 5 TWh of generation annually for each gigawatt of capacity, what is proposed for generating capacity from renewable resources would provide only about

2 TWh/GW. This is because the latter kind of capacity is less available. The sun does not shine at night, and the wind does not always blow.

89. The estimate is in Table 1 of Cherni and Kentish (2007); the quotation is from p. 3,627 of the same source. The estimate did not include the huge potential for electricity generation from wind turbines located off-shore on the continental shelf off China's east coast, said to be 750 GW (Feller, 2006).
90. For example, in June 2007 the US Energy Department announced $175 million for three new biofuel research centers in Tennessee, Wisconsin and California. See Wald, M., "US is creating 3 centers for research on biofuels," *New York Times,* 26 June 2007, select.nytimes.com/gst/abstract.html?res=F40816FC345B0C758EDDAF0894DF404482.
91. Eberhart (2007) highlights the potential of grid-connected transport on p. 252 of his book on energy futures. We also described such a transition in Gilbert and Perl (2007).

References

ACORE (2007) *The Outlook on Renewable Energy in America,* American Council for Renewable Energy, Washington DC, 64 pp., acore.org/files/RECAP/docs/OutlookonRenewableEnergy2007.pdf.

APTA (2007) *Public Transportation Factbook,* American Public Transportation Association, Washington DC, apta.com/research/stats/factbook/index.cfm.

Bardach, E. (1976) "Policy termination as a political process," *Policy Sciences,* vol 7, no 2, pp. 123–131.

Bel Geddes, N. (1940) *Magic Motorways,* Random House, New York, NY, 297 pp.

Bisnow, M. (1983) *Diary of a Dark Horse,* Southern Illinois University Press, Carbondale, IL, 329 pp.

BP (2009) *Statistical Review of World Energy June 2009,* BP plc, London, 48 pp., bp.com/statisticalreview.

Brown, L. R. (2006) *Plan B 2.0: Rescuing Planet Under Stress and a Civilization in Trouble,* W W Norton, New York, NY, 365 pp.

Cherni, J. A. and Kentish, J. (2007) "Renewable energy policy and electricity market reforms in China," *Energy Policy,* vol 35, pp. 3,616–3,629.

Cherry, C. R. (2006) "Implications of electric bicycle use in China: Analysis of costs and benefits," Presentation at the Volvo Center Workshop, University of California Berkeley Center for Future Urban Transport, 24 July 2006, its.berkeley.edu/volvocenter/July2006workshop/Presentations/03b%20Chris.ppt.

Delucchi, M. A., and Murphy, J. (2006) *US Military Expenditures to Protect the Use of Persian-Gulf Oil for Motor Vehicles,* Institute of Transportation Studies, University of California, Davis, CA, its.ucdavis.edu/publications/2004/UCD-ITS-RR-96-03(15)_rev2.pdf.

Diamond, J. (2005) *Collapse: How Societies Choose to Fail or Succeed,* Viking Penguin, New York, NY.

Eberhart, M. (2007) *Feeding the Fire: The Long History and Uncertain Future of Mankind's Energy Addiction,* Harmony Books, New York, NY, 304 pp.

Engler, Y. and Mugyenyi, B. (2005) "Red road rising," *The Ecologist,* vol 35, no 2, pp. 60–63.

Feller, G. (2006) "China's great potential: Wind power," *Power Engineering International,* vol 14, no 7, pp. 30–32.

Feng, L. (2006) "Status & problems of transportation related socio-economic issues in China," *World Transport Policy & Practice,* vol 12, no 4, pp. 35–40, ecoplan.org/library/wt12-4.pdf.

Fogel, R. W. (1960) *The Union Pacific Railroad: A Case in Premature Enterprise,* Johns Hopkins Press, Baltimore, MD, 129 pp.

Fournier, D. F. and Westervelt, E. T. (2005) *Energy Trends and Their Implications for US Army Installations,* US Army Corps of Engineers, Washington DC, 86 pp., stinet.dtic.mil/oai/oai?&verb=getRecord&metadataPrefix=html&identifier=ADA440265.

Francis, A. and Wheeler, J. (2006) *One Planet Living in the Suburbs,* World Wildlife Federation, Goldaming, Surrey, UK, 116 pp., assets.panda.org/downloads/opl_uk_suburbs_fullrpt.pdf.

Friedmann, J. (2006) "Four theses in the study of China's urbanization," *International Journal of Urban and Regional Research,* vol 30, pp. 440–451.

Gelernter, D. (1995) *1939: The Lost World of the Fair,* Free Press, New York, NY.

Gellings, C. W., Wikler, G. and Ghosh, D. (2006) "Assessment of US electric end-use energy efficiency potential," *The Electricity Journal,* vol 19, no 9, pp. 55–69.

Gilbert, R. and Perl, A. (2007) "Grid-connected vehicles as the core of future land-based transport systems," *Energy Policy,* vol 35, no 5, pp. 3,053–3,060.

Goodrich, C. (1960) *Government Promotion of American Canals and Railroads: 1800–1890,* Columbia University Press, New York, NY, 382 pp.

Gustavsson, E. (1995) "Energy efficiency of personal rapid transit," *ECEEE Summer Study Proceedings 1995,* European Council for an Energy-Efficient Economy, Stockholm, Sweden, eceee.org/conference_proceedings/eceee/1995/Panel_6/page 6_1/Paper/.

Hartz, L. (1948) *Economic Policy and Democratic Thought: Pennsylvania 1776–1860,* Harvard University Press, Cambridge, MA, 366 pp.

Homer-Dixon, T. (2006) *The Upside of Down: Catastrophe, Creativity, and the Renewal of Civilization,* Island Press, Washington DC, 429 pp.

IEA (2006) *World Energy Outlook 2006,* International Energy Agency, Paris, France, 596 pp.

IEA (2008) *World Energy Outlook 2008,* International Energy Agency, Paris, France, 578 pp.

Jacobs, J. (2004) *Dark Age Ahead,* Random House, New York, NY, 256 pp.

Kenworthy, J. and Hu, G. (2002) "Transport and urban form in Chinese cities," *DISP,* Issue 151, pp. 4–14, disp.ethz.ch.

Kunstler, J. H. (1993) *Geography of Nowhere: The Rise and Decline of America's Man-Made Landscape,* Free Press, New York, NY, 304 pp.

Kunstler, J. H. (2005) *The Long Emergency: Surviving the Converging Catastrophes of the Twenty-First Century,* Grove/Atlantic, New York, NY, 320 pp.

Leicester, A. (2006) *The UK Tax System and the Environment,* Institute for Fiscal Studies, London, UK, ifs.org.uk/comms/r68.pdf.

Levinson, D., Mathieu, J. M., Gillen, D. and Kanafani, A. (1997) "The full cost of high-speed rail: An engineering approach," *Annals of Regional Science,* vol 31, pp. 189–215.

Lieberthal, K. (1997) "China's governing system and its impact on policy implementation," *China Environment Series 1,* Woodrow Wilson International Center for Scholars, Washington DC, pp. 3–8, wilsoncenter.org/topics/pubs/ACF4CF.PDF.

Lin, J. (2007) "Energy conservation investments: A comparison between China and the US," *Energy Policy,* vol 35, pp. 916–924.

Liu, J. and Diamond, J. (2005) "China's environment in a globalizing world," *Nature,* vol 435, pp. 1,179–1,186.

Molotch, H. (1976) "The city as a growth machine: Toward a political economy of place," *American Review of Sociology,* vol 82, pp. 309–332.

Nakićenović, N. (2007) "Endogenous technological change," in Greene, D. L. (ed) *Modeling the Oil Transition: A Summary of the Proceedings of the DOE/EPA Workshop on the Economic and Environmental Implications of Global Energy Transitions,* Oak Ridge National Laboratory, Oak Ridge, TN, pp. 157–159, www-cta.ornl.gov/cta/Publications/Reports/ORNL_TM_2007_014_EnergyTransitionsWorkshopSummary.pdf.

NBSC (2006) *China Statistical Yearbook 2005,* National Bureau of Statistics of China, China Statistics Press, Beijing Info Press, Beijing, China, stats.gov.cn/tjsj/ndsj/2005/indexeh.htm.

Normile, D. (2008) "China's living laboratory in urbanization," *Science,* vol. 319, pp. 740–743.

O'Rourke, D. and Connolly, S. (2003) "Just oil? The distribution of environmental and social impacts of oil production and consumption," *Annual Review of Environment and Resources,* vol 28, pp. 587–617.

Oliver, H. He (2006) "Reducing China's thirst for foreign oil: Moving towards a less oil-dependent road transport system," *China Environment Series 8,* Woodrow Wilson International Center for Scholars, Washington DC, pp. 41–60, wilsoncenter.org/topics/pubs/CEF_Feature.3.pdf.

Pearce, F. (2006) "Master plan: A suburb of Shanghai could become the blueprint for cities worldwide," *New Scientist,* 17–23 June, vol 190, no 2556, pp. 43–46.

Perl, A. (2002) *New Departures: Rethinking Rail Passenger Policy for the Twenty-First Century,* University Press of Kentucky, Lexington, KY, 334 pp.

Rose, M. (1990) *Interstate: Express Highway Politics 1939–1989,* University of Tennessee Press, Knoxville, TN, 192 pp.

Sadownik, B. and Jaccard, M. (2002) "Shaping sustainable use in Chinese cities," *DISP,* Issue 151, pp. 15–22, disp.ethz.ch.

Seely, B. (1987) *Building the American Highway System: Engineers as Policy Makers,* Temple University Press, Philadelphia, PA, 312 pp.

UNCTAD (2006) *Review of Maritime Transport, 2006,* United Nations Conference on Trade and Development, New York and Geneva, Switzerland, unctad.org/en/docs/rmt2006_en.pdf .

US CIA (2009) *The World Factbook,* United States Central Intelligence Agency, Washington DC, https://cia.gov/library/publications/the-world-factbook/index.html.

US DOT (2007a) *National Transportation Statistics 2006*, Bureau of Transportation Statistics, US Department of Transportation, Washington DC, bts.gov/publications/national_transportation_statistics/index.html.

US DOT (2007b) *Fiscal Year 2008 Budget in Brief,* US Department of Transportation, Washington DC, 70 pp., dot.gov/bib2008/pdf/bib2008.pdf.

US EIA (2007a), *Annual Review of Energy 2006,*United States Energy Information Administration, Washington DC, eia.doe.gov/emeu/aer/contents.html.

US EIA (2007b) *International Energy Outlook 2007,* US Energy Information Administration, Washington DC, 230 pp., eia.doe.gov/oiaf/ieo/index.html.

Wang, R. and Ye, W. (2004) "Eco-city development in China," *Ambio,* vol 33, pp. 341–342.

Weinert, J., Ma, C. and Cherry, C. (2007) "The transition to electric bikes in China: History and key reasons for rapid growth," *Transportation,* vol 34, pp. 301–318.

WTO (2006) *International Trade Statistics 2006,* World Trade Organization, Geneva, Switzerland, wto.org/english/res_e/statis_e/statis_e.htm.

Zhou, W. and Szyliowicz, J. S. (2006) "The development and current status of China's transportation system," *World Transport Policy and Practice,* vol 12, no 4, pp. 10–16, ecoplan.org/library/wt12-4.pdf.

Zhu, J. (2004) "Local developmental state and order in China's urban development during transition," *International Journal of Urban and Regional Research,* vol 28, pp. 424–447.

Zi, M. (2001) "Should Chinese households have cars?," *Beijing Review,* vol 44, no 8, pp. 15–16.

CHAPTER 6: Leading the Way Forward

1. Data on Saudi Arabia's production are from Table 3 in several issues of the monthly *Oil Market Report* produced by the International Energy Agency, available at omrpublic.iea.org/.
2. Information about the Khurais field is from "Saudi Aramco Starts Production From Khurais Oil Field In Riyadh, Saudi Arabia," *Energy Business Re-*

view, June 11, 2009, at energy-business-review.com/news/saudi_aramco_starts_production_from_khurais_oil_field_in_riyadh_saudi_arabia_090611.

3. The quotation is from p. 9 of Gladwell (2002).
4. The article is Maynard M., "Industry Fears Americans May Quit New Car Habit," *New York Times,* May 31, 2009, at nytimes.com/2009/05/31/business/31car.html.
5. For fewer licensed young people, see Johnston L., "More teens waiting to get driver's license," *Cleveland Plain Dealer,* March 28, 2008, at blog.cleveland.com/metro/2008/03/more_teens_waiting_to_get_driv.html. But also see the letter from Anne McCartt, Vice-President Research, Insurance Institute for Highway Safety, to Richard Capka, Acting Administrator, US Federal Highway Administration, March 9, 2006, at iihs.org/laws/comments/pdf/fhwa_ds_atm_030906.pdf. Among many other things, Ms. McCartt noted, "Especially for 16 year-olds, large year-to-year anomalies in some states' data were found.... For example, the reported number of 16-year-old licensed drivers in Illinois was 79,391 in 1998, 8,159 in 1999, and 88,872 in 2000. In Louisiana, the reported number was 25,675 in 2001 but only 2 in 2002 and 3 in 2003."
6. For car driving by young people in south-central Ontario, see Gilbert, R., O'Brien C., *Child- and Youth-Friendly Land-Use and Transport Planning Guidelines for Ontario,* Centre for Sustainable Transportation, University of Winnipeg, 2009, at kidsonthemove.ca/uploads/Ontario%20Guidelines,%20v2.6.pdf.
7. The vehicle sales data for the US in Figure 6.1 are from the US Bureau of Economic Analysis, National Economic Accounts (motor vehicles), at bea.gov/national/index.htm. The data on vehicle-kilometers traveled are from US Federal Highway Administration, Traffic Volume Trends (Historical VMT from 1970, plus monthly supplements), at fhwa.dot.gov/ohim/tvtw/tvtpage.cfm. Note that the latter applies to all vehicle movement, but well over 90% of this comprises movement of light-duty vehicles.
8. Because of the scaling used in Figure 6.1, the fall in travel appears steeper than the fall in sales, but the reverse is true.
9. If the 6 percent decline in 16 months were to continue, it would take 16 years for the amount of driving to fall by half. However, if the decline is considered to be 9 percent, because the amount of driving was previously rising by about 3 percent (before 2004), it would take only 10 years to fall by half.
10. There is a minor wrinkle in the argument being developed here that rises in oil price caused falls in vehicle sales: first gas-guzzling SUVs etc. and then regular automobiles. It is that seasonally adjusted sales of hybrid gasoline-electric vehicles in the US peaked late in 2007 and fell during 2008 even more steeply than sales of other light-duty vehicles. (For hybrid sales, see greencarcongress.com/2009/06/sales-20090603.html. For sales of other light-duty vehicles, see the first source detailed in Note 7.) If high oil prices

were the direct cause of the falls in sales, sales of the more fuel-conserving hybrids might have been expected to remain high, at least until sales of regular automobiles began to fall during the second half of 2008. The late-2007 peak in sales of the higher-priced hybrid vehicles suggests that the higher prices of SUVs etc. could also have been a factor in their earlier decline. However, hybrid sales were such a small share of total auto sales (2.8 percent in November 2007) and so regionally limited (they were disproportionately in California) that their sales may not be reflect any overall trend.

11. US gasoline price data are from the US Energy Information Administration at tonto.eia.doe.gov/dnav/pet/pet_pri_gnd_dcus_nus_m.htm.
12. By mid 2009, according to the results of a survey, over 60 percent of a sample of chief executive officers of Canadian companies believed that world oil production would soon reach a peak. See Castaldo J., "The CEO Poll: On black gold," *Canadian Business Online*, June 8, 2009, at canadianbusiness.com/managing/ceo-poll/article.jsp?content=20090608_115156_8760.
13. UK gasoline prices are from the Excel file behind Figure 12-14 of UK ONC (2009).
14. Data on car sales for Europe are from the website of the European Automobile Manufacturers Association at acea.be/images/uploads/files/20090505_03_PC_90-09_By_Country_Enlarged_Europe.xls. They are actually new-vehicle registrations in 18 western European countries: the 15 countries of the European Union as it was in 2003 plus Iceland, Norway and Switzerland.
15. The average pump price of liter of regular gasoline in Japan fell from ¥153 to ¥134 between March 31 and April 3, 2008 as a long-time tax expired (see uk.reuters.com/article/idUKTKG00302420080404). By May 1 the price was back to ¥153 and by July 7 it was ¥182 (uk.reuters.com/article/idUKTKG00317720080709).
16. Data for the US and Europe are from the first source detailed in Note 7 and the source detailed in Note 14, respectively. Japanese car sales and export data are from the Japanese Automobile Manufacturers Association's *Active Matrix Database System* at jamaserv.jama.or.jp/newdb/eng/index.html.
17. Buttonwood, the finance and economics columnist for *The Economist*, wrote on June 4, 2009, "...the role of oil at $140 a barrel in sparking this recession has probably been underestimated." James Surowiecki, whose columns serve a similar role for the *New Yorker*, appeared in his June 22, 2009, column to endorse the analysis of James Hamilton discussed here in Chapter 3. Hamilton concluded, in Surowiecki's words, "...given the already weak state of the U.S. economy in 2007, the sharp increase in oil prices might have been enough, on its own, to tip the economy into recession, even without that year's blowup in the credit markets."
18. For an example of the conventional wisdom about the current global recession, see the analysis of Paul Krugman in his three June 2009 Robbins

lectures at the London School of Economics under the general title "The Return of Depression Economics." Links to the podcast, video and slides for the first lecture are at lse.ac.uk/collections/LSEPublicLecturesAndEvents/events/2009/20090311t1955z001.htm, where there is also a link to these items for the second lecture, where there can be found a corresponding link for the third lecture. For further examples, see the sources below in Notes 28, 34, 35 and 36. For suggestions that there may have been corruption as well as risky lending behavior, see Rich, F., "Bernie Madoff is no John Dillinger," *New York Times,* July 5, 2009, at nytimes.com/2009/07/05/opinion/05rich.html?_r=1&emc=eta1, and the sources noted therein.

19. Home price data are from *S&P/Case-Shiller Home Price Indexes. 2008, A Year in Review,* Standard & Poor's, New York, January 2009, at www2.standardandpoors.com/spf/pdf/index/Case-Shiller_Housing_Whitepaper_YearinReview.pdf.
20. Data on borrowing are from *Flow of Funds Accounts of the United States,* Federal Reserve Board, Washington DC, June 2009, at federalreserve.gov/RELEASES/z1/20030605/z1.pdf.
21. Data in Figure 6.2 are from the US Bureau of Labor Statistics at bls.gov/data/#employment.
22. The quotation is from Molvig D., *Tough Times Series: A Closer Look at Home Equity Loans,* Credit Union National Association, Madison, WI, June 18, 2009, at hffo.cuna.org/12433/article/2324/html.
23. Unemployment data are from the US Bureau of Labor Statistics at bls.gov/data/#unemployment.
24. GDP data are from the US Bureau of Economic Analysis at bea.gov/national/index.htm (Current account and "real" GDP).
25. Earnings in the vehicle production and home building industries are at the source in Note 21.
26. An increasingly favored index of financial market distress is what is known as the LIBOR-OIS spread, explained by Sengupta and Tam (2008), who noted, "The LIBOR-OIS spread has been a closely watched barometer of distress in money markets for more than a year.... The 3-month London Interbank Offered Rate (LIBOR) is the interest rate at which banks borrow unsecured funds from other banks in the London wholesale money market for a period of 3 months. Alternatively, if a bank enters into an overnight indexed swap (OIS), it is entitled to receive a fixed rate of interest on a notional amount called the OIS rate. In exchange, the bank agrees to pay a (compound) interest payment on the notional amount to be determined by a reference floating rate (in the United States, this is the effective federal funds rate) to the counterparty at maturity." A chart in the article indicates that the LIBOR-OIS spread rose above its historic level of close to zero in August 2007. It remained at about the new level until September 2008 and then rose steeply by a factor of five within a few weeks. By June 2009,

the LIBOR-OIS spread had fallen to just below the August 2007 level (see bloomberg.com/apps/quote?ticker=.LOIS3%3AIND).

27. For US borrowing, see the source in Note 20.
28. For examples of commentary in early 2008, see Reinhart and Rogoff (2008a, 2008b)
29. For the indication of extreme distress in financial markets at this time, see Note 26.
30. These approximate shares are based on Tables 3-1a, 3-2a, and 3-20b of US BTS (2009). Direct employment by the auto industry in the US in 2007 was about 350,000. The total of direct and indirect employment was about 4.2 million, or about 2.7 percent of the labor force; see Cole et al. (2008) and US CIA (2008). In Europe in 2007 the corresponding totals were about 2.2 million and 12 million, or about 5.4 percent of the labor force; see the European Commission's website at ec.europa.eu/enterprise/automotive/pagesbackground/sectoralanalysis/index.htm, and also indexmundi.com/european_union/labor_force.html.
31. Changes in the value of world trade are from the website of the World Trade Organization at wto.org/english/res_e/statis_e/quarterly_world_exp_e.htm. Oil represented only 6.8 percent of world trade by value in 2005—according to another part of the same website—and so, even with the large rises and falls in oil price in 2007–2008, the effect on the rest of the world's trade would remain remarkable. Oil comprised 38 percent of the world's seaborne freight movement in 2007, according to Table 5 of UNCTAD (2009).
32. For a discussion of underinvestment in the oil industry, see "Bust and boom," *The Economist*, May 21, 2009. The article notes that oil firms "must find another Saudi Arabia's worth of oil every two years just to maintain their production at today's levels. Yet the oil industry is short of equipment and manpower, thanks to decades of underinvestment in the 1980s and 1990s." It notes too that investment is down 15–20% in 2009 compared with 2008.
33. The term "giddy leap" was used in the article detailed in Note 32.
34. The 2009 paper by Eichengreen and O'Rourke, "A tale of two depressions," with an update is at voxeu.org/index.php?q=node/3421.
35. See World Bank (2009).
36. See OECD (2009).
37. According to "The big sweat," *The Economist*, June 13, 2009, "The sheer scale of their fiscal burdens may tempt governments to lighten their loads by inflation or even outright default. Inflation seems increasing plausible because many central banks are already printing money to buy government bonds."
38. The simultaneous occurrence of high rates of inflation and economic stagnation is sometimes known as "stagflation," a term most often applied to the period after the first oil shock of 1973.
39. The quotation is from p. 213 of Keynes (1937).
40. Funding has been made available through two subsidiary programs, the

Automotive Industry Financing Program ($80.3 billion) and the Automotive Supplier Support Program ($5.0 billion). See financialstability.gov/impact/DataTables/additionaltransactions.html.

41. The quotation is from a document at the White House website, "Obama Administration New Path to Viability for GM & Chrysler," at whitehouse.gov/assets/documents/Fact_Sheet_GM_Chrysler.pdf, March 30, 2009. This is one of the few items of publically available information produced by the Presidential Task Force on the Auto Industry.
42. For Robert Nardelli on electric vehicles, see Casey Williams, Chrysler Electric Vehicles, *Boulder Daily Camera,* January 20, 2009, at dailycamera.com/news/2009/jan/20/chrysler-electric-vehicles/. For Nardelli's farewell letter to Chrysler employees, see scoop.chrysler.com/2009/04/30/message-from-bob-nardelli-6/.
43. For Fiat's partnership with Micro-Vett S.p.A, see micro-vett.it/english/company.html. For Fiat's unveiling of its own electric car, see zoomilife.com/2008/11/06/fiat-electric-car-unveiled-at-sao-paulo-motor-show/.
44. For GM's anticipation of a smaller auto market, see "For U.S. auto sales, a long hangover awaits," at reuters.com/article/ousivMolt/idUSTRE5546PO20090605?sp=true.
45. US sales data are from the source for sales in Note 7.
46. For the Toyota statement, see "Toyota's new boss warns of two more tough years," at reuters.com/article/businessNews/idUSTRE55O0YB20090625?feedType=RSS&feedName=businessNews. Whether the reference applies to Japan only or to all Toyota's operations is unclear.
47. The quotation is from the report of a seminar held in April 2009 by the EU-Japan Centre for Industrial Cooperation entitled, "Economic and Environmental Challenges for the Automotive Industry in the EU and Japan," at eu-japan.gr.jp/data/current/newsobj-268-datafile.pdf.
48. The quotation concerning GM's proposed Volt automobile is from Presidential Task force on the Auto Industry, *GM February 17 Plan: Viability Determination, Summary,* March 30, 2009, at whitehouse.gov/assets/documents/GM_Viability_Assessment.pdf. The Volt is discussed in Chapter 3.
49. The remark is by George Iny, president of the Automobile Protection Association, a consumer group, quoted in Van Praet, N., "GM: Go green or go profitable?" *Financial Post,* June 20, 2009.
50. For the Canadian analysis of the rescue of GM, see Coyne, A., "We'll pay for this bailout for years," *Maclean's,* June 9, 2009, at www2.macleans.ca/2009/06/09/we%E2%80%99ll-pay-for-this-bailout-for-years/.
51. A sixth, potential measure is the American Clean Energy and Security Act of 2009, which was passed by the US House of Representatives on June 26, 2009, and is to be considered by the US Senate. The proposed legislation chiefly concerns implementation of a "cap-and-trade" system to reduce emissions of greenhouse gases. It contains numerous other measures,

including many concerning deployment of electric vehicles, enhancement of the electrical grid, and production of electricity from renewable sources. The current version of the legislation can be reached from thomas.loc.gov/cgi-bin/query/z?c111:H.R.2454:. This measure is not discussed here, partly because this complex, 1,428-page bill was approved by the House late in our writing, and partly because, at the time of writing, Senate approval of the proposed legislation seemed uncertain.

52. For the US Federal Railroad Administration's *Overview, Highlights and Summary of the Passenger Rail Investment and Improvement Act of*, 2008, March 2009, see fra.dot.gov/downloads/PRIIA%20Overview%20031009.pdf.
53. For the loan to Nissan, see PR Newswire, "Nissan's Plan for Zero-Emissions Vehicles Advances with U.S. Department of Energy Loan," at sev.prnewswire.com/null/20090623/CL3681823062009-1.html.
54. Information about the American Recovery and Reinvestment Act of 2009 is based on the Act itself, which is at gpo.gov/fdsys/pkg/PLAW-111publ5/content-detail.html, and the US Government website explaining the Act, which is at recovery.gov/.
55. President Obama's proposed 2010 federal budget, entitled *A New Era of Responsibility: Renewing America's Promise,* is at whitehouse.gov/omb/assets/fy2010_new_era/A_New_Era_of_Responsibility2.pdf.
56. See Bullis K., "Q & A: Steven Chu," *Technology Review,* May 14, 2009, at technologyreview.com/business/22651/.
57. For the May 18, 2009, announcement on fuel economy, see whitehouse.gov/the_press_office/Background-Briefing-on-Auto-Emissions-and-Efficiency-Standards/.
58. On p. 24009 of *Notice of Upcoming Joint Rulemaking To Establish Vehicle GHG Emissions and CAFE Standards,* at epa.gov/fedrgstr/EPA-AIR/2009/May/Day-22/a12009.pdf, the US Environmental Protection Agency and Department of Transportation wrote "...the emissions reductions and fuel economy improvements under consideration for the proposal would be expected to involve more widespread use of [electric vehicles and other technologies] across the fleet." On p. 24011 there is the further point, "...EPA is currently considering proposing additional credit opportunities to encourage the commercialization of advanced GHG/ fuel economy control technology such as electric vehicles and plug-in hybrid electric vehicles."
59. For discussion of the replacement transport legislation, see MacGallis, A., "White House Says Transportation System Overhaul Must Wait," *Washington Post,* June 26, 2009, at washingtonpost.com/wp-dyn/content/article/2009/06/25/AR2009062504055_pf.html.
60. A mostly complimentary article about the US president—"...at home and abroad, he invariably appears to be the only adult in the room..."—nevertheless included the following: "Only miniscule portions of the stimulus

bill or his budget proposals were dedicated to mass transit, and his indifference to the issue—what must be a major component of any serious effort to go green—was reflected in his appointment of a mediocre Republican time-server, Ray LaHood, as his transportation secretary." (See Baker, K., "Barack Hoover Obama: The best and the brightest blow it again," *Harper's Magazine,* July 2009, pp. 29–37.)

61. See Gertner, J., "Getting up to speed," *New York Times Magazine,* June 10, 2009, at nytimes.com/2009/06/14/magazine/14Train-t.html?_r=1.
62. Costing of the first phase is from Figure 21 of the *California High Speed Train Business Plan,* November 2008, at cahighspeedrail.ca.gov/images/chsr/20081107134320_CHSRABusinessPlan2008.pdf. Costs for the remainder are derived from information at cahighspeedrail.ca.gov/faqs/financing.htm.
63. For benefits of the proposed rail system see the first source detailed in Note 62.
64. For an analysis of California's condition, see "UCLA Anderson Forecast: Economy Healing But Not Out of Hospital Yet: California Budget Crisis Will Impact State's Ability to Recover from Recession," uclaforecast.com/contents/archive/2009/media_61609_1.asp.
65. Information about China's rail expansion comes chiefly from an interview with Zheng Jian, chief planner with China's Ministry of Railways, in "High-speed trains to take the strain," *China Daily,* June 24, 2008, at chinadaily.cn/cndy/2009-06/24/content_8315454.htm. The quotation is from Wong, J., and Light, A., *China Begins Its Transition to a Clean-Energy Economy,* Center for American Progress, Washington DC, June 4, 2008, at americanprogress.org/issues/2009/06/china_energy_numbers.html. The number of cities over 200,000 population is from Table 10-1 of the *China Statistical Yearbook 2008,* at sei.gov.cn/try/hgjj/yearbook/2008/indexeh.htm.
66. See "IBM rides the high-speed rail: The tech giant looks for ways to make high-speed trains smarter, safer and more efficient," June 22, 2009, at money.cnn.com/2009/06/22/technology/ibm_china_fast_trains.fortune/index.htm.
67. See Mingtai, L., "Rail investment promises express delivery for stimulus plan," *China Daily,* March 20, 2009, at chinadaily.com.cn/cndy/2009-03/20/content_7597900.htm.
68. See Bradsher, K., "Clash of subways and car culture in Chinese cities," *New York Times,* March 26, 2009, at nytimes.com/2009/03/27/business/worldbusiness/27transit.html. Bradsher suggested, "The question is whether the burrowing machines [digging the subway tunnels] can outrace China's growing love affair with the automobile...."
69. The quotation reflects the view of Goldman Sachs Asia-Pacific Chief economist, Michael Buchanan, as reported in Stutchbury, M., "China's stimulus package helps us," *The Australian,* June 27, 2009, at theaustralian.news.com.au/story/0,25197,25696230-7583,00.html.

70. Data on vehicle production are from the National Bureau of Statistics of China's database at stats.gov.cn/english/statisticaldata/monthlydata/index.htm (Output and growth of major industrial products).
71. For the National Bureau of Statistics of China's Consumer Confidence Index, see the source detailed in Note 70 (Consumer Confidence Index).
72. For data on trade between China and the US, see the website of the US Census Bureau at census.gov/foreign-trade/top/index.html#2009.
73. For discussion of the 2009 burst in China car sales, see Ying T., "China car sales jump 'beyond imagination,' bring two-month wait," at bloomberg.com/apps/news?pid=newsarchive&sid=ayyzWK6yOnlY.
74. Beginning in 2009, China has a new system of setting the state-controlled prices for transport fuels, which in the past have often had little relation to the costs of production, including the price of crude oil. Now, retail prices may be changed whenever the crude oil price changes at a rate of more than 4 percent per month. The latest change at the time of writing was on June 30, 2009, the third increase this year. Gasoline pump prices rose 11 percent. A survey suggests that 90 percent of users would now drive less. See "China's fuel price hike hurts drivers, economic impact limited," July 2, 2009, at news.tradingcharts.com/futures/5/1/126295115.html.
75. According to Bradsher, K., "China Vies to Be World's Leader in Electric Cars," *New York Times,* April 1, 2009, at nytimes.com/2009/04/02/business/global/02electric.html, "Chinese leaders have adopted a plan aimed at turning the country into one of the leading producers of hybrid and all-electric vehicles within three years, and making it the world leader in electric cars and buses after that. The goal, which radiates from the very top of the Chinese government, suggests that Detroit's Big Three, already struggling to stay alive, will face even stiffer foreign competition on the next field of automotive technology than they do today."
76. For the quotation and the other information in this paragraph, see *Medium and Long-Term Development Plan for Renewable Energy in China,* National Research and Development Council, September 2007, at chinaenvironmentallaw.com/wp-content/uploads/2008/04/medium-and-long-term-development-plan-for-renewable-energy.pdf.
77. The information and quotation about the electrical grid are from "China becomes world leader in UHV power technology," *People's Daily Online,* April 16, 2009, at english.peopledaily.com.cn/90001/90776/90881/6638814.html.
78. The quotation and most of the other information in this paragraph has come from Hayward, D., "China's oil supply dependence," *Journal of Energy Security,* June 2009, at ensec.org/index.php?option=com_content&view=article&id=197:chinas-oil-supply-dependence&catid=96:content&Itemid=345.
79. For data on oil imports, and for the link to vehicle sales, see Mohr P.,

"Scotiabank's Commodity Price Index begins to recover in May," Scotiabank Group, Toronto, ON, June 29, 2009, at scotiacapital.com/English/bns_econ/bnscomod.pdf. According to this author, China's crude oil imports moved over four million barrels a day for the first time.

80. The quotation is from a June 25, 2009, press release of the International Air Transport Association at iata.org/pressroom/pr/2009-06-25-01.htm. Passenger and freight data are at iata.org/pressroom/facts_figures/traffic_results/2009-06-25-01.htm. Jet fuel price data are at iata.org/whatwedo/economics/fuel_monitor/index.htm.
81. See Higgins, M. and Shih, G., "Airlines, already suffering, brace for further woes," *New York Times*, July 14, 2009, at nytimes.com/2009/07/14/business/14airlines.html.
82. The Baltic Dry Index is one of several indices of marine activity issued daily by London's Baltic Exchange, whose website is at balticexchange.com/. For how the index has changed, visit bloomberg.com/apps/cbuilder?ticker1=BDIY%3AIND.
83. For information on the Hamburg Index, see the website of the Hamburg Shipowners' Association at vhss.de/objectives_and_services.php.
84. For Rahm's comment, see Zeleny, J., "Obama weighs quick undoing of Bush policy," *New York Times*, November 9, 2008, at nytimes.com/2008/11/10/us/politics/10obama.html.
85. The quotation is from McKenna, B., "Rocky Road to recovery," *Globe and Mail*, July 7, 2009, at theglobeandmail.com/report-on-business/crash-and-recovery/rocky-road-to-recovery/article1208835/.
86. The quotation is from Krugman, P., "That '30s show," *New York Times*, July 2, 2009, at nytimes.com/2009/07/03/opinion/03krugman.html.
87. The quotation is from Baker, K., "Barack Hoover Obama: The best and the brightest blow it again," *Harper's Magazine*, July 2009, pp. 29–37.
88. The remarks of the president and vice-president at the launch of the US high-speed rail initiative can be found at whitehouse.gov/the_press_office/Remarks-by-the-President-and-the-Vice-President-on-High-Speed-Rail/. The April 2009 document *Vision for High-Speed Rail in America*, produced by the US Federal Railroad Administration, is at fra.dot.gov/Downloads/Final%20FRA%20HSR%20Strat%20Plan.pdf
89. In December 2007, Mr. Edward Hamberger, President of the Association of American Railroads, condemned a proposal by the Passenger Rail Working Group of the National Surface Transportation Policy and Revenue Study Commission to expand the shared use of rail corridors by freight and passenger trains by stating "Piggy-backing on privately owned and operated freight railroad assets will give America a third-rate passenger rail system, one that is not attractive to passengers or competitive with automobile or air travel." See his full statement, available at aar.org/NewsAndEvents/PressReleases/2007.aspx.

90. For the GAO's assessment, see the statement of Susan A. Fleming, Director, Physical Infrastructure Issues, "High Speed Passenger Rail: Effectively Using Recovery Act Funds for High Speed Rail Projects," June 23, 2009, at gao .gov/new.items/d09786t.pdf.
91. GDP projections are from *World Economic Outlook Update*, International Monetary Fund, Washington DC, July 8, 2009, at imf.org/external/pubs/ft/ weo/2009/update/02/index.htm.
92. See Harper's Index, *Harper's Magazine*, July 2009, and indicated sources.
93. James Madison famously wrote that "ambition must be made to counter ambition" in US political institutions. For the complete articulation of this philosophy of institutional design, see Hamilton et al. (1982).
94. For litanies of continuing US economic ills, and talk of a second stimulus package, see the sources detailed in Notes 85 and 86.
95. Among those calling for a second US stimulus package are Laura Tyson, an external economic adviser to the president (see Coy, P., "Stimulus Gets a Second Wind in Washington," *Business Week*, July 7, 2009, at businessweek .com/investor/content/jul2009/pi2009077_215057.htm) and Warren Buffet, said to be the world's second richest person (see Frye, A., et al., "Buffett says US may need a second stimulus package," June 24, 2009, at bloomberg .com/apps/news?pid=20601087&sid=aHaggIeAoZjk). The administration's initial response has been to say that the February 2009 package will prove to be adequate (see Travers K., "Labor leaders push Obama for second stimulus package," July 13, 2009, at blogs.abcnews.com/politicalpunch/2009/07/ labor-leaders-push-obama-for-second-stimulus-package.html).

References

Cole, D., McAlinden, S., Dziczek, K. and Menk, D. M. (2008) *The Impact on the U.S. Economy of a Major Contraction of the Detroit Three Automakers*, Center for Automotive Research, Ann Arbor, MI, November, cargroup.org/ documents/FINALDetroitThreeContractionImpact_3__001.pdf.

Gladwell, M. (2002) *The Tipping Point*, Little, Brown & Co, New York, NY, 301 pp.

Hamilton, A., Madison, J. and Jay, J. (1982) *The Federalist Papers*, New York: Bantam Dell Publishing Group, 621 pp.

Keynes, J. M. (1937) "The General Theory of Employment," *Quarterly Journal of Economics*, 51(2), pp. 209–223.

OECD (2009) *Economic Outlook*, Organization for Economic Cooperation and Development, Paris, France, 265 pp.

Reinhart, C. M. and Rogoff, K. S. (2008a) *Is the 2007 US sub-prime financial crisis so different? An international historical comparison*, Working Paper 13761, National Bureau of Economic Research, Cambridge, MA, nber.org/papers/ w13761.

Reinhart, C. M. and Rogoff, K. S. (2008b) *This time is different, A panoramic view of eight centuries of financial crises,* Working Paper 13882, National Bureau of Economic Research, Cambridge, MA, nber.org/papers/w13882.

Sengupta, R. and Tam, Y. M. (2008) "The LIBOR-OIS Spread as a Summary Indicator," *Economic Synopses,* Federal Reserve Bank of St. Louis, MI, No. 25, research.stlouisfed.org/publications/es/08/ES0825.pdf.

UK ONC (2009) *Social Trends 39,* Office of National Statistics, London, UK, 283 pp, statistics.gov.uk/socialtrends39/.

UNCTAD (2009) *Review of Maritime Transport 2008,* United Nations Conference on Trade and Development, Geneva, Switzerland, unctad.org/en/docs/rmt2008_en.pdf.

US BTS (2009), *National Transportation Statistics,* US Bureau of Transportation, Washington DC, 499pp, bts.gov/publications/national_transportation_statistics/pdf/entire.pdf.

US CIA (2008) *The World Fact Book,* US Central Intelligence Agency, Washington DC, 755 pp, https://cia.gov/library/publications/the-world-factbook/.

World Bank (2009) *Global Development Finance: Charting a Global Recovery,* World Bank, Washington DC, 167 pp., econ.worldbank.org/WBSITE/EXTERNAL/EXTDEC/0,,contentMDK:22216733~pagePK:64165401~piPK:64165026~theSitePK:469372,00.html.

Index

S

T

About the Authors

Anthony Perl usually walks to work at Simon Fraser University's Vancouver, British Columbia campus, where he directs the Urban Studies Program and is Professor of Political Science and Urban Studies. He previously rode light rail to teach at the University of Calgary, took the subway to direct the City University of New York's Aviation Institute, and caught the metro to work at the Laboratoire d'Economie des Transports in Lyon, France. Anthony earned an AB (Hon.) in Government from Harvard University, and MA and Ph.D. degrees in Political Science from the University of Toronto. He currently serves on the Board of VIA Rail, Canada's national passenger rail carrier. For more about Anthony and his work, see sfu.ca/urban/faculty.html.

Richard Gilbert is an independent consultant with current and recent private- and public-sector clients in Asia, Europe, and North America. His work focuses on transport, energy, and urban governance. In earlier careers, he taught experimental psychology and pharmacology at universities in the UK, Mexico, the US and Canada, and then, from 1976–1991, served as a full-time "big-city" politician, elected six times with increasing majorities. He and Rosalind have lived in the same house in downtown Toronto for 40 years, where they enjoy visits from a daughter and three sons and numerous, mostly teenage, grandchildren. More information about Richard is at richardgilbert.ca.

If you have enjoyed *Transport Revolutions*,
you might also enjoy other

Books to Build a New Society

Our books provide positive solutions for people who
want to make a difference. We specialize in:

Sustainable Living • Ecological Design and Planning
Natural Building & Appropriate Technology • New Forestry
Environment and Justice • Conscientious Commerce
Progressive Leadership • Resistance and Community
Nonviolence • Educational and Parenting Resources

New Society Publishers

ENVIRONMENTAL BENEFITS STATEMENT

New Society Publishers has chosen to produce this book on recycled paper made with 100% post consumer waste, processed chlorine free, and old growth free.

For every 5,000 books printed, New Society saves the following resources:[1]

46	Trees
4,132	Pounds of Solid Waste
4,546	Gallons of Water
5,930	Kilowatt Hours of Electricity
7,511	Pounds of Greenhouse Gases
32	Pounds of HAPs, VOCs, and AOX Combined
11	Cubic Yards of Landfill Space

[1]Environmental benefits are calculated based on research done by the Environmental Defense Fund and other members of the Paper Task Force who study the environmental impacts of the paper industry.

For a full list of NSP's titles, please call 1-800-567-6772 or check out our web site at:

www.newsociety.com

NEW SOCIETY PUBLISHERS